Also in the Variorum Collected Studies Series:

RHODA RAPPAPORT
Studies on Eighteenth-Century Geology

STEPHEN CLUCAS
Magic, Memory and Natural Philosophy in the Sixteenth and Seventeenth Centuries

SONJA BRENTJES
Travellers from Europe in the Ottoman and Safavid Empires, 16th–17th Centuries
Seeking, Transforming, Discarding Knowledge

JOHN GASCOIGNE
Science, Philosophy and Religion in the Age of the Enlightenment
British and Global Contexts

KENNETH L. TAYLOR
The Earth Sciences in the Enlightenment
Studies on the Early Development of Geology

WILLIAM A. WALLACE
Domingo de Soto and the Early Galileo
Essays on Intellectual History

W.G.L. RANDLES
Geography, Cartography and Nautical Science in the Renaissance
The Impact of the Great Discoveries

JOHN GASCOIGNE
Science, Politics and Universities in Europe, 1600–1800

P.M. HARMAN
After Newton: Essays on Natural Philosophy

WILLIAM A. WALLACE
Galileo, the Jesuits, and the Medieval Aristotle

CHARLES B. SCHMITT
Reappraisals in Renaissance Thought

VARIORUM COLLECTED STUDIES SERIES

Religion, Magic, and the Origins of Science in Early Modern England

I dedicate this volume, with love,
to my daughter Isla, who was born and grew up
while I was working on these studies.

John Henry

Religion, Magic, and the Origins of Science in Early Modern England

Routledge
Taylor & Francis Group
LONDON AND NEW YORK

First published 2012 by Ashgate Publishing

Published 2016 by Routledge
2 Park Square, Milton Park, Abingdon, Oxon OX14 4RN
711 Third Avenue, New York, NY 10017, USA

First issued in paperback 2017

Routledge is an imprint of the Taylor & Francis Group, an informa business

British Library Cataloguing in Publication Data
Henry, John, 1950–
Religion, magic, and the origins of science in early modern England.
– (Variorum collected studies series ; CS999)
1. Religion and science – England – History – 16th century. 2. Religion and science – England – History – 17th century. 3. Physics – Religious aspects – Christianity. 4. Science and magic – History – 16th century. 5. Science and magic – History – 17th century. 6. Philosophy and religion – England – History –16th century. 7. Philosophy and religion – England – History – 17th century. 8. Philosophy and science – England – History – 16th century. 9. Philosophy and science – England – History – 17th century.
I. Title II. Series
215.3'0942'09031–dc23

ISBN 978–1–4094–4458–9

Library of Congress Control Number: 2011944852

VARIORUM COLLECTED STUDIES SERIES CS999

ISBN 13: 978-1-138-11707-5 (pbk)
ISBN 13: 978-1-4094-4458-9 (hbk)

CONTENTS

This volume contains xii + 314 pages

PUBLISHER'S NOTE

The articles in this volume, as in all others in the Variorum Collected Studies Series, have not been given a new, continuous pagination. In order to avoid confusion, and to facilitate their use where these same studies have been referred to elsewhere, the original pagination has been maintained wherever possible.

Each article has been given a Roman number in order of appearance, as listed in the Contents. This number is repeated on each page and is quoted in the index entries.

PREFACE

Arguments about the relationship between science and religion, and between science and magic have proved to be enduring aspects of the historiography of early modern science. The essays collected here contribute to one or another of these on-going debates; or more often to both, since historical realities seldom divide themselves along the neat lines that historians would wish to impose upon them. At a time when the newly conceived mechanical philosophy seemed to offer the best hope of replacing the increasingly moribund Aristotelianism of the universities, the nature of matter inevitably became a major focus of philosophical concern. By proposing to replace the hylomorphism of scholasticism with atomistic or corpuscularist theories, the new philosophers pinned their hopes on explaining all physical phenomena in terms of 'matter in motion'. There were those, such as Galileo, Descartes, and Hobbes, who hoped to confine their physics purely to kinematics – so that the motions of bodies, and the interactions arising from their motions, were all that was required to explain any set of circumstances.[1] For many other thinkers, however, especially those with a greater knowledge of, and practical expertise in, what we would now call chemical phenomena, a strictly mechanistic kinematics seemed unworkable. This led to an alternative view, in which bodies were invested with 'active principles' which enabled them to interact in ways that could not simply be reduced to their motions.

Broadly speaking, it was the occult nature of these supposed 'active principles' which ensured that the traditions of natural magic came to the fore in numerous versions of the new philosophy. But it is also evident that these attempts to develop new theories about the nature of matter and body mark one of the major aspects of thc new sciences which Amos Funkenstein has seen as leading early modern natural philosophers to believe that traditional modes of theologizing had been rendered obsolete. Consequently, Funkenstein has persuasively argued, natural philosophers developed a 'new and unique approach to matters divine', and began to treat 'theological issues at length'.[2] Matter theory became in this period a major site for assessing the nature of

---

[1] See, for example, John Henry, 'Galileo and the Scientific Revolution: The Importance of His Kinematics', *Galilaeana* 8 (2011), pp. 3–36; Stephen Gaukroger, *Descartes' System of Natural Philosophy* (Cambridge: Cambridge University Press, 2002); Frithiof Brandt, *Thomas Hobbes' Mechanical Conception of Nature* (Copenhagen: Levin and Munksgaard, 1928).

[2] Amos Funkenstein, *Theology and the Scientific Imagination from the Middle Ages to the Seventeenth Century* (Princeton: Princeton University Press, 1986), p. 3.

God's Providence. All natural philosophers seemed to be agreed that matter, according to its own nature, was passive and inert. If it could be shown that bodies had inherent principles of activity, therefore, it had to be supposed either that God had bestowed active principles upon these bodies, or that God had chosen some other way to make bodies appear to be active. It is the contention of a number of the papers gathered here that most thinkers simply assumed that God endowed active principles on matter, but that the Cambridge Platonists opted for an alternative view, which strictly maintained the passivity of matter and supposed that activity could not originate from bodies but derived from an all-pervasive immaterial Spirit of Nature – supposedly acting as the vicarious power of God in the world.[3] In most of the articles that follow, therefore, a study of the role of occult qualities and other ideas from the natural magical tradition, leads us simultaneously into a study of the relationship between religion and natural philosophy.

The papers are presented in roughly chronological order with regard to their historical content. We begin with a study of William Gilbert's pioneering study of magnetism, widely renowned as one of the first works of natural philosophy to rely heavily on the experimental method, the *De magnete* of 1600. Although it had been noticed before that Gilbert's main aim in this work was to establish that the Earth itself was a giant magnet and therefore might be capable of spontaneously performing the kind of rotatory movements required by Copernican theory, I believe this was the first study to really emphasise this, and in so doing it was the first to bring out the importance of animism for Gilbert's enterprise. Gilbert evidently did not feel the need to explore the religious implications of his new theory of an animate Earth, being content with rhetorical claims about the absurdity of denying the Earth a soul while granting souls to lowly creatures like insects and worms. The second essay features the interaction between religion and the new philosophy, in this case without any need to consider occult or magical notions. Sir Kenelm Digby

[3] I have argued this point in a number of other places: 'A Cambridge Platonist's Materialism: Henry More and the Concept of Soul', *Journal of the Warburg and Courtauld Institutes* 49 (1986), pp. 172–95; 'England', in Roy Porter and Mikuláš Teich (eds), *The Scientific Revolution in National Context* (Cambridge: Cambridge University Press, 1992), pp. 178–210; 'Henry More (1614–1687)', *Stanford Encyclopaedia of Philosophy* (August, 2007): http://plato.stanford.edu/entries/henry-more/; 'Omnipotence and Thinking Matter: John Locke and the Use of Reason in Religion', in Delfina Giovannozzi and Marco Veneziani (eds), *Materia: Atti del XIII Colloquio Internazionale del Lessico Intellettuale Europeo (Roma, 7–9 gennaio 2010)* (Florence: Leo S. Olschki, 2012), in press; and 'The Reception of Cartesianism', in Peter Anstey (ed.), *The Oxford Handbook of British Philosophy in the Seventeenth Century* (Oxford: Oxford University Press, 2012), in press. See also, Alan Gabbey, Daniel Garber, John Henry and Lynn Joy, 'New Doctrines of Body and Its Powers, Place and Space', in D. Garber and M. Ayers (eds), *The Cambridge History of Seventeenth-Century Philosophy*, 2 vols (Cambridge: Cambridge University Press, 1998), i, pp. 553–623.

and his mentor, the secular priest Thomas White, made a remarkable but little known contribution to the Catholic Counter-Reformation during the English Interregnum, in which their version of the mechanical philosophy, and its implications for Christian dualism, was made to support a new version of Catholicism which made many concessions to Protestant belief, and which they hoped would win Protestants back to the Roman Church.

The third essay sets the agenda for the rest of the volume. It was in my paper on 'Occult qualities and the experimental philosophy' that I first tried to demonstrate that, contrary to the standard accounts, it was not Newton who re-introduced occult qualities back into natural philosophy after they had been vigorously excluded by Galileo, Descartes, and their respective followers. In a survey of English natural philosophers I showed that the strict mechanistic and kinematic approach of Galileo and Descartes (in which matter is completely passive) was only really adopted by Thomas Hobbes, and by the Cambridge Platonists, Henry More and Ralph Cudworth. Every other English natural philosopher of note accepted to a greater or lesser extent the presence of active principles in matter. It was while doing the research for this paper that I noticed the attack on Francis Glisson's theory of active matter by the Cambridge Platonist Henry More. I expanded this into a separate study, which is included here ('Medicine and pneumatology'). I was able to indicate here that the divide was not simply between natural philosophers on the one hand and theologians on the other, because I was able to show that the Presbyterian divine Richard Baxter supported Glisson. I followed this up with a broader survey of attitudes to the soul, in which it again became clear that the Cambridge Platonists differed in significant respects from the majority of their contemporary theologians. It emerged from this work that Henry More was opposed to the voluntarist theology that I had been led to believe was favoured by the leading experimental philosopher Robert Boyle.[4] Accordingly, I examined the relations between More and Boyle and found that, far from being fellow-travellers helping to promote the mechanical philosophy in late seventeenth-century England (as was usually supposed), they were in fact seriously at odds with one another. The result was my essay on 'Henry More versus Robert Boyle'.

---

4 Voluntarist theology plays a major role in a number of the studies here. Readers may be aware of Peter Harrison's attempts to dismiss the historical significance of theological voluntarism on the grounds that it is nothing more than a misconception of historians. Peter Harrison, 'Voluntarism and Early Modern Science', *History of Science* 40 (2002), pp. 63–89; and 'Was Newton a Voluntarist?' in J.E. Force and Sarah Hutton (eds), *Newton and Newtonianism* (Dordrecht: Kluwer Academic, 2004), pp. 39–63. I have responded to these claims, and I hope re-asserted the historical importance of voluntarist theology to our understanding of early modern natural philosophy, in my 'Voluntarist Theology at the Origins of Modern Science: A Response to Peter Harrison', *History of Science* 47 (2009), pp. 79–113.

Robert Boyle and Robert Hooke are portrayed in the standard historiography of early modern science as dyed-in-the-wool mechanical philosophers. Having offered a somewhat different, but necessarily brief, treatment of them in my survey article on 'Occult qualities and the experimental philosophy', I felt that I should try to reinforce my claims at greater length.[5] The articles on Boyle and his notion of 'cosmical qualities', and on Robert Hooke are intended, therefore, to show that the two leading natural philosophers in England before Newton both allowed principles of activity into their theories of matter. Although Newton has been recognised (by contemporaries such as Christiaan Huygens and G.W. Leibniz, and subsequently by R.S. Westfall and other Newton scholars) as re-introducing occult qualities into post-Cartesian natural philosophy, there has been a tendency to deny that he believed gravity acted at a distance. For modern historians and philosophers of science alike, action at a distance has been regarded as too occult, indeed too absurd, a notion for a great mind like Newton's to accept. Given that Newton freely accepts action at a distance as a reality throughout his later work, I decided to try to understand how modern scholars could have so badly misunderstood him. The result was my article, reaffirming Newton's acceptance of this quintessential occult notion, '"Pray do not ascribe that notion to me"'. Resistance to my argument still persists, however, and I have recently supplemented this early article with a more detailed study of Newton's belief in action at a distance.[6]

While working for so long on the use of natural magical traditions by early modern natural philosophers, it was inevitable that my thoughts should turn to one of the major problems in the historiography of magic – why did belief in magic among the learned go into decline towards the end of the seventeenth century, and the beginning of the eighteenth? The final essay in this collection offers my own contribution to this on-going debate.

JOHN HENRY

University of Edinburgh
October 2011

---

[5] With regard to Boyle, at least, this is perhaps no longer such a prominent aspect of the historiography, thanks to the work of Antonio Clericuzio, Lawrence Principe, William R. Newman and Michael Hunter, all of whom have since emphasised non-mechanistic aspects of Boyle's work. Consider, for example, A. Clericuzio, 'A Redefinition of Boyle's Chemistry and Corpuscular Philosophy', *Annals of Science* 47 (1990), pp. 561–89; L. Principe, *The Aspiring Adept: Robert Boyle and his Alchemical Quest* (Princeton: Princeton University Press, 1998); William R. Newman, *Atoms and Alchemy: Chymistry and the Experimental Origins of the Scientific Revolution* (Chicago: University of Chicago Press, 2006); and Michael Hunter, *Boyle: Between God and Science* (New Haven and London: Yale University Press, 2009).

[6] 'Gravity and *De gravitatione*: The Development of Newton's Ideas on Action at a Distance', *Studies in History and Philosophy of Science* 42 (2011), pp. 11–27.

ACKNOWLEDGEMENTS

I am grateful to the following publishers, institutions and journals for permission to reproduce the studies included in this volume: The University of Pennsylvania Press (I); Cambridge University Press (II, V, and VII); Michael Hoskin and Science History Publications Ltd (III, and X); The Trustees, The Wellcome Trust (IV); Springer, Dordrecht (VI, and IX); and Boydell & Brewer Ltd (VIII). Full bibliographic details of the original publications are provided in the list of Contents.

I

Animism and Empiricism: Copernican Physics and the Origins of William Gilbert's Experimental Method

In the second year of this journal's run, way back in 1941, appeared Edgar Zilsel's classic and still widely cited paper on *The Origins of William Gilbert's Experimental Method*.[1] Focusing on Gilbert's *De magnete* of 1600, undoubtedly a seminal text in the history of the experimental method,[2] Zilsel argued that Gilbert borrowed his methodology from elite craftsmen, artisans, and other manual workers involved in mining, smelting, smithing, compass making, navigation, sailing, and other activities which involved working with iron or with magnets.[3] On its appearance the paper contributed to a growing and still continuing debate in the history of science about the relative importance of scholars and craftsmen in the origins of modern science. Although Zilsel's general thesis has been critically evaluated in terms of that wider debate,[4] his very specific claims about the

Thanks to Drs. Stephen Pumfrey and Katherine Neal for their help and encouragement.

[1] Edgar Zilsel, "The Origins of William Gilbert's Experimental Method," *JHI*, 2 (1941), 1-32 (henceforward cited as "Zilsel").

[2] William Gilbert, *De magnete, magnetisque corporibus, et de magno magnete tellure; Physiologia nova, plurimis & argumentis, & experimentis demonstrata* (London, 1600); English edition: *De magnete*, tr. P. Fleury Mottelay (New York, 1958), cited hereafter as "De magnete"; and see D. H. D. Roller, *The De Magnete of William Gilbert* (Amsterdam, 1959), 50-91.

[3] See Zilsel, 24, but also 12-24. Zilsel makes the more general claim that modern science had its origin in "the experimenting manual workers of early capitalism" (24, 26-29), and in "The Sociological Roots of Science," *American Journal of Sociology*, 47 (1942), 544-62.

[4] See also R. K. Merton, *Science, Technology and Society in Seventeenth-Century England* (New York, 1970), first published in *Osiris*, 4 (1938); and B. Hessen, "The Social and Economic Roots of Newton's 'Principia,' " in *Science at the Cross Roads: Papers Presented to the International Congress of the History of Science and Technology held in London from June 29th to July 3rd, 1931, by the Delegates of the USSR* (London, 1931), 149-212; also G. N. Clark,

role of elite craftsmen in providing Gilbert with a ready-made experimental method, have never been properly considered. This paper attempts to reassess Zilsel's claims about the origin of Gilbert's experimental method and, in so doing, to arrive at a richer understanding of what Gilbert was trying to do in *De magnete* and to suggest an alternative source for his experimentalism.

The first thing to note about Zilsel's argument is that it had two strands.[5] One of these pointed to actual experimental techniques used by Gilbert and showed that they can all be found in earlier writings on magnets. To be more specific, the techniques in question are to be found either in the thirteenth-century "Letter on the Magnet" written by Petrus Pergrinus or Pierre de Maricourt, first published in 1558, or in *The Newe Attractive*, a work on the magnetic compass written by a retired mariner, Robert Norman, first published in 1581 but reprinted three times before 1600.[6] The second strand did not involve experiments at all, with one notable exception, but simply implied or suggested that Gilbert's experimental techniques must have been inspired by miners and foundrymen, or navigators and instrument makers. The problem here, as others have objected,[7] is that it is by no means clear that elite craftsmen were engaged in performing experiments in anything like the sense of the word required for the argument to cut any ice. For Zilsel, however, "we cannot doubt that many of [the miners and foundrymen of the period], stimulated to improvements by economic competition, were wont to try new techniques and to observe natural processes."[8] A few sentences later we are told that it is "obvious" that "among such manual labourers there were experimentalists, though experimentalists with practical aims only and without theoretical knowledge."[9]

Science and Social Welfare in the Age of Newton (Oxford, 1949); A. R. Hall, "The Scholar and the Craftsman in the Scientific Revolution," and A. C. Crombie "The Significance of Medieval Discussions of Scientific Method for the Scientific Revolution," both in M. Clagett (ed.), *Critical Problems in the History of Science* (Madison, 1959), 3-23 and 79-101; P. Rossi, *Philosophy, Technology and the Arts in the Early Modern Era* (New York, 1970); J. A. Bennett, "The Mechanics' Philosophy and the Mechanical Philosophy," *History of Science*, 24 (1986), 1-28; and H. F. Cohen, *The Scientific Revolution: A Historiographical Inquiry* (Chicago, 1994), 328-50.

[5] Zilsel, 27.

[6] Petrus Pergrinus Maricurtensis, *De magnete, seu rota perpetui motus, libellus' nunc primum promulgatus* (Augsburg, 1558), and *The Epistle of Peter Peregrinus de Maricourt to Sygerus de Foucaucourt, Soldier, Concerning the Magnet*, tr. S. P. Thompson (London, 1902); and H. D. Harradon, "Some Early Contributions to the History of Geomagnetism—I: The Letter of Peter Peregrinus de Maricourt to Sygerus de Foucaucourt, Soldier, Concerning the Magnet," *Terrestrial Magnetism and Atmospheric Electricity*, 48 (1943), 3-17. Cited below as "Harradon"; letter dated 8 August 1269. Robert Norman, *The Newe Attractive. Containing a short discourse of the Magnes or Loadstone* (London, 1581), reprinted in 1585, 1592, and 1596.

[7] See Cohen, *op. cit.*, 340-41.

[8] Zilsel, 14.

[9] *Ibid.*, 15.

The philosopher of science at this point might legitimately want to suggest that an experiment is not an experiment unless it is testing a theory, but we do not have to enter into such arguable niceties.[10] We can see that there is something profoundly tendentious in Zilsel's claims simply by reading on. At this point Zilsel mentioned the only experiment specifically referred to in this strand of his argumentation. Quoting Gilbert, Zilsel showed that Gilbert himself must have descended into a mine in order, as Zilsel says, "to verify the hypothesis" that the polarity of a magnet derives from the polarity of the earth:

> We had a twenty pounds' heavy loadstone dug and hauled out after having first observed and marked its ends in its vein. Then we put the stone in a wooden tub on water, so that it could turn freely. Immediately the surface which had looked to the North in the mine turned itself to the North on the water.[11]

Zilsel did not make the absurd claim that Gilbert must have seen miners doing this kind of experiment, but he did all he could to imply that Gilbert's experiment owed a great deal to his acquaintance with miners:

> It is almost symbolic that Gilbert performed a laboratory experiment just after having left a pit and talked to miners. Of course Gilbert's experiments were not plain copies of the trials of the miners and foundrymen. But his spirit of observing and experimenting was taken over not from scholars but from manual workers.[12]

The suggestion that miners themselves might have taken it into their heads to see whether an extracted loadstone oriented itself in the same way as it had lain in its seam is not just unlikely but almost ludicrous, and so we have to wonder what other possible "trials of the miners" inspired Gilbert. If we cannot think of any, then we have to doubt the claim that Gilbert's spirit of experimenting was taken from these manual workers.

Zilsel tried to bolster his rather weak position by claiming that in some cases Gilbert's experiments "simply repeated the working processes of contemporary iron manufacture." Referring to chapters nine to eleven of Book I of *De magnete*, Zilsel claimed that part of Gilbert's "laboratory must have looked like a smithy," because Gilbert tells us that "he heats the ore for hours in a furnace and melts it; then he hammers the product, puts it into a second furnace, and so on." In fact Zilsel was conflating two different parts of Gilbert's account here. In chapter

[10] See M. B. Hesse, "Gilbert and the Historians," *British Journal for the Philosophy of Science*, 11 (1960), 1-10, 130-42.

[11] Zilsel, 15, translation differing slightly from Mottelay's (*De magnete*, Bk. III, ch. 4, 184).

[12] *Ibid.*

nine Gilbert told the reader that if he were to roast some pieces of stony iron ore "in a moderate fire" for some ten or twelve hours and then allow them to cool in a north-south orientation, the stones would now be attracted to a loadstone, even though they were not before (due to their excessive stoniness). In chapter eleven Gilbert described the process of making wrought iron, which "as everyone knows" is attracted to a loadstone, but he did not suggest that he had done this himself. During smelting, we are told, "the metal flows out ... and forms a great long mass, which under the blows of a large hammer is cut into pieces: from these, after being reduced in another furnace and again put on the anvil, the workmen form cubical masses, or more usually bars, which are sold to merchants and blacksmiths: from these blocks or bars are everywhere made in smiths' shops various implements."[13] It is perfectly evident that Gilbert was here telling his scholarly readership how wrought iron is made (by workmen); there is no suggestion that he has done it himself. It is one thing to roast some stones in a moderate fire, and quite another to have a furnace capable of heating iron ore to "very great heat" for eight to twelve hours and then to be able to contrive for the metal to pour out to form "a great long mass," and so on.

If Zilsel's attempts to suggest in fairly general terms that Gilbert borrowed his experimental method from artisans and craftsmen remain unsatisfactory, maybe we should look to the other strand of his argument, where he considered the very specific influences of Pierre De Maricourt and Robert Norman. There can be no doubt of the influence of Pierre's "Epistle on the Magnet" on Gilbert. As Zilsel pointed out, Gilbert's method of determining the poles of magnets and his technique of floating magnets in water on wooden (or in Gilbert's case, cork) supports to enable them to move freely, are identical to Pierre's techniques.[14] But Zilsel was aware that this thirteenth-century philosopher presented a difficulty for his case. The fact that Pierre was already performing experiments centuries before sits uneasily with Zilsel's claim that "experimental science could not come into existence so long as liberal and mechanical arts were kept separate by the contempt for manual labor."[15] Zilsel did not allow this to detain him long. Pierre de Maricourt is simply seen as a great precursor who did not allow his interests to be hampered by the "social prejudices" of his day. "We do not know," Zilsel wrote, "where Petrus's freedom from bias [against manual labor] originates," but his attitude merely serves to confirm "the importance of manual labor and the mechanical arts for the history of science."[16] But as we shall see, the significance of Pierre lies elsewhere.

[13] See *De magnete*, Bk. I, ch. 10, 48; also 47. Cf. Zilsel, 15, and 14 where, citing the same passage, he writes that Gilbert "describes how he himself manufactures the wrought iron he needs for his experiments."

[14] Zilsel, 12; and see also Roller, *op. cit.*, 39-42.

[15] Zilsel, 29.

[16] Zilsel, 30.

Zilsel discusses at some length the other undeniable influence upon Gilbert, the retired navigator, Robert Norman, and he does seem to provide an undeniable case of an elite craftsman who really was performing experiments in an effort to understand the behavior of magnets. Like Pierre de Maricourt, Norman floated magnets on water in specially prepared experimental vessels or suspended them on threads. Gilbert used both these techniques. Norman weighed iron before and after magnetizing it to determine that the magnetic virtue is imponderable, and he showed experimentally that a floating magnet merely turns toward the pole but is not attracted towards it. Again, Gilbert repeated both of these experiments and gave due acknowledgment to Norman for the second. Gilbert clearly knew Norman's *Newe Attractive*, therefore, and recognized its importance. In the opening chapter of *De magnete* Gilbert mentioned Norman as a "skilled navigator and ingenious artificer, who first discovered the dip of the magnetic needle."[17] Since the phenomenon of magnetic dip was to play a major role in Gilbert's own theory of magnetism, it is fitting that he should acknowledge Norman in this way.

If Gilbert was indebted to Norman, it would seem that Zilsel was even more beholden. "Robert Norman," Zilsel wrote, "is of great importance for our problem." Even if the supposed influence of miners and foundrymen on Gilbert's experimental method is discounted and even if the relevance of Pierre de Maricourt to an account about the role of economic forces in stimulating the Scientific Revolution is rescinded, Norman remains as an undeniable element in Zilsel's case. It was crucially important to Zilsel that his readers should recognize that "This remarkable man who, twenty five years before Galileo's first publication, speaks of the 'incredible delight' of experimental discovery, was a craftsman."[18] For Zilsel, Norman, "the experimenting compass-maker," was "even nearer than Gilbert to the sober objectivity of modern natural science."

While Zilsel believed that there was a significant historical "problem" in trying to understand how a learned scholar like Gilbert might have come up with the experimental method, he perceived no such problem arising from the fact that Norman relied upon the experimental method. After all, Norman was a manual worker, one who was used to the hands-on approach of the experimentalist. He was a practical man trying to earn his living and as such can be seen, quite unproblematically, as one of those "experimenting manual workers of early capitalism."[19]

Those of us who do not share Zilsel's Marxist economic determinism and his naïve view of the nature of experimentation might wish to differ. After all, as

[17] *De magnete*, Bk. I, ch. 1, 15. See Zilsel, 19.

[18] Zilsel, 23. The "incredible delight" is attributed by Norman (*Newe Attractive*, Epistle Dedicatory, sig. Aii^{v}) not to himself but to Archimedes, Pythagoras and "sundrie others, that are mentioned in aucthours."

[19] Zilsel, 24; and 6, 12, 23, on the origins of Gilbert's method, and therefore of modern science in general.

we have already suggested, it would be completely unwarranted to regard Norman as just one of many artisan experimenters. The experimental research he carried out for his *Newe Attractive* was highly unusual, not the routine kind of thing that any compass maker or navigator might have been expected to do. Where then might Norman have received a stimulus to develop his experimental techniques? In fact Robert Norman informs us himself (although Zilsel seems to have overlooked it). It arises from Norman's accidental discovery of the phenomenon of magnetic dip:

> Havying made many and divers compasses and using alwaies to finish and ende them, before I touched the needle, I found continually, that after I had touched the Irons with the Stone, that presently the North poinct thereof would bende or *Decline* downwards under the Horizon in some quantitie: in so much that to the Flye of the Compasse, which before was made equall, I was stil constrained to put some small peece of Ware in the South part thereof, to counterpoise this *Declining*, and to make it equall againe.[20]

What this means is that a carefully balanced iron needle, made for the compass, would no longer pivot on the level once it had been magnetized by rubbing with a loadstone. After magnetization the needle would now dip down, as though unbalanced by the act of magnetizing.

Norman tells us that, for a long time, he paid no special attention to this effect: "Whiche effect havyng many tymes passed my handes, without any great regard therunto, as ignorant of any suche propertie in the Stone, and not before havyng heard nor read of any such matter."[21] The day came, however, when he was commissioned to make a compass with a needle of "sixe ynches long." This time, he tried to correct the unbalancing by cutting some of the North end of the needle away, but he cut it too short and spoiled the whole needle. Somewhat enraged by this ("stroken into some choler"), Norman tells us that he now decided to try to understand this effect. We can see here, then, that Zilsel is partly right: Norman certainly was stimulated to understand this natural effect by economic factors. But if economic necessity was the Mother of Norman's invention, it also had some potential fathers from (for Zilsel anyway) an unexpected source:

> I applied my self to seeke further into this effect, and makyng certaine learned and expert men, my freendes, acquainted in this matter, they advised me to frame some Instrument, to make some exacte triall, how much the Needle touched with the Stone would Decline, or what great-

[20] *New Attractive*, 8.
[21] *Ibid.*

est Angle it would make with the plane of the Horizon. Whereupon I made diligent proofes, the maner whereof is shewed in the Chapter followyng.[22]

The artisan Robert Norman developed his experimental method, therefore, because some learned experts told him that was the right way to go about it.

In keeping with his thesis that modern experimental science owes its origins to the breakdown of the barriers between craftsmen and scholars, Zilsel sees Gilbert as the "first academically trained scholar who dared to adopt the experimental method from the superior craftsmen and to communicate the results in a book not to helmsmen and to mechanics but to the learned public."[23] Because of his understanding of the scientific importance of experiment (which he supposedly gleaned from his personal knowledge of various craftsmen and their work), Gilbert was able to help make manual experimental work respectable among the ranks of the educated. Gilbert was seen by Zilsel, therefore, to be in the vanguard of those intellectuals who, "with the advancement of early capitalistic society," rebelled against both Scholasticism and humanism and began to make contacts with elite craftsmen.[24]

Zilsel's slant on Gilbert and what he was trying to do seriously affected the way Zilsel presented the *De magnete* and its purpose. Zilsel gave the distinct impression that the *De magnete* was a compendium of knowledge about iron, loadstones, and magnetism, presented to intellectuals for entirely pragmatic reasons. For Zilsel the *De magnete* was "nearer to a technological than to a physical work of our time," and he famously broke down the subject matter of the *De magnete* into its constituent parts: only 40% of the book deals with physical experiments, he wrote, while 25% deals with nautical instruments and navigation, 10% is concerned with mining, smelting, and fashioning of iron, and so on.[25] After showing Gilbert's knowledge of mining and metallurgy, Zilsel insisted that these are the "mining and metallurgy of rapidly advancing early capitalism." Similarly, Gilbert's interest in navigation was linked to the circumnavigations of the globe and European expansion.[26]

Zilsel was so bound up with his own account of what *De magnete* was all about that the real and very obvious purpose of Gilbert's book was hardly alluded to at all. Indeed, difficult as it is to believe of anyone who had read the book so closely, it looks very much as though Zilsel completely missed the point.

[22] *New Attractive*, 9.
[23] Zilsel, 28.
[24] Zilsel, 27.
[25] Zilsel, 12-13.
[26] Zilsel, 14, and 25, where Zilsel (note 34) suggests the importance of military engineering to Gilbert as well.

But Gilbert's point is clearly spelled out, as was usual in those days, by the full title of his book: *On the magnet and magnetic bodies, and on the great earthly magnet*. A glance at the contents pages will show that the whole argument of the book culminates in the sixth and final part, where the first of nine chapters speaks "Of the globe of the earth as a great loadstone." Certainly, Zilsel was aware of this aspect of Gilbert's achievement, he knew that "Gilbert was the first to conceive of the earth as a large magnet," but he dismissed this aspect of Gilbert's work as a lucky guess: "the resemblance of his magnetic philosophy to modern science is merely a matter of chance."[27]

Zilsel's failure to see the importance of Gilbert's discovery seems to be confirmed by the fact that he failed to notice its connection with the Copernican theory about the motion of the earth. Zilsel noted Gilbert's support for the Copernican theory, but only incidentally as he detailed Gilbert's knowledge of contemporary authors.[28] The mathematician Edward Wright, who wrote a "Laudatory address concerning these books on magnetism" prefixed to Gilbert's book, made it perfectly plain, however: "no one, I do believe, ever gave more certain demonstrations" of the spherical motion of the earth.[29]

In fact the whole point of *De magnete* was to offer a solution to a crucial problem for Copernican theory. In his *De revolutionibus orbium coelestium* of 1543 Copernicus effectively left unanswered the question as to how the earth might be able to move and keep on moving. According to Aristotelian physics, of course, the continued motion of the earth was as absurd a proposition as it was possible to imagine. Although Copernicus tried to show the inadequacy of the standard arguments against the motion of the earth, enabling him to conclude that "nothing prevents the earth from moving," he did not follow this up with an argument as to what made the earth move.[30]

As Robert S. Westman has shown, Copernicus tried to introduce a new epistemology of natural science in which the dictates of mathematics were taken to establish the way things are. According to the prevailing tradition, it was the natural philosopher who established the truth of things, and he did so by explaining how the phenomena in question took place. In the case of astronomical phenomena that explanation would be couched in terms of Aristotelian physics. Mathematics was regarded, like logic, as an instrument by which certain kinds

[27] Zilsel, 5.

[28] Zilsel, 9. See *De magnete*, Bk. VI, ch. 3, 318.

[29] *De magnete*, "Address by Edward Wright," xliii. See F. R. Johnson, *Astronomical Thought in Renaissance England: A Study of the English Scientific Writings from 1500 to 1645* (Baltimore, 1937), 216; G. Freudenthal, "Theory of Matter and Cosmology in William Gilbert's *De magnete*," *Isis*, 74 (1983), 22-37; and Stephen Pumfrey, "Magnetical Philosophy and Astronomy, 1600-1650," *Planetary Astronomy from the Renaissance to the Rise of Astrophysics, Part A: Tycho Brahe to Newton*, eds. R. Taton and C. Wilson (Cambridge, 1989), 45-53.

[30] Nicholas Copernicus, *Complete Works: On the Revolutions*, tr. E. Rosen (Baltimore, 1992), 17 (Bk. I, ch. 9).

of truth claims could be analyzed and tested, but it was not held to explain anything. In what Westman has seen as an attempt to bring about a major change in both the intellectual and social role of the astronomer, Copernicus professed to be content with the fact that his mathematics told him that the earth must be in motion, as one of the planets, around the stationary sun.[31]

We can see this, for example, in his discussion of the order of the planets. In the Ptolemaic system the order of the planets (outward from the earth) could be established only by convention. Copernicus's theory changed this, making it possible to establish the order of planets according to the geometry of the system. For Copernicus this was sufficient proof that his system was true. Noticing that the "grand circle" between the convexity of the sphere of Venus and the concavity of the sphere of Mars, all the planets being centered on the sun, was sufficient to contain the earth and the moon, Copernicus boldly declared: "Hence I feel no shame in asserting that this whole region engirdled by the moon, and the centre of the earth, traverse this grand circle amid the rest of the planets in an annual revolution around the sun."[32] Copernicus's lack of embarrassment, however, was not sufficient to convince natural philosophers, who continued to demand a physical explanation of the earth's putative motion. Since Copernicus died the same year that his book appeared on the public stage, this problem was left as a legacy to all would-be Copernicans.[33]

William Gilbert, evidently one of the earliest committed Copernicans, believed he had the answer. Magnets have the ability to rotate spontaneously, and the earth is a giant magnet; therefore, the earth has the power to rotate spontaneously. The whole of *De magnete* was dedicated to showing the truth of this.

Although the first part, Book I, sets out to provide what, for its day, was a typically encyclopaedic natural history of the loadstone and of iron, it ends with a chapter declaring "That the terrestrial globe is magnetic and is a loadstone; and, just as in our hands the loadstone possesses all the primary powers (forces) of the earth, so the earth by reason of the same potencies lies ever in the same direction in the universe."[34] Book II begins with a series of chapters on magnetic

[31] R. S. Westman, "The Astronomer's Role in the Sixteenth Century: A Preliminary Study," *History of Science*, 18 (1980), 105-47; also N. Jardine, *The Birth of History and Philosophy of Science: Kepler's* A Defence of Tycho against Ursus, *with Essays on Its Provenance and Significance* (Cambridge, 1984), 225-57; *idem*, "Epistemology of the Sciences," in *The Cambridge History of Renaissance Philosophy*, ed. C. B. Schmitt and Q. Skinner (Cambridge, 1988), 697-702; J. M. Lattis, *Between Copernicus and Galileo: Christoph Clavius and the Collapse of Ptolemaic Astronomy* (Chicago, 1994); and P. Dear, *Discipline and Experience: The Mathematical Way in the Scientific Revolution* (Chicago, 1995).

[32] Copernicus, Bk. I, ch. 10, 20.

[33] Such thinkers were called "philosophizing astronomers" by Johannes Kepler (see Westman, *op. cit.*, p. 117), and included Thomas Digges, Christopher Rothmann, Michael Maestlin, Kepler, and Galileo; Gilbert's name should also be added to this list.

[34] *De magnete*, Bk. I, ch. 17, 64-71. See W. B. Ashworth Jr., "Natural History and the Emblematic World View," in *Reappraisals of the Scientific Revolution*, eds. D. C. Lindberg and R. S. Westman (Cambridge, 1990), 303-32.

movements, but soon compares the earth to magnets, with chapters on "the earth's magnetic meridians" (ch. 10), "the magnetic horizon" (ch. 12), "Of the magnetic axis and poles" (ch. 13) and so on. In this book the reason for Gilbert's choice of spherical magnets in his experiments becomes apparent. By showing how a spherical loadstone behaves, or how small iron needles behave in relation to the spherical surface of the loadstone, Gilbert is easily able to draw analogies with the giant magnet of the earth and with the behavior of iron needles, or compass needles, in relation to the earth. "All the experiments that are made on the terrella, to show how magnetic bodies conform themselves to it," Gilbert wrote, "may ... be shown on the body of the earth."[35]

This, incidentally, is something else which Zilsel failed to understand. "In all his experiments," Zilsel pointed out, "he [Gilbert] uses spherical loadstones, although he himself knows that bar-like magnets are more effective." Zilsel's bafflement is evident from his subsequent, inconsistent, speculation: "He calls them 'little earths' (*terrellae* [Bk] I, [ch.] 3) and presumably clings to the medieval shape of his magnets because he believes in a metaphysical connection of spherical form and magnetism."[36] It is difficult to imagine how Gilbert might have recognized the greater attracting power of bar magnets and yet believe that magnetism was dependent upon a spherical form. Later Zilsel blamed the influence of Pierre De Maricourt for Gilbert's "rather inexpedient" choice of shape for his experimental magnets.[37] In fact Pierre deliberately used spherical magnets to imitate the shape of the starry vault, since he believed that magnets orientated themselves to the heavenly poles. Given that Gilbert rejected this belief and wanted to establish empirically that magnets orientated themselves with respect to the magnetic poles of the earth, Gilbert's choice of spherical magnets, far from being inexpedient, was even more appropriate than it was for Pierre.

Zilsel must have been in the same uncomprehending mood at the point where he came closest to stumbling over what the *De magnete* was all about:

> How near his [Gilbert's] "magnetic philosophy" still is to medieval vitalism is revealed by the fact that he believes in a metaphysical correlation of magnetism and rotation. He speaks of the "magnetic rotation" of the Terrestrial globe ([Bk] VI, [ch.] 3, p. 214 [317]), and would like to accept the statement of Pierre de Maricourt that a spherical magnet rotates continuously by itself, were it not for his conscience as a cautious experimentalist. He reproduces Pierre's statement and adds (VI, 4, p. 223 [332]): "until now we have not succeeded in seeing this. We even doubt this movement because of the stone's weight and because the whole

[35] *De magnete*, Bk. VI, ch. 1, 313.
[36] Zilsel, 4.
[37] Zilsel, 12.

> earth moves by itself, as it is moved by the other stars also. That does not hold proportionally of some part [the *terrella*]."[38]

Ironically, in the accompanying footnote, Zilsel wrote that "In order to understand Gilbert's argument we have to realize that he was among the earliest adherents of Copernicus in England and was already convinced of the rotation of the earth." Certainly Gilbert himself was already convinced of the rotation of the earth, but he was trying to convince his contemporaries. According to Zilsel, Pierre's claim that a *terrella* undergoes a diurnal rotation was rejected by Gilbert because he, unlike Pierre, believed in the rotation of the earth. In fact Gilbert was trying to offer a reason why his experimental efforts to reproduce diurnal motion of a *terrella* had repeatedly failed. Zilsel thought that Gilbert wanted this experiment of Pierre's to be repeatable simply because of his belief in a "metaphysical correlation of magnetism and rotation," but his Copernicanism overcomes his naïve medieval vitalism and led him to reject it. Zilsel failed to realize that Gilbert wanted the experiment to work because it would have clinched his argument that the earth rotates because it is a giant loadstone.

Books I and II of *De magnete* make it quite clear that the major concern is to draw an analogy between the loadstone and the earth itself, and so it is with the rest of the text. Book III begins with a consideration of the causes of magnetic motions and goes on to say that "Earth, the mother of all, hath these causes shut up in her recesses: all magnetic movements are to be considered with respect to her law, position, constitution, verticity, poles, equator, horizon, meridians, centre, periphery, diameter, and to the form of her whole inward substance."[39] Book IV is concerned with magnetic "variation," which is Gilbert's term for the deviation of the compass needle from true north and so is entirely concerned with discussion of the earth and the behavior of compasses in different parts of the globe. Book V is concerned with magnetic dip, the phenomenon discovered accidentally by Robert Norman, and again is largely concerned with the behavior of compass needles in different parts of the globe. Book VI finally spells out the inescapable conclusion of the foregoing books, that the earth is a giant loadstone and that, like magnets, it is capable of spontaneous circular movement about its axis and of maintaining itself in a particular orientation.

In many respects it is the material in Book V, explaining the phenomenon of magnetic dip, which completes Gilbert's argument that the earth is a giant magnetic sphere. It is also the area where Gilbert approaches closest to the work of

[38] Zilsel, 5-6, translation from *De magnete*, differing slightly from that of Mottelay, *De magnete*, Bk. VI, ch. 4, 332.

[39] *De magnete*, Bk. III, ch. 1, 179-80; and see Freudenthal, *op. cit.*

Robert Norman and where Zilsel believed the image of Gilbert's originality was most "impaired."[40]

We have already heard Norman's account in the *Newe Attractive* as to how he came to investigate magnetic dip—the mysterious declination from the level position of the carefully balanced iron needle, after it had been magnetized. Having been advised by some learned friends to make an instrument capable of making an "exacte triall" of the extent to which the needle would decline, Norman readily put his compass-making skills to good use. Pivoting an iron pointer so that it could move in a vertical instead of a horizontal plane, against a vertical round plate "like an Astrolabe," divided exactly into 360 parts, Norman was quickly able to establish that at London the needle would dip to 71 degrees and 50 minutes below the horizon.[41]

After showing that a floating magnetized needle (buoyed up by cork) merely altered its inclination but did not move from its place, Norman concluded that the needle was orientating itself to what he called a "point Respective" and was not physically drawn to a "point attractive." The implication of this, as Norman recognized, was that the behavior of the magnetized needle could not easily be attributed to some external drawing power but must be in the needle itself (presumably bestowed upon it by the magnet).[42] As for the "point Respective" itself, Norman was only able to say that it must lie along a line declining at 71° 50' below the horizon at London. It was impossible to say precisely where along that line the respective point was until observations of magnetic dip in other parts of the globe allowed a triangulation. Nevertheless, Norman clearly states his belief that the point respective is in the body of the earth and not in heaven.[43]

In his "Epistle on the Magnet" Pierre de Maricourt had already described something very like the kind of triangulation exercise which Norman advocated, but on the surface of a magnetic terrella. In order to determine the poles of a terrella, Pierre wrote, one should allow slender iron needles to be picked up by the stone, and then draw a line along the length of the needle. After repeating this procedure a number of times, the various lines will be seen to converge in two points: "Know then that one is north and the other south." According to Pierre, however, the accuracy of this can be checked:

> In order, therefore, that you may determine one point [pole] on the stone exactly, break from the needle or iron a little piece which shall be oblong, and about as long as two finger nails and place it on the spot at

[40] Zilsel, 25.

[41] *Newe Attractive*, 9-10. In fact the text says the astrolabe-like plate should be divided into 160 parts, but this is obviously a typographical error.

[42] *Newe Attractive*, 12-15, also 18.

[43] *Newe Attractive*, 17, 18.

> which, as already stated, the point has been found, and if it stands perpendicular to the stone, there is no doubt that the place sought is there, if not move it about until it does stand perpendicular.[44]

Pierre's technique relies upon the fact that a small pointer on the surface of the magnetic sphere will point straight to the magnetic pole (always located not on the surface but at a point just below the surface of the sphere) at an appropriately sloping angle until it is directly over the pole, in which case it will point straight in toward the center of the sphere. At this point, in Norman's terms, the angle of dip is 90° from the horizontal.

We know that Gilbert read both Norman and Pierre de Maricourt, and so it is hardly surprising that we see him putting them both together in the way that I have here. Whereas Pierre was concerned only to determine the position of the pole, Gilbert used his *terrellae* and small needles to demonstrate how the dip of the needle in Norman's "Dip Instrument" can be imitated by moving small needles over the surface of a magnetic sphere (see illustration).[45] The conclusion then seems hard to deny: the dip of a compass needle, or a needle in a "dip instrument," is due to the fact that the earth itself is a giant spherical loadstone. At London the needle points directly to the north magnetic pole somewhere beneath the surface of the earth near the geographic north pole. It does so by pointing straight into the earth at an angle of 71° 50'. Take the instrument further north and the inclination will be steeper until, at the magnetic pole, it will be pointing straight down towards the center of the earth.

Norman freely admitted that he did not know why the magnetic needle pointed to the point respective, and although Pierre de Maricourt knew why his needles pointed straight down at the poles of his *terrellae*, he had no awareness of a corresponding geomagnetic dip. It took Gilbert to put the two together and to realize that the earth was a great magnet. Questions of originality are endlessly debatable, but we should be wary of allowing hindsight to distort our judgment. It seems to me that Gilbert's originality lay not just in the fact that he realized the magnetic nature of the earth but also that he saw that it could be clearly demonstrated using the experimental techniques of Pierre de Maricourt and the phenomenon of dip. Furthermore, he had the boldness to believe that this could lead to a solution to the major problem of Copernican physics—how the earth moves. Although this last aspect of Gilbert's thinking was mistaken, it was a powerful and justly influential idea, which was taken up and extended by other early

[44] Harradon, 7-8.
[45] See *De magnete*, 275-304.

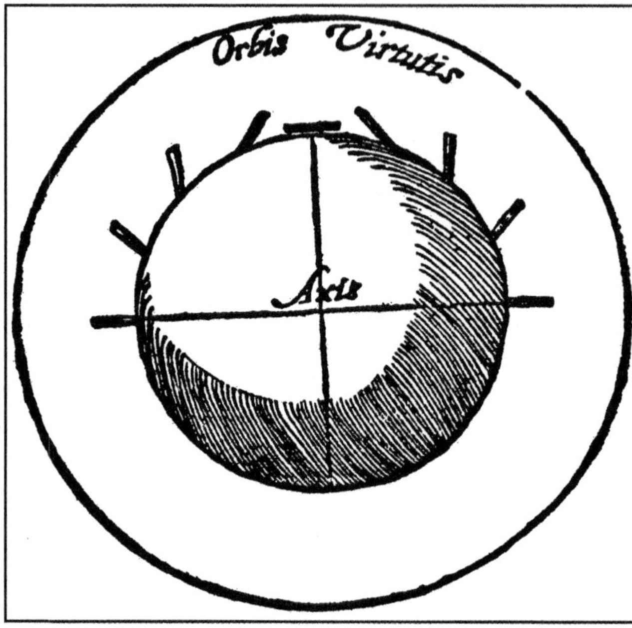

Gilbert's diagram showing how the "dip" of a small magnetized needle varies over the surface of a *terrella*, *De magnete*, Bk. V, ch. 2, 285. Since the convention of orienting maps with the north/south axis vertical was only just being established at this time (J. B. Harley and David Woodward, "Concluding Remarks," *The History of Cartography*, ed. J. B. Harley and D. Woodward [Chicago, 1987], I, 506), and there was no corresponding convention on depictions of magnets, Gilbert, like the printers of Pierre De Maricourt's *De magnete* (Augsburg, 1558), chose a horizontal depiction of the north/south axis.

Copernicans, including even Galileo and Kepler.[46] To deny Gilbert's originality is surely to misunderstand what he was trying to do.[47]

There can be no doubt of the importance of magnetic dip for Gilbert's philosophy. As he wrote in the opening words of Book V, "We come at last to that fine experiment, that wonderful movement of magnetic bodies as they dip be-

[46] See Pumfrey, *op. cit.*, and J. A. Bennett, "Magnetical Philosophy and Astronomy from Wilkins to Hooke," *Planetary Astronomy from the Renaissance to the Rise of Astrophysics, Part A: Tycho Brahe to Newton*, eds. R. Taton and C. Wilson (Cambridge, 1989), 222-30; and Johnson, *op. cit.*, 235-43.

[47] Zilsel has not been alone in denying Gilbert's originality. For J. L. Heilbron, *Electricity in the 17th and 18th Centuries: A Study of Early Modern Physics* (Berkeley, 1979), 169, Gilbert is best seen as a "moderate peripatetic and not above plagiarizing those he criticizes" (a view opposed by Freudenthal, *op. cit.*).

neath the horizon in virtue of their natural verticity; after we have mastered this, the wondrous combination, harmony and concordant interaction of the earth and the loadstone (or magnetized iron), being made manifest by our theory, stand revealed."[48] It should be noted, then, that Gilbert's primary concern was to demonstrate that the earth is a giant loadstone: "This motion [of dip] we have so illustrated and demonstrated with many experiments, and purpose in what follows so to point out the causes and reasons, that no one endowed with reason and intelligence may justly contemn, or refute, or dispute our chief magnetic principles."[49] Certainly, Gilbert took the trouble to show the pragmatic relevance of the phenomenon of dip for navigators. Unfortunately, the use of the dip instrument together with proposed maritime tables of dip merely provided an alternative way of determining one's latitude; they provided no help with longitude, the real navigational problem at this time.[50] In any case this concern is merely incidental. It is perfectly clear that the magnetic principles Gilbert mentioned referred to the nature of the earth and its inherent motions.[51]

Having established the magnetic nature of the earth by his experiments, Gilbert feels justified in claiming that the earth's ability to move itself with a circular motion "is proved by its parts, which, when separated from the whole, do not simply travel in a right line, as the Peripatetics taught, but rotate also." The natural movements of the whole and the parts are alike, Gilbert claims, "hence, since the parts move in a circle, the whole, too, hath the power of circular motion."[52] The conclusion is offered as undeniable: "The earth therefore rotates."[53] As Edward Wright declared in his "Laudatory Address," no one else "ever gave more certain demonstrations" of the motion of the earth. In 1600 it was certainly a reasonable claim to make.

I hope I have said enough to throw serious doubt upon Zilsel's claim that Gilbert took the experimental method from superior craftsmen. The question remains, where did Gilbert get the inspiration for his experimental approach?

In a very real sense I do not think that we have to look much farther than Pierre de Maricourt's "Epistle on the Magnet," published for the first time not many years before, in 1558. Pierre's epistle is short and does not contain many experiments, but all the basic techniques which Gilbert uses can be found there.

[48] *De magnete*, Bk. V, ch. 1, 275. See also Bk. VI, ch. 1, 314: "one experiment [magnetic dip] reveals plainly the grand magnetic nature of the earth..."

[49] *Ibid.*

[50] See, e.g., C. H. Cotter, *A History of Nautical Astronomy* (London, 1968).

[51] Freudenthal's analysis of Gilbert's attempts to develop a new theory of matter in support of his new physics and cosmology supports the argument that if Gilbert's main concern was the practical uses of iron and magnets, he would not need to bother himself about such niceties.

[52] *De magnete*, Bk. VI, ch. 4, 331.

[53] *De magnete*, Bk. VI, ch. 4, 333.

Moreover, Pierre's epistle is very much concerned with showing the cosmic significance of magnets and magnetism. Although Gilbert was to differ from Pierre, the Frenchman's suggestions that the loadstone "bears in itself the similitude of the heavens" and that "it is from the poles of the world that the poles of the magnet receive their virtue," can easily be seen as the kind of thinking about magnets that Gilbert might have found stimulating and inspirational.[54]

Zilsel, as we have seen, tried to present Pierre as a thinker ahead of his time, one who rejected the contemporary bias against manual labor and who, like Gilbert, paid attention to the experimenting of craftsmen. Certainly Pierre de Maricourt had a reputation for being, as even a renowned experimentalist like Roger Bacon acknowledged, "a master of experiment."[55] Pierre himself extolled the importance of the hands-on approach in his epistle:

> the investigator in this subject ... must himself be very diligent in handicraft also, in order that through the operation of this stone he may know wonderful effects. For by his carefulness he will be able to correct an error which in an age he could not possibly do by any means of his knowledge of nature and mathematics, if he lacked carefulness in use of hands.[56]

But Pierre was not writing here as a man out of his time. He was simply writing as a natural magician. "For in occult operations," Pierre explained to his correspondent, "we investigate many things by manual industry, and in general without it we are unable to bring anything to completion."[57] Indeed, Pierre wrote in the opening chapter of his letter that he might have dealt with even more occult aspects of the loadstone but that here he intended to confine himself to the magnet's more manifest properties, "on the ground that this teaching will form part of a Tract, in which we shall show how to construct physical instruments." The more occult properties Pierre had in mind required knowledge of "the art of engraving stones," by which, presumably, he meant some sort of talismanic knowledge. But he immediately went on to say that even the comparatively manifest properties of magnets which he intends to consider in this letter are only mani-

[54] Harradon, 7 and 11.

[55] Harradon, 4.

[56] Harradon, 6.

[57] Harradon, 7. See Lynn Thorndike, *History of Magic and Experimental Science* (8 vols.; New York, 1923-58), especially VI, ch. 43, 390-436; and VII, ch. 10, 272-322; S. Clark, *Thinking with Demons: The Idea of Witchcraft in Early Modern Europe* (Oxford, 1997), 214-32; "The Scientific Status of Demonology," in *Occult and Scientific Mentalities in the Renaissance*, ed. B. Vickers (Cambridge, 1984), 351-74; B. P. Copenhaver, "Astrology and Magic," in *The Cambridge History of Renaissance Philosophy*, 264-300.; and J. Henry, "Magic and Science in the Sixteenth and Seventeenth Centuries," in *A Companion to the History of Modern Science*, ed. G. N. Cantor et al. (London, 1990), 583-96.

fest to "astrologers" and "naturalists" but will be regarded by the common people as "illusions and phantasms, because they are secrets to the layman."[58]

The point is, then, that Pierre de Maricourt was a natural magician, and the use of the experimental method was a commonplace within that tradition. Although many historians of science seem to prefer to ignore what they anachronistically regard as an embarrassing aspect of science's origins, the links between magic and experimentalism have been impossible to deny ever since the appearance of Lynn Thorndike's magisterial study of the history of magic. Devoted to establishing the proposition "that magic and experimental science have been connected in their development; that magicians were perhaps the first to experiment; and that the history of both magic and experimental science can be better understood by studying them together," Thorndike's enterprise can hardly be said to have suffered from a lack of evidence.[59] Even Edgar Zilsel acknowledged the links between magic and experiment, though he mentioned it only to dismiss it. The prejudice against manual labor, he wrote, "did not prevent the experiments of the alchemists." Alchemy was made "respectable by the charm of both magic and gold," so that even well-bred people may practice it "as a hobby."[60]

There would seem to be a prima facie case for suggesting that Gilbert derived his experimental method not from elite craftsmen but from the well-known magical tradition. This is certainly the impression that Gilbert himself gives in the opening words to *De magnete* in his preface:

> Since in the discovery of secret things and in the investigation of hidden causes, stronger reasons are obtained from sure experiments and demonstrated arguments than from probable conjectures and the opinions of philosophical speculators of the common sort; therefore to the end that the noble substance of the great loadstone, our common mother (the earth), still quite unknown, and also the forces extraordinary and exalted of this globe may the better be understood, we have decided first to begin with the common stony and ferruginous matter, and magnetic bodies, and the parts of the earth that we may handle and perceive with the senses; then to proceed with plain magnetic experiments, and to penetrate to the inner parts of the earth.[61]

[58] Harradon, 6.

[59] Thorndike, *op. cit.*, I, 2; see also P. Rossi, *Francis Bacon: From Magic to Science* (London, 1968), W. Eamon, *Science and the Secrets of Nature: Books of Secrets in Medieval and Early Modern Culture* (Princeton, 1994); and T. L. Hankins and R. J. Silverman, *Instruments and the Imagination* (Princeton, 1995).

[60] Zilsel, 26.

[61] *De magnete*, "Author's Preface," xlvii.

These do not seem to be the words of a man introducing an unheard-of procedure to his scholarly readership. Just as he seemed to expect his readers to know what he meant by "speculators of the common sort," so he anticipated no difficulty in their seeing that, if you are studying occult phenomena (and magnets were always regarded as the supreme occult objects), then you use experiments. Later on in his preface Gilbert referred to his book as presenting "a new style of philosophizing," but it is clear that what Gilbert claimed as new is the content—"these foundations of magnetic science." Similarly, two paragraphs later Gilbert refers to his natural philosophy of magnetic forces as "almost a new thing, unheard-of before."[62] Nowhere does Gilbert ever make any corresponding claims about the originality of his experimental method. Given this and the fact that Robert Norman told us he was advised to proceed experimentally by some learned friends, it seems unwise to assent to Zilsel's claim that experimental science could not have come into existence before the barrier between scholars and craftsmen was demolished.[63]

There are other reasons for seeing Gilbert as a writer in the magical tradition. Thanks to the work of Gad Freudenthal we now know of Gilbert's awareness of Paracelsian alchemical theory and his eclectic use of it in developing his own mineralogical theory to support his magnetic philosophy.[64] But we do not need to rely on such penetrating scholarly excavations, for if there is a patently obvious feature of Gilbert's work apart from his experimentalism, it is his animism. For Gilbert the earth is not only a giant loadstone, it is a giant animate creature. Indeed, it is its magnetic ability to move itself which testifies to its animate nature.[65]

Rejecting the traditional Aristotelian position, in which the earth was held to be "imperfect, dead, inanimate, and subject to decay," Gilbert insisted that just as the motions of the heavenly bodies testified to their animate nature, so the magnetic motions of the earth revealed it to be animated. Gilbert even suggested that the earth's magnetic soul was in many ways superior to the human soul, which can be deceived by the senses:

> The earth's magnetic force and the formate soul or animate form of the globes, that are without senses, but without error and without the injuries of ills and diseases, exert an unending action, quick, definite, con-

[62] *De magnete*, "Author's Preface," xlix, and l. J. L. Heilbron, *op. cit.*, 169, has noted with some contempt what he calls Gilbert's advocacy of his own originality, but in no case does Gilbert refer to his experimental method in these terms.

[63] Zilsel, 28.

[64] Freudenthal, *op. cit.*, 24-28.

[65] *De magnete*, especially Bk. V, ch. 12, 308-12. See *Self-Motion, From Aristotle to Newton*, ed. M. L. Gill and J. G. Lennox (Princeton, 1994).

stant, directive, motive, imperant, harmonious, through the whole mass of matter.... Yet these movements in nature's founts are not produced by thoughts or reasonings or conjectures like human acts, which are contingent, imperfect and indeterminate, but connate in them are reason, knowledge, science, judgment....[66]

It is the very obvious combination of animism and empiricism in Gilbert's *De magnete* which has caused different historians to make conflicting claims as to his significance in the history of science. Because of the subsequent extreme divergence of experimentalism and animism in western culture, historians seem to have had difficulty in recognizing the simple fact that Gilbert worked at a time before that divergence had begun, when animistic thinking and the experimental method were both to be found in the tradition of natural magic. In the accounts of different historians, as Mary Hesse has observed, different pictures of Gilbert have been made to take shape.[67] But they are distorted pictures, caused by a failure properly to locate Gilbert's thought in its own context. Wishing to see Gilbert as a modern thinker, Zilsel was forced to accept that it was "not easy to draw the picture of Gilbert's scientific attitude correctly." Although Gilbert was "usually as critical minded as a modern experimentalist," his animism, Zilsel believed, showed "the strongly emotional background of the idea of the maternal earth."[68] Zilsel evidently could not see the possibility of critical animism, even though, in studying Gilbert, he was studying one of the greatest critical animists who ever lived.

What allows us to call Gilbert's animism "critical" is the very fact that he performed experiments in order to establish the extent and the precise nature of the magnet's animation. In so doing, Gilbert should not be said to have been applying an inappropriately "scientific" method to a figment of the "occult mentality." He was simply drawing upon the common procedures, "manual industry" as Pierre de Maricourt called it, of the tradition of natural magic for understanding "occult operations."[69]

Far from being contradictory strands in his work, Gilbert's animism and empiricism were entirely complementary. His experiments enabled him to establish, at least to his own satisfaction, that there is a magnetic soul which enables magnets to move themselves; and this newly established animism in turn allowed Gilbert to clinch his solution to the Copernican problem of how the earth moves. It is important to note, however, that, without the animistic strand to the argument, Gilbert could not satisfactorily explain the rotation of the earth. The magnetic nature of the earth will not suffice for Gilbert's purposes.

[66] *De magnete*, Bk. V, ch. 12, 311-12. See Roller, *op. cit.*, 153.
[67] Hesse, *op. cit.*, 1-2.
[68] Zilsel, 3 and 5.
[69] Harradon, 7.

It has long been recognized that Gilbert's magnetic philosophy did not allow him to make any claims at all about the annual revolution of the earth around the sun. Johnson and Freudenthal see this as the reason for Gilbert's pronouncement that he passes by "the earth's other movements," even though he was in fact a fully committed Copernican.[70] The implication of Johnson's and Freudenthal's account, however, is that Gilbert's magnetic philosophy allowed him to establish the diurnal rotation of the earth. But this is not strictly true; even his account of the diurnal motion is problematic. After all, magnets do not continually rotate on their polar axes. As Gilbert points out,

> The whole earth regards Cynosura by its steadfast nature; and similarly each true part of the earth seeks a like place in the world, and turns with circular motion to that position. The natural movements of the whole and of the parts are alike: hence, since the parts move in a circle, the whole, too, hath the power of circular motion.[71]

So, the earth has the power, like a magnet, to turn along part of a circle to face in a particular direction. But this is not the same as saying it can completely rotate once every twenty-four hours.

Gilbert lays claim to having established the diurnal rotation in what can only be described as a roundabout way. Having previously rejected the suggestion that the fixed stars continually rotate around the earth once every twenty-four hours, Gilbert declares that "Bodies that by nature move with a motion circular" must have in their parts "various metes and bounds," namely "poles that are not merely mathematical expressions, an equator that is not a mere fiction, meridians, too, and parallels." The earth fulfills this requirement, as Gilbert has shown, and therefore we know "that the earth rotates on its own poles, proved by many magnetical demonstrations to exist"—notice that it is the existence of the real (not mathematical) poles which has been demonstrated by magnetism, not the rotation.[72] To justify the claim that the earth must rotate about these poles Gilbert effectively invokes the nature-does-nothing-in-vain gambit: why else would God have created real physical poles in the globe of the earth, if not to provide a fixed axis about which it could rotate?

> For not in virtue only of its stability and its fixed permanent position does the earth possess poles and verticity; it might have had another

[70] *De magnete*, Bk. VI, ch. 3, 327. Freudenthal, *op. cit.*, 33; also, Johnson, *op. cit.*, 216. Others have preferred to deny that Gilbert was a fully-fledged Copernican (Freudenthal, *op. cit.*, 33, note 44, and Roller, *op. cit.*, 162 and 173).

[71] *De magnete*, Bk. VI, ch. 4, 331.

[72] *De magnete*, Bk. VI, ch. 4, 328.

> direction, as eastward or westward, or toward any other quarter. By the wonderful wisdom of the Creator, therefore, forces were implanted in the earth, forces primarily animate, to the end the globe might, with steadfastness, take direction, and that the poles might be opposite, so that on them, as at the extremities of an axis, the movement of diurnal rotation might be performed.[73]

In the end, then, the daily rotation of the earth was explained not by an experimentally established earthly magnetism but by the animate nature of the earth. It was that animate nature which was established by magnetic experiments.

[73] *De magnete*, Bk. VI, ch. 4, 328-29.

II

ATOMISM AND ESCHATOLOGY: CATHOLICISM AND NATURAL PHILOSOPHY IN THE INTERREGNUM

Introduction

In spite of vigorous opposition by a number of historians it has now become a commonplace that the rapid development of the 'new philosophy' sprang from the ideology of Puritanism.[1] What began its career as the 'Merton thesis'[2] has now been refined, developed, and so often repeated that it seems to be almost unassailable.[3] However, the two foremost historians in the entrenchment of this new orthodoxy are willing, in principle, to concede that 'in reality things were very mixed up',[4] and that non-Puritan natural philosophers at the time were operating 'in a precisely similar manner'[5] to their Puritan contemporaries. Indeed, it would be impossible not to concede this in the face of the many critiques launched against the Merton thesis.

Unfortunately, because of its undoubted successes in helping the historian to make sense of the period, there is now a very real danger that natural philosophers who cannot be assimilated into this picture will be ignored as irrelevant to the broader issues (whatever they may be taken to be). For example, although he has not said so in print, it would seem that Charles Webster has dismissed the relevance of the Catholic natural philosophers working with Richard Towneley in Lancashire during this period. In his brief but pioneering work on the predominantly Catholic Towneley group, Webster was able to raise his voice against the Merton thesis and to criticize Christopher Hill's *Intellectual origins of the English*

An earlier version of this paper was read at the Open University in December 1980 and I would like to thank those present for their helpful comments. In particular my thanks are due to Drs David Goodman and Colin Russell. The paper has also benefited from my discussions with a number of others whom I should now like to thank. These are: Drs David Grylls, Michael Hunter, H. W. Jones, B. C. Southgate, Simon Schaffer, and Mr Stephen Pumfrey.

[1] For some examples of this opposition see the articles aimed against Christopher Hill published in Charles Webster (ed), *The intellectual revolution of the seventeenth century*, London, 1974. These are by H. F. Kearney, 218–42 and 254–61; T. K. Rabb, 262–79 and 284–85; and B. J. Shapiro, 286–316. See also Lotte Mulligan, 'Puritans and English science: a critique of Webster', *Isis*, 1980, *71*, 456–69.

[2] R. K. Merton, *Science, technology and society in seventeenth-century England*, Hassocks and Atlantic Highlands, N.J., 1970. This is a re-issue of the 1938 paper which appeared in *Osiris*, 1938, *4*, 414–565.

[3] The major instruments of this entrenchment are Christopher Hill, *The intellectual origins of the English revolution*, Oxford, 1965; and Charles Webster, *The Great Instauration, science, medicine and reform, 1626–1660*, London, 1975.

[4] Christopher Hill, 'Puritanism, capitalism and the scientific revolution', in C. Webster, op. cit. (1), pp. 243–53, 252.

[5] C. Webster, op. cit. (3), p. 504.

revolution for taking too simplistic a view of things.[6] By the time he came to write *The Great Instauration*, however, the importance of this group was all but forgotten and Webster denied them any serious part in his considerations. Webster was now assuming that the Towneley group, along with other 'dissidents' merely turned to natural philosophy because 'for those excluded from civil office or political participation, the sciences were an unexceptionable form of recreation'.[7] The Puritan movement towards science, however, was much more interesting and important, Webster argues, because 'it can be shown that their scientific beliefs were framed with conscious reference to their religious views, each section within the Puritan sphere of influence developing a scientific outlook consistent with its doctrinal position'.[8]

In this paper I wish to take up the challenge implicit in the work of Hill and Webster and show that there was indeed a small group of thinkers who operated in a similar manner to the Puritans and of whom we could say (as Webster has of the Puritans) that they provide 'an invaluable case study' by virtue of 'their pronounced attention to scientific matters and the explicit manner in which they articulated their ideas about religion, society and the natural world'.[9]

The Blackloists were a small group of English Roman Catholic thinkers who made a bold effort to establish a new counter-reforming Catholic theology in England. They took their name from the alias, Blacklo, of their leader, Thomas White (1593–1676), a secular priest. In spite of the small size of the group and the ultimate (inevitable?) failure of all their efforts, they deserve far more attention than historians have granted them so far.[10] This is especially true in the field of history of science

[6] C. Webster, 'Richard Towneley (1629–1707), the Towneley group and seventeenth-century science', *Transactions of the Historic Society of Lancashire and Cheshire*, 1966, *118*, 51–76, on p. 75–6.

[7] C. Webster, op. cit. (3), p. 485.

[8] Ibid., p. 503–4. [9] Ibid., p. 504.

[10] The Blackloists played an important part in the history of English Catholicism but, in spite of this, they have largely been ignored by historians. Even their leader, Thomas White, is surprisingly little studied in view of his stature among his contemporaries. There are a number of brief treatments of his life and work in published sources. These include: Charles Dodd, *The Church history of England*, Brussels, 1742, iii, 285–8 and 350–6; H. W. Jones, 'Thomas White (or Blacklo), 1593–1676: new data', *Notes and queries*, 1973, *10*, 381–8; G. Anstruther, *The seminary priests: a dictionary of the secular clergy of England and Wales, 1588–1850*, Great Wakering, 1975, ii, 348–54; and the *Dictionary of national biography*. However, these should be supplemented by three unpublished theses which go a long way towards establishing the full significance of White and his followers. These are: R. I. Bradley, 'Blacklo: an essay in counter-reform', Ph.D. thesis, Columbia University, 1963; J. M. Lewis, 'Hobbes and the Blackloists: a study in the eschatology of the English revolution', Ph.D. thesis, Harvard University, 1976; and B. C. Southgate, 'The life and work of Thomas White, 1593–1676', Ph.D. thesis, University of London, 1979. I must acknowledge an intellectual debt to the last two of these in particular: I have drawn heavily upon them in this paper.

The other known members of the Blackloists are, with one exception, even more obscure: Henry Holden (1596–1662), Peter Fitton (1602–1657), E. Tyrrel (1621–76), and H. Ellis (d. 1676). There is some information about these thinkers in Anstruther's dictionary cited above. The only other member of the group to make a name for himself is also little studied: John Sergeant (1621–1707). John Sergeant was an ardent controversialist who came to the fore after the Restoration. I hope to examine the later fortunes of the Blackloists and the ideas of Sergeant in a forthcoming paper. Meanwhile, the best treatment of his thought can be found in G. H. Tavard, *The seventeenth-century tradition: a study in recusant thought*, Leiden, 1978, 219–45 (see also pp. 180–96 for a discussion of Henry Holden).

since the two main protagonists of this group, Sir Kenelm Digby and Thomas White, were extremely important in the early development of the mechanical philosophy in England. It is the intention of this paper to give a brief account of the religious and ideological concerns of Digby and White and to show how their scientific ideas were developed in order to promote their counter-reforming scheme. In so doing the paper will not simply provide another critique of Webster and Hill but will enrich and expand the prevailing historiography beyond the bounds of the 'Puritanism-and-science' thesis.

Digby, White and the mechanical philosophy

Sir Kenelm Digby and his mentor, Thomas White, are important in the history of science because they were the first Englishmen to produce a fully developed system of mechanical philosophy. Thomas White's *De mundo* published in Paris in 1642, has still not received much attention but the recent discovery and publication of Thomas Hobbes' long and detailed examination and refutation of the work goes some way to indicating its contemporary significance.[11] It is immediately clear that the *De mundo* was greatly influenced by the work of Galileo and, in its turn, profoundly influenced Digby's major contribution to natural philosophy.[12] Sir Kenelm Digby's *Two treatises*[13], which appeared in 1644, shares with Descartes' *Principles of philosophy*[14] the honour of being the first fully considered attempt at a mechanistic system. The fact that it was written in English can only have enhanced its importance among contemporary English thinkers.

Fairly detailed expositions of Digby's *Treatise on body*, the first part of the *Two treatises*, have been provided elsewhere[15] so we need not cover the same ground here. Suffice it to say that, while working almost entirely within an Aristotelian framework, Digby develops a heuristic system which has justly been described as 'impeccable' according to the precepts of the mechanical philosophy.[16] In essence, Digby develops what can only

[11] Thomas White, *De mundo dialoghi tres,* Paris, 1642. Thomas Hobbes, *Critique du* De mundo *de Thomas White,* edited by J. Jacquot and H. W. Jones, Paris, 1973. There is an English translation by H. W. Jones: Thomas Hobbes, *Thomas White's* De mundo *examined,* London, 1976.

[12] For a discussion of White's mechanical philosophy see Southgate, op. cit. (10), 346–403. On White's influence upon Digby see R. T. Petersson, *Sir Kenelm Digby: the ornament of England, 1603–1665,* London, 1956, p. 342.

[13] Sir Kenelm Digby, *Two treatises. In the one of which the nature of bodies; in the other the nature of mans soule is looked into: in way of the discovery of the immortality of reasonable soules,* Paris, 1644; reprinted in facsimile, New York and London, 1978. Hereafter cited as: Digby, *Treatises.*

[14] René Descartes, *Principia philosophiae,* Amsterdam, 1644. This can be consulted in *Oeuvres,* edited by A. Tannery, Paris, 1905, vol. viii.

[15] K. Lasswitz, *Geschichte der Atomistik vom Mittel alter bis Newton,* 2 vols., Hamburg and Leipzig, 1890, ii, 188–207; R. H. Kargon, *Atomism in England from Hariot to Newton,* Oxford, 1966, pp. 70–3; and B. J. T. Dobbs, 'Studies in the natural philosophy of Sir Kenelm Digby', *Ambix,* 1971, *18,* 1–25; 1973, *20,* 143–63; 1974, *21,* 1–28.

[16] B. J. T. Dobbs, op. cit. (15), Part I, p.6.

be described as 'Aristotelian atomism'. This may seem a peculiar hybrid but Digby himself tried to pass it off as a thoroughbred:

> Let any man read his [Aristotle's] books of Generation and Corruption, and say whether he doth not expressly teach, that mixtion (which he delivereth to be the generation or making of a mixt body) is done *per minima,* that is in our language and in one word by atomes.[17]

According to Digby's exegesis Aristotle professed that the natural qualities which arose from the combination of elements were brought about

> by the mingling of the least partes or atomes of the said Elements, which is in effect to say that all the nature of bodies their qualities and their operations, are compassed by the mingling of atomes: the showing and explicating of which hath been our labour in this whole Treatise.[18]

Unfortunately, the fact that Digby chose to develop his atomistic mechanical philosophy within the framework of Aristotelianism has tended to obscure his undoubted significance. Digby has often been regarded as a predominantly old-fashioned thinker, fighting a rear-guard action at a time when all the major natural philosophers were trying to undermine and overthrow Aristotle. Such an attitude involves a rather simplistic assessment of the so-called scientific revolution even in general terms,[19] and with regard to the revival of atomism in the early modern period it is completely misconceived. Kenelm Digby's *Treatise on body* must be seen as the culmination of the Aristotelian *minima naturalia* tradition. Some of Aristotle's early commentators, notably Averroes, interpreted certain of his pronouncements to imply the existence of *minima naturalia,* which were held to be the smallest particles of a substance which could take part in physical or chemical operations. The importance of this tradition for the revival of atomistic philosophies in the sixteenth and seventeenth centuries should not be underestimated.[20] Digby (and White) took this tradition as far as it was to go and invested it with renewed vigour. When seen from this perspective, the *Treatise on body* would seem to show that the new mechanical philosophy, in its early stages, owed as much to Aristotelianism as it did to Descartes.[21]

[17] Digby, *Treatises,* p. 343. [18] Ibid.

[19] For general indications of the vitality of Aristotelianism during the 'scientific revolution' see C. B. Schmitt, 'Towards a re-assessment of renaissance Aristotelianism', *History of Science,* 1973, *11,* 159–93. See also W. Pagel, 'The reaction to Aristotle in seventeenth-century biological thought', in E. A. Underwood (ed.), *Science, medicine and history: essays on the evolution of scientific thought and medical practice written in honour of Charles Singer,* 2 vols., London, 1953, i, 489–509.

[20] A. G. van Melsen, *From atomos to atom,* New York, 1960, pp. 41–4, 46–8 and 58–76; E. J. Dijksterhuis, *The mechanisation of the world picture,* Oxford, 1961, pp. 200–9, 277–9; and V. P. Zubov, 'Zur Geschichte des Kampfes Zwischen dem Atomismus und dem Aristotelismus im 17 Jahrhundert (*minima naturalia* und *mixtio*)', in G. Harig (ed.), *Sowjetische Beitrage zur Geschichte der Naturwissenschaft,* Berlin, 1960, 161–91.

[21] It is worth remembering in this context that F. Brandt concluded that the biggest influence on Hobbes' early mechanism was Aristotle. See F. Brandt, *Thomas Hobbes' mechanical conception of nature,* Copenhagen and London, 1928, pp. 55–72.

Even so, there is more to Digby's work than this. The *Treatise on body* is so different from the work of the other Aristotelians in the *minima* tradition as to be almost in a class of its own. Digby did not merely tack on bits of atomist doctrine in order to shore up the crumbling structure of Aristotelianism, rather he re-built the whole edifice using atomism or mechanism as his mortar. Furthermore, Digby and his philosophical partner Thomas White were well-respected members of Mersenne's circle and so were in the vanguard of philosophical developments. They were deeply influenced by the work of Galileo and Descartes, and were fully acquainted with the developing ideas of Hobbes and Gassendi.[22] All of these thinkers, as is well known, were vigorous in their rejection of Aristotle, and we must ask ourselves why Digby and White did not follow suit. The main reason for their continued adherence to Aristotle will emerge as we turn to consider the ambitious motives that lay behind their work.

The origins of the Blackloist religio-political programme

Although he was brought up as a Roman Catholic, Digby abandoned his mother church in 1630 and became an Anglican. It may be that he became disenchanted with the Catholic Church at this time because he recognized that 'Rome fetters reason', but for the most part his conversion seems to have been an understandable case of opportunism, in order to avoid the disadvantages of being Catholic at a time when Catholicism was anathema to most Englishmen.[23] However, by 1635 he had returned to the church of his birth, and shortly after this (1638) he embarked upon his career as a philosophical writer. Undoubtedly Digby reached this turning point and was given new direction by the aid and influence of the now little-remembered Catholic priest, Thomas White. Very quickly after their first meeting White's influence upon Digby became more and more profound. Digby seldom failed to recommend the philosophy of this priest who, thereby, was initiated into Mersenne's circle and attracted the interest of Digby's friends, Descartes and Hobbes. Digby repeatedly referred to White as his master, mentor, and teacher. He once wrote to White saying that his whole philosophy had been shaped by the priest, even going so far as to say

[22] For their connections with Mersenne's circle consult the index to M. Mersenne, *Correspondance*, edited by P. Tannery and C. de Waard, 13 vols., Paris, 1955–77. On the influence of Galileo see note 12 above. Indications of their close association with Hobbes have been given above in note 11. Digby discusses an atomist argument of Gassendi's which was unpublished at the time in *Two treatises* pp. 154–5. For Descartes' reactions to Digby and White consult the index of R. Descartes, *Correspondance*, edited by C. Adam and G. Milhaud, 8 vols., Paris, 1936–63.

[23] There are a number of sources for Digby's biography besides the useful *Biographia britannica* and *Dictionary of national biography* articles. These include E. W. Bligh, *Sir Kenelm Digby and his Venetia*, London, 1932; J. F. Fulton, *Sir Kenelm Digby: writer, bibliophile and protagonist of William Harvey*, New York, 1937; V. Gabrieli, *Sir Kenelm Digby: un inglese italianato nell 'eta della controriforma*, Rome, 1957; but still the fullest is R. T. Petersson, op. cit. (12). I will restrict my references on bibliographical details to this last work. On Digby's conversion to Anglicanism and return to Catholicism see idem, p. 94.

I should not lie if I said that your redeeming me out of vulgar ignorance hath been in some regard a misfortune to me, as the cure of madness was to the poor wretch who then saw his misery.[24]

The congruence between the ideas of these two men can be seen from even a cursory comparison of their published works, and yet the full significance of this intellectual partnership has yet to be fully assessed.[25]

Unfortunately it is not possible to give an exact date for their meeting, but the circumstances which brought them together are fairly evident. It is reasonable to assume that during the few years which Digby spent as an uneasy Anglican he would have been interested in ecumenical movements. During this period he was a close friend of Archbishop Laud whose 'high-flying' Anglicanism was often thought to be dangerously close to Catholicism.[26] Quite quickly after Digby's conversion to Anglicanism there was a softening in attitude towards Catholicism among many Englishmen, and there were even a number of converts to the Church of Rome. Perhaps the most surprising of these was the temporary conversion of William Chillingworth, a thinker more usually known for his vociferous opposition to Catholicism.[27] It began to seem to the Papal authorities that some sort of reconciliation between Rome and Canterbury was possible and in 1634 Pope Urban VIII sent Gregorio Panzani over to England to negotiate along ecumenical lines.[28]

In 1633 a former colleague of Thomas White at the English College in Douai, Christopher Davenport, published *Deus, Natura, Gratia*[29] which argued that no fundamental dogmas separated Anglicans and Catholics. White indicated his own ecumenism by writing a commendatory preface to the work. During Panzani's visit, White played a large role in his attempts to reconcile rival factions among English Catholic clergy. By 1636 when Panzani left England, White had emerged as a powerful figure in the secular clergy and had even been nominated as a possible candidate

[24] Letter from Digby to White, 25 February 1650, reprinted in R. Pugh, *Blacklo's cabal* (full citation given in note 91 below, p. 93.

[25] See J. M. Lewis, op. cit. (10) for the only attempt to assess this partnership so far.

[26] Nicholas Tyacke, 'Puritanism, Arminianism and counter-revolution', in Conrad Russell (ed.), *The origins of the English civil war*, London, 1973, 119–43, on p. 130, 134 and 136.

[27] William Chillingworth, *An account of what moved the author to turn Papist, with his confutation of the arguments that persuaded him thereto*, in *Works*, London, 1719. See also R. R. Orr, *Reason and authority: the thought of William Chillingworth*, London, 1967, especially chapter 1, pp. 1–22. The reasons for this temporary softening of attitude are no doubt complex and, as far as I am aware, have not received sufficient analysis from historians. No doubt the rise of Arminianism played some role; see Tyacke, op. cit. (26) and J. B. Dockery, *Christopher Davenport, friar and diplomat*, London, 1960, pp. 47–61.

[28] Gregorio Panzani, *The memoirs of G. Panzani giving an account of his agency in England in the years 1634, 1635, 1636*. Translated and edited by J. Berington, Birmingham, 1793. See especially pp. 165–6, 172, 176.

[29] Franciscus, a Sancta Clara [Christopher Davenport], *Deus, Natura, Gratia, sive tractatus de praedestinatione, de meritis & peccatorum remissione, seu de iustificatione & denique de sanctorum in vocatione, Ubi ad trutinam fidei catholicae examinatur confessio anglicana etc.*, Lyon, 1634. For a complete study of Christopher Davenport see J. B. Dockery, op. cit. (27). Chapters 4 and 5 deal with *Deus, Natura, Gratia*, pp. 62–93, and see also Appendix D, pp. 146–49. For a brief contemporary account of Davenport's work see Richard Baxter, *A key for catholics to open the juggling of the Jesuits* . . . a new edition revised by J. Allport, London, 1839, pp. 307–16.

for the Bishopric of Chalcedon—the Catholic Bishopric so-called in order not to 'prejudice any right his majesty hath to the nomination of English bishops'.[30] It seems quite likely that Digby became acquainted with White during these hopeful times for Catholics, and that White was one of the Catholic friends who, we are told by Laud, guided Digby back to his mother church in 1635.[31] Certainly, by 1637 both men were well acquainted.[32]

The ecumenical intentions of the Blackloists had their beginnings during these times, but the hoped for reconciliation between Rome and England did not materialize. Digby and White, however, despaired of nothing. From this time onwards they dedicated all their efforts to gaining toleration for English Catholics. Thomas White's most recent biographer[33] has argued forcefully that, as far as White was concerned, all his writings, in whatever field of human knowledge, were part of a unified whole: a comprehensive system of religion, politics and philosophy. When one of White's most trenchant critics objected that it was not possible to know where his philosophy ended and where his theology began, White merely replied: 'I see not how you could give a schollar a greater praise'.[34] On another occasion White declared that all his endeavours were directed towards showing the conformity of faith with 'inferior sciences'.[35] It is my contention that the one guiding principle which directed all of White's unified efforts was his belief in the importance of ecumenism.

White and his faithful follower and accomplice, Digby (along with other followers), tried to accomplish this (as we shall see) by direct political means: negotiating with members of the English Government or with Papal authorities in Rome. But what is more significant from our point of view is that they aspired to gain toleration for Catholics by making a deliberate effort to develop a reformed Catholic theology which was close to, or amenable to, English Protestant thought. Their writings, therefore, were all carefully composed with a view to removing those doctrinal differences between Catholics and English Protestants which stood in the way of reconciliation. In order to see this we must briefly examine the way their works developed from the late thirties to the early sixties.

In their earliest published works Digby and White tried to defend the

[30] John Sergeant, *An account of the Chapter erected by William, titular Bishop of Chalcedon, and ordinary in England and Scotland,* with preface and notes by W. Turnbull, London, 1853.

[31] Letter from Laud to Digby, March 27 1636. Reprinted in William Laud, *The history of the troubles and tryal of the most reverend Father in God, and blessed martyr, William Laud, Lord Archbishop of Canterbury . . .* edited by Henry Wharton, London, 1695, p. 610.

[32] In that year, or perhaps earlier, White and Chillingworth debated Catholic infallibility in Digby's lodgings. See George Digby, *Letters between the Lord George Digby and Sir Kenelm Digby kt. concerning religion,* London, 1651, p. 85. Thomas Birch in his 'Life' of Chillingworth dates this meeting in 1635, Chillingworth, *Works,* London, 1820, pp. 4–7, but Lewis, op. cit. (10) argues that it must have been between July 1637 and January 1638, p. 60.

[33] B. C. Southgate, op. cit. (10), p. 1 and *passim.*

[34] Thomas White, *Religion and reason, mutually corresponding and assisting each other,* Paris, 1660, p. 25.

[35] Thomas White, *A letter to a person of honour . . . in vindication of himself and his doctrine,* 1659 unpaginated [p. 18].

Catholic emphasis on tradition as an authority equal to the scriptures. For Catholics, the oral traditions of the early Church seemed to be every bit as reliable as the written testaments because they, so Digby and White claimed, could all be traced back to the direct teachings of Christ. While study of the scriptures alone could lead a man to the truths of religion, more often than not the difficulties of interpretation will lead to error and controversy. The proliferation of different sects within Protestantism testify to this. It is necessary, therefore, to be guided by the use of reason and by the traditional interpretations given to those controversial parts of scripture by the early members of the church. These early fathers, of course, had every advantage over the moderns in being able to determine Christ's original meanings. Those to whom Jesus preached, they argued, ought to be believed as firmly as Jesus himself. The Roman Catholic Church was held to be unique in having an unbroken line of masters and disciples which could be traced back to Christ and his original chosen disciples. This was sufficient to guarantee that no false doctrines or errors could have crept in: only what Christ himself taught has been handed down. As White put it: 'If we look into the immediate progresse and joints of the descent we cannot find where it can misse.'[36] Digby concluded that it was safe to allow oneself to be

> guided by the unanimous consent of the wisest, the learnedest, and the piousest Men of the whole world, that have bene instructed in what they believe by men of the like quality living in the age before them, and soe from age to age untill the Apostles and Christ,—and that in this manner have derived from that fountayne, both a perfect and a full knowledge of all.[37]

Now, White first developed these ideas in the context of a polemic with the Anglican apologist, Lucius Cary, Viscount Falkland (1610–43). However, White's *An answer to the Lord Faulklands discourse of infallibility* was not the last word on the matter. Falkland himself was quick to reply and his close associate, William Chillingworth (1602–44) also entered the lists by composing a refutation of White's ideas which was published posthumously.[38] We also know that Chillingworth and White debated this topic face to face in Digby's lodgings sometime between 1635 and 1637.[39] The arguments of Falkland and Chillingworth were

[36] Thomas White, *An answer to the Lord Faulklands discourse of infallibility,* reprinted in Lucius Cary, Viscount Falkland, *A discourse of infallibility, with Mr. White's answer to it, and a reply to him,* second edition, London, 1660, p. 4. White's answer was first published in 1651 but must have been written prior to 1637, see Lewis, op. cit. (10), 59.

[37] Kenelm Digby, *Conference with a Lady about choice of religion,* Paris, 1638, reprinted London, 1969, pp. 109–10.

[38] See note 36 above and: William Chillingworth, *A discourse concerning tradition (i.e. an answer to some passages in Rushworth's dialogues. Beginning at the third dialogue, 12 p. 181 . . . about traditions),* in *Works,* London, 1719. *The dialogues of William Richworth* [Rushworth] *or the judgement of common sense in the choice of religion,* Paris, 1640, were seen through the press by Thomas White. See G. H. Tavard, op. cit. (10), 158–79 and B. C. Southgate, 'A note on the authorship of Rushworth's *Dialogues*', *Notes and queries,* 1981, *28,* pp. 207–8.

[39] See note 32 above.

essentially the same. They sought to show that the doctrinal development of the Roman Church was by no means as smooth and indeflectable as White and Digby wanted their readers to believe. They were able to find a number of doctrines which separated Protestant and Catholic and which they could show had equally divided Catholic from Catholic in the past. The major doctrinal issues were the immaculate conception of the Blessed Virgin, infants receiving the Eucharist, transubstantiation, Papal infallibility and eschatological beliefs.[40] From this moment on the theology of Digby and White underwent an astonishing transformation. They recognised that the arguments of Falkland and Chillingworth were sufficient to undermine completely their own arguments about the infallibility of tradition and the historical coherence of orthodox Catholic theology. From then on White and his followers became convinced of a genuine need to reform the Catholic Church from within. White and the Blackloists now began to reject certain Catholic dogmas in the hope of making their own reformed Catholicism not only more amenable to English Protestants, but also more comfortable to the *true* faith.

When White wrote his own mechanical system of philosophy, an expansion of Digby's *Two treatises,* he showed no hesitation in denying that an accident could exist out of its own subject. As White himself pointed out he thereby denied Catholic teaching upon transubstantiation; nevertheless he was adamant:

> 'Tis answered there's neither Authority nor Demonstration in Theologie, which convinces that an Accident may be preserv'd out of a Subject . . .[41]

Similarly, he indicated that he held no brief for Catholic Mariology.[42] However, by far the most important aspect of the new theology of White and Digby was their rejection of the concept of purgatory and their development of a new Catholic eschatology.

Much recent research has shown the prevalence and importance of apocalyptical beliefs and related eschatological concepts in seventeenth-century England.[43] The work of White and Digby, completely overlooked

[40] Lucius Cary, *Reply,* p. 65, 82, 187. Another important aspect of the debate between Falkland and Chillingworth on the one hand, and White and Digby on the other, arose from the rational scepticism of the two Protestants, as opposed to the firm belief in the possibility of absolute knowledge professed by the two Catholics. I do not intend to deal with this here but see Southgate, op. cit. (10), chapters 5, 6, 9 and *passim.* For the background to the scepticism of Falkland and Chillingworth see H. R. McAdoo, *The spirit of Anglicanism: a survey of Anglican theological method in the seventeenth century,* London, 1965, chapter 1, pp. 1–23; R. H. Popkin, *The history of scepticism from Erasmus to Descartes,* revised edition, New York, 1968; and H. G. van Leeuwen, *The problem of certainty in English thought, 1630–1690,* The Hague, 1963.

[41] Thomas White, *Peripatetical institutions: in the way of . . . Sr. Kenelm Digby: the theoricall part,* London, 1656, p. 196. This work first appeared in Latin: *Institutionum peripateticarum,* Lyon, 1646. As the title indicates, White draws heavily upon Digby's *Treatise* but he expands the scope of his own work to include cosmological, meteorological and geographical phenomena.

[42] Thomas White, *Apology for Rushworth's dialogues,* Paris, 1654, pp. 64–6.

[43] Interest in this topic seems to have been inspired by the pioneering work of Norman Cohn in 1957, now re-issued as Norman Cohn, *The pursuit of the millennium: revolutionary millenarians and mystical anarchists of the middle ages,* revised and expanded edition, London, 1970. Works which concentrate on

in all of this research, is highly significant in this respect and adds greatly to our understanding of this historical development. In view of the apocalyptic excitement which we now know was stirred up in those times, it is hardly surprising that White and Digby recognized its importance in English religious thought. Furthermore, Digby's friend Thomas Hobbes, as is well known, was developing his own eschatology which emphasised the resurrection of the body and the Last Judgement.[44] And then, in those works of Falkland and Chillingworth directed against White, millenarianism appeared as one of the major examples of a piece of early Catholic tradition now rejected as heresy.

Chillingworth rightly insisted that the 'doctrine of the millenaries' was 'by the present Roman Church held false and heretical'.[45] However, bearing White's criteria in mind it ought to be part of the true faith because it was 'believed and taught by the Eminent Fathers of the Age next after the Apostles, and by none of that age opposed or condemned'. Irenaeus, for example, was said to have derived it 'from Priests which saw John, the Disciple of the Lord'.[46] Similarly, Justin Martyr is quoted as saying that those who denied the millennium also denied the Resurrection.[47] Furthermore, the doctrine of millenarianism is to be found in the Revelations to St John and was an accepted part of Christian belief within thirty years of his death, 'when in all probability there were many alive, that had heard him expound his own words and teach this doctrine'.[48] It remained part of the orthodoxy, according to Chillingworth, for two hundred and fifty years until contradicted by Dionysius Alexandrinus.[49]

Although Falkland and Chillingworth invoke millenarians in this way it should not be assumed that they were thoroughgoing millenarians themselves. The evidence tends to suggest that they were simply using these arguments to demonstrate the untenability of the Catholic claim to a theology based on the earliest traditions of the Church. In fact the recent proliferation of studies on millenarianism has rather tended to obscure the

these themes in seventeenth-century England include: J. F. Wilson, *Pulpit in Parliament: Puritanism during the English civil wars, 1640–1648,* Princeton, 1969; W. M. Lamont, *Godly rule: politics and religion 1603–1660,* London, 1969; Peter Toon (ed.), *Puritans, the millennium, and the future of Israel: Puritan eschatology 1600 to 1660,* Cambridge and London, 1970; Christopher Hill, *Antichrist in seventeenth-century England,* London, 1971; B. S. Capp, *The Fifth Monarchy men: a study in seventeenth-century English millenarianism,* London, 1972; P. Christianson, *Reformers and Babylon: English apocalyptic visions from the reformation to the eve of the civil war,* Toronto, 1978; and K. R. Firth, *The apocalyptic tradition in reformation Britain, 1530–1645,* Oxford, 1979. See also the debate between B. S. Capp and W. M. Lamont in Charles Webster (ed.), op. cit. (1), pp. 386–434. For the relevance of these ideas to the history of science see Webster, op. cit. (3), chapter 1, pp. 1–31. In this paper I hope to show that Webster's view is somewhat one-sided. See below, pp. 234–36.

[44] J. G. A. Pocock, 'Time, history and eschatology in the thought of Thomas Hobbes', in J. H. Elliot and H. G. Koenigsberger (eds.), *The diversity of history: essays in honour of Sir Herbert Butterfield,* London, 1970, 149–98, now reprinted in J. G. A. Pocock, *Politics, language and time: essays on political thought and history,* London, 1972, 148–201.

[45] Chillingworth, *Discourse concerning tradition* (full citation in note 38), p. 33.

[46] Ibid., 33, 34. [47] Ibid., 34.

[48] Ibid., 36. [49] Ibid., 37.

fact that it was regarded by most English thinkers as an unorthodox view, which could lead to heresy in religion and political upheaval in everyday life.[50] This is not to say that orthodox Anglicans denied the general resurrection and the Last Judgement, but for the most part they concentrated more on the very clear scriptual statements that the blessed would enter heaven and the damned would be punished immediately after death. This emphasis straight away raises questions about the importance of the Last Judgement—most obviously: why should it be necessary at all? Anglicans for the most part followed Calvin's teaching in the *Institutes of the Christian religion*:

> . . . as the Scripture uniformly commands us to look forward with eager expectation to the coming of Christ, and defers the crown of glory which awaits us till that period, let us be content within these limits which God prescribes to us—that the souls of pious men, after finishing their laborious warfare, depart into a state of blessed rest, where they wait with joy and pleasure for the fruition of the promised glory; and so, that all things remain in suspense till Christ appears as the Redeemer. And there is no doubt that the condition of the reprobate is the same as Jude assigns to the devils, who are confined and bound in chains till they are brought forth to the punishment to which they are doomed.[51]

The Last Judgement, therefore, is still anticipated but, in the mean time, the departed souls are kept at some sort of half-way stage. It is essential to think in such terms in order to account for Jesus's promise to the blessed thief: 'Today shalt thou be with me in paradise',[52] and for the fact that Dives, after his death, could see Lazarus comforted in 'Abraham's bosom'.[53] Jeremy Taylor, writing in 1649, explained the quandary by relying on a distinction made by St Paul between paradise and the highest heaven:

> for paradise is distinguished from the heaven of the Blessed, being itself a receptacle of holy souls, made illustrious with visitation of Angels and happy by being a repository for such spirits, who at the first day of Judgement shall go forth into eternall glory.[54]

[50] See, for example, Jeremy Taylor, *The liberty of prophesying*, London, 1648, pp. 84–5, where he denies the authority of Papias and other early chiliasts. The unorthodoxy of these ideas may also explain why Chillingworth's refutation of White remained unpublished during his life-time. See also below, pp 234—36.

[51] Jean Calvin, *Institutes of the Christian religion*, translated by John Allen, 3 vols., London, 1813. Book III, chapter 25, section 6, p. 478. It is not necessary to distinguish here between the Calvinist and Arminian or Laudian Anglicans who have been delineated by Nicholas Tyacke, op. cit. (26). Tyacke points out that Anglicanism in the early part of the seventeenth century was doctrinally Calvinist, 119–29 (see also McAdoo, op. cit. (40), pp. 21, 24–30), but this was superseded by Arminianism with the rise of the Laudian faction, after Charles I came to the throne. The Arminian rejection of Calvinist doctrine was centred on predestination, however, and it seems that Laudian Anglicans did not differ from Calvin on the issues of eschatology. This is certainly the case with Jeremy Taylor, cited below, and is also true of Henry Hammond, one of the leading Laudians. H. Hammond, *The Works*, 4 volumes edited by W. Fulman, London, 1684, i, 703; iii, 363, 855–61. On Hammond see J. W. Packer, *The transformation of Anglicanism, 1643–1660, with special reference to Henry Hammond*, Manchester, 1969.

[52] Luke, 23, 43. [53] Luke, 16, 22.

[54] Jeremy Taylor, *The great exemplar of sanctity and holy life according to the Christian institution*, London, 1649, p. 170.

From this point of view, however, it seems that the Last Judgement has been reduced to a mere formality: those in Abraham's bosom know their blessedness will be ratified,while those 'bound in chains' must know there will be no reprieve.[55] The only way that orthodox theologians could make the Last Judgement relevant was to argue that complete justice could not be meted out to the whole person, body *and* soul, without the resurrection of the body. In life the bodies of the innocents suffered as much as their souls and bodies too deserved their rewards, and conversely it was often bodily pleasures which led the wicked astray and so it is fitting that their bodies should be punished.[56]

In spite of these efforts it should still be recognized that, in general, the Day of Judgement was played down by representatives of orthodoxy and more emphasis was given to the immediacy of punishment or reward after death. This was partly in order to avoid encouraging the political and social upheavals that seemed inherent in the millenarian views, but orthodox thinkers were also determined to reject the heresy of mortalism which was becoming, from their point of view, alarmingly prevalent. Mortalism, the belief that the soul dies with the body or, in a less extreme version, remains sleeping in total oblivion until the resurrection (when it is resuscitated along with the body), was feared to have social implications as bad as those of millenarianism. If the soul died with the body then the normal religious constraints towards good morals counted for nothing, and immorality of the worst kinds would ensue. Implicit in these fears is the assumption that reward or punishment deferred indefinitely to the time of the Last Judgement would simply be ignored by most people.[57]

Be that as it may, the more radical Protestant thinkers were not satisfied with the comparatively unimportant Last Judgement of orthodoxy. For them, the bodily resurrection and the Last Judgement were the major aspects of their belief; the culmination of the history of the world.[58] The state of souls between death and the eventual Judgement Day, therefore, required more careful consideration than it was granted by conservative theologians. It was clearly for this very reason that mortalism became so prevalent.

It is only in the light of all this that we can see the ideological purpose behind the philosophical works of Digby and White. In order to reunite all Christians: Puritan, Anglican and Catholic, in a reformed Ecumenical Church, Digby and White would have to rediscover the *true* faith.

[55] The possibility of a reprieve would have to be ruled out to avoid similarities with the Catholic concept of purgatory.

[56] Jeremy Taylor, *Great examplar,* p. 171. This notion was regarded as heretical, however, in Thomas Edwards, *The third part of Gangraena, or a new and higher discovery of the errors, heresies, blasphemies, and insolent proceedings of the sectaries of these times . . .*, London, 1646, p. 8.

[57] See Norman T. Burns, *Christian mortalism from Tyndale to Milton,* Cambridge, Mass., 1972, p. 99, where he quotes, for example, Thomas More, *A dyalogue of Syr Thomas More Knyghte,* London, 1529, pp. 276, 279.

[58] Charles Webster refers to the tradition of 'providential history', op. cit. (3), 1–6.

Furthermore, they had to be able to demonstrate that it really *was* the one true faith in order to persuade all Christians to join them. In general terms they sought to do this by developing a unified philosophical theology based on what they took to be the three major sources for incontrovertible knowledge: scripture, tradition and reason. Throughout their work Digby and White return again and again to these three sources as the only infallible guides to the doctrines of the true faith, and it is this which accounts for the striking coherence of all White's works which has already been noted.[59] More specifically, what the Blackloists required was a reformed theology which would include the early traditional belief in the general resurrection and the Last Judgement, but would also avoid the mortalist heresy and prove the immortality of the human soul.

The mechanical philosophy in the service of eschatology

The first major step towards this counter-reforming theology was made in Digby's *Two treatises*. Although it is the *Treatise on body* which has attracted most attention from historians of science, for Digby the really important part of the work was its development of a comprehensive physiology of the human soul, in the second treatise. As he wrote in the 'Dedicatory epistle', he intended to study corporeal agents only in so far as 'the knowledge of them serveth to the knowledge of the soul'.[60] That the *Two treatises* were actually written with this religious purpose in mind is also clear from what Digby wrote in 1642 in his *Observations upon Religio medici*.[61] In this short work Digby insists that the immortality of the soul can be demonstrated by philosophy 'as well as faith delivereth it':

> I take the immortality of the Soule (under his [Browne's] favour) to bee of that nature that to them onely that are not versed in the wayes of proving it by reason, it is article of faith; to others it is an evident conclusion of demonstrative Science.[62]

[59] See above p. 217 and Southgate, op. cit. (10), p. 1 and *passim*. White continually emphasized the use of reason in order to get at the truths of scripture and tradition. This can be seen in the titles of two of his works: *Religion and reason mutually corresponding and assisting each other*, Paris, 1660?; and *Devotion and reason. First essay wherein modern devotion for the dead is brought to solid principles and made rational*, Paris, 1661. White, like others in the seventeenth century (notably Hobbes and Descartes) believed that rigorous reasoning could be ensured by following the axiomatic method of Euclidean geometry. There are numerous examples of this method in the work of White. Perhaps the most interesting example for the historian of science is his *Appendicula tentans solutionem problematis Toricelliani de subsistentia hydrargyri in tubo superne sigillato*, London?, 1663 (appended to *Scirri*, see note 114 below). But see also his *Euclides physicus, sive de principiis naturae, stoecheida etc* . . ., London 1657; and *Euclides metaphysicus, sive de principiis sapientiae, stoecheida etc* . . ., London, 1658. For a fuller discussion of this aspect of White's work see H. W. Jones, 'Leibniz' cosmology and Thomas White's *Euclides physicus*', *Archives internationales d'histoire des sciences*, 1975, *25*, pp. 277–303. For a general survey of this important topic see H. Schüling, *Die Geschichte der axiomatischen Methode in 16 und beginnenden 17 Jahrhundert*, Hildesheim, 1969.

[60] Digby, *Treatises*, sig. ã iv.

[61] Sir Kenelm Digby, *Observations upon Religio medici*, London, 1643. The work is dated December 23, 1642. It is, of course, a response to Sir Thomas Browne, *Religio medici*, London, 1643. See J. H. Wise, *Sir Thomas Browne's 'Religio medici' and two seventeenth-century critics*, Columbia, Missouri, 1973, pp. 57–121.

[62] Digby, *Observations*, pp. 10, 11–12.

He then gives advance notice that he has demonstrated the immortality of the soul by rational means in a long philosophical work. It would seem from this that the *Two treatises* were actually written sometime before the end of 1642. Digby also makes it abundantly clear that his philosophy can be used to refute mortalism:

> I shall observe how if hee had traced the nature of the soule from its first principles, hee could not have suspected it should sleepe in the grave till the Resurrection of the body.[63]

In order to trace the nature of the soul, however, Digby felt it 'an unavoidable necessity' to deal in an exhaustive way with the nature of bodies. Furthermore, he recognizes the need to explain all the properties of bodies in mechanical terms. The current doctrines of the schools make no real distinction between material and spiritual things whose operations are alike attributed to 'certain powers or qualities' and so, as Digby complains:

> What hope could I have, out of the actions of the soule to convince the nature of it to be incorporeall if I could give no other account of bodies operations, then that they were performed by qualities occult, specificall or incomprehensible?[64]

If, on the other hand, he could banish occult qualities and powers from the realms of physics he would be at liberty to explain spiritual phenomena in terms of occult qualities. An exhaustive account of the phenomena of bodies which relied only on the principles of the mechanical philosophy would enable Digby to show that the operations of the soul

> are such as cannot proceed from those principles, which being adequate and common to all bodies we may rest assured, that what cannot issue from them, cannot have a body for its source.[65]

The essence of Digby's thesis is to show that the soul is incorporeal and to go on from there to demonstrate its immortality. Mortality, involving change, decay, dissolution and so forth, can only apply to material entities since all change can ultimately be explained in terms of the rearrangement in space of the constituent parts of bodies. A separated or 'unbodyed' soul, therefore, must be completely incapable of change. The very last chapter of the *Two treatises* treats 'Of the perseverance of a soul in the state she findeth her selfe in, at her first separation from her

[63] Ibid., 12. For a discussion of the extent of Browne's mortalism see Burns, op. cit. (57), 150–53. There is not much to go on, other than Browne's own brief statement in *Religio medici*. See Thomas Browne, *Works*, edited by G. Keynes, London, 1928, i, 16. Digby was not alone in attributing mortalism to Browne, however. Henry Power, in a letter to Browne (February 10, 1647) regarded Browne's talk of the 'reindividualling of an incinerated plant' (cf. *Works*, i, 59) as a 'notable illustration of that psychopanuchy which antiquity so generally received, how these forms of ours may be lulled and ly asleepe after the separation untill that great and generall day . . .' See J. O. Halliwell (ed.), *A collection of letters illustrative of the progress of science in England*, London, 1841, reprinted Hamden, Conneticut, 1965, p. 92.

[64] Digby, *Treatises*, sig. õ vi.

[65] Ibid., sig. ũ.

body'.[66] Here, Digby reaches the conclusion that 'no change at all can happen to an abstracted soul'.[67] Once again, there are intimations of these ideas presented in his *Observations upon Religio medici*. For example, he tells us that 'in the state of eternity there is no succession, no change, no variety. Soules or Angells in that condition, doe not so much as change a thought'.[68] He also takes Browne to task for implying a contradiction 'in his consideration of the activity of glorified eyes (which shall be in a state of rest; whereas motion is required to seeing)'.[69]

Implicit in Digby's conclusion to the *Two treatises* is a denial of the Catholic concept of purgatory. If the soul is unchanging after death then it is ludicrous to suppose that souls can be purged of their sins before the resurrection on the Day of Judgement.[70] Digby makes it rather more explicit in his shorter *Observations:*

> to be in such a condition as maketh us understand damned soules miserable; is a necessary effect of the temper it is in, when it goeth out of the Body, and must necessarily (out of its owne nature) remain in, invariably for all eternity.[71]

It is quite evident that Digby's *Treatise on body,* the earliest fully worked-out system of mechanical philosophy in English was written to provide a philosophical basis on which to erect a new eschatology.

The details of this new eschatology were set out clearly and fully for the first time in Thomas White's *De medio animarum statu,* which was published in 1653.[72] This work marks the culmination of White's attempts to arrive at a reformed Catholicism with no fundamental doctrinal differences from English Protestantism. With this new theology he believed he could bring together the more radical Protestant sects which emphasized the apocalyptic and eschatological features of Christianity and the more conservative Anglicans who tended to deny or play down the apocalyptic aspects. By paying sufficient attention to the 'middle state' of souls, he believed he could avoid the heresy of mortalism while also placing great emphasis on the importance of the resurrection and Last Judgement. As he himself insisted: 'whether we cast our eyes on the old or new Testament, we shall find our faith founded and rooted in the resurrection.'[73]

From the outset of White's *Middle state of souls* it is clear that the

[66] Ibid., 443–45. [67] Ibid., 445.

[68] Digby, *Observations,* p. 17. [69] Ibid., 41.

[70] Unfortunately, Digby never explicitly considers how a soul can be purged when it is conjoined to the body. It is fairly clear, however, that he believes the soul to become capable of change in some way while established in the body.

[71] Digby, *Observations,* p. 13.

[72] Thomas White, *Villicationis suae de medio animarum statu ratio episcopo Chalcedonensi reddita a Thoma Anglo . . .,* Paris, 1653. This appeared in English as: *The middle state of souls from the hour of death to the day of Judgement,* 1659, and this edition has been used here. The ideas expressed in this work are re-iterated throughout White's subsequent work. See especially: *The state of the future life, and the presents order to it,* London, 1654, and *Notes on Mr. F. D.'s result of a dialogue concerning the middle state of souls,* Paris, 1660.

[73] White, *Middle state,* p. 37.

author has thoroughly absorbed the criticisms of Falkland and Chillingworth. However, this does not mean that he must rescind his views on the infallibility of tradition. On the contrary, he reaffirms it by following the paths indicated by the two Protestants. He categorically repudiates the concept of purgatory as something 'new in the Church' and so not part of true tradition.[74] Furthermore, from the stand-point of his own natural philosophy he is able to argue that the very idea of purgation of sins after death and prior to a bodily resurrection is completely untenable.

Following Digby, he argues that only corporeal entities can suffer from external corporeal agents such as fire or instruments of torture. It is 'a fiction vain and altogether impossible', therefore, to suppose that a spiritual, 'indivisible subject' like the human soul could be affected by such external agents.[75] But, if it is true that purgatory cannot purge the soul, it must also be true that the soul cannot suffer in hell-fire. It would seem that White's theology is leading him towards mortalism.[76] White manages to avoid this by introducing a concept of *internal* gladness or, when appropriate, suffering. In other words he introduces a strictly psychological interpretation of Abraham's bosom, and the torments of the soul in the interval between death and the resurrection. So, just as 'the Embrio or seminal concreation delineats the future man', similarly the whole course of a man's life will determine the state of his soul at the moment of his death. As White himself puts it: 'to have had, in the course of his whole life, these and these thoughts and affections, designes and points out, by the impressions left, the future condition of his soul'.[77] However, as soon as the soul becomes separated from the body it is no longer capable of or susceptible to change—since change is a physical phenomenon which it is nonsensical to apply to spiritual entities.[78] Therefore, the disembodied soul must remain in the state it was in at the

[74] Ibid., p. 260. [75] Ibid., p. 97.

[76] In fact, White and Digby have been accused of mortalism from time to time. Alexander Ross wrote against Digby and Richard Overton, a notorious mortalist, in the same work: *The philosophical touchstone: or observations upon Sir Kenelm Digbie's discourse of the nature of bodies and of the reasonable soul . . . and the weak fortifications of a late Amsterdam ingeneer* [Richard Overton] *patronising the soules mortality briefly slighted*, London, 1645. The fullest treatment of White's eschatology before Lewis, op. cit. (10), saw him as a mortalist: Francis Blackburne, *An historical view of the controversy concerning an intermediate state and the separate existence of the soul between the death and the general resurrection, deduced from the beginning of the Protestant reformation to the present times*. The second edition, corrected and greatly enlarged, London, 1772, p. 115. See also C. A. Patrides, *Milton and the Christian tradition*, Oxford, 1966, pp. 264–5.

[77] White, *Middle state*, p. 153. Cf. note 71 above. White's psychological interpretation of the suffering endured by the soul after death is very similar to the view expressed by the ranter Jacob Bauthumley in *The light and dark sides of God*, London, 1650. See N. T. Burns, op. cit. (57), p. 84; and Cohn, op. cit. (43), pp. 305–6. Such ideas were fairly common among the radical protestant sects according to Christopher Hill, *The world turned upside down: radical ideas during the English revolution*, Harmondsworth, 1975, pp. 175, 177. This gives further confirmation of my argument that White's theology was capable of absorbing radical ideas and integrating them with more orthodox beliefs. It should be remarked, however, that these ideas were not totally revolutionary but can be traced back to Origen and other early Fathers. See D. P. Walker, *The decline of hell: seventeenth-century discussions of eternal torment*, London, 1964, 61–3.

[78] White, *Middle state*, 103–4.

very moment of death: 'the Entity so made continueth such', White says, 'till it be, as it were, new moulded, which is the worke of the Resurrection'.[79]

The soul neither dies nor sleeps, then, but remains fully self-conscious, indefinitely, in a state of self-inflicted guilty torments or tranquil joy, until the resurrection when it is once again conjoined to the body. The whole person, body and soul, can then suffer hell-fire, or be 'admitted *locally* to Heaven'.[80] White sums it up like this:

> Being therefore by the operation of death, as it were new moulded and minted into a purely spiritual substance, he carries inseparably with him the matter of his torment in the like manner as he also doth who takes leave of the body with his affections only venially disordered. We have no occasion here to employ infernal Architects to invent strange racks and dungeons, since the innate and intimately inhering strife and fury of the affections bent against reason, perform alone that execution; which is therefore proportional to the sins, because springing and resulting from them, nor ever otherwise possibly capable to cease and determine, unless the soul by a new conjunction with the body, became again susceptible of contrary impressions.[81]

Clearly, White's carefully-thought-out natural philosophy, leading him to his new psychology of the disembodied soul, enabled him to develop a theology which he believed could resolve the exegetical problems of eschatology which divided English Protestants. Furthermore, if Catholics would accept, as he had, the arguments of Falkland, Chillingworth and other Protestants, that purgatory was not part of the true tradition of the Church, then White's theology could be accepted by all as the true faith.[82]

It should also be noticed that White's new eschatology carried with it some important corollaries which could only have helped his ecumenical designs. Keith Thomas has pointed out that the belief in ghosts—souls revisiting their haunts from purgatory—was seen as a Catholic superstition which tended to be denied by Protestants.[83] For this very reason, White was concerned to deny the reality of ghostly appearances.[84] More importantly, White's denial of purgatory undermined the validity of indulgences. The idea of buying one's way out of purgatory, to which the concept of indulgences so often amounted, was anathema to Protestants. Besides, the abuses of indulgences were still remembered as one of the major incentives to the initial Lutheran Reformation.[85]

[79] Ibid., 154. [80] Ibid., 260. [81] Ibid., 5–6.

[82] That White hoped to convince his fellow Catholics as much as Protestants is evident from his dedication of the book to Richard Smith, the Bishop of Chalcedon.

[83] Keith Thomas, *Religion and the decline of magic,* Harmondsworth, 1973, p. 703. Also Southgate, op. cit. (10), 251–2.

[84] White, *Middle state,* 196.

[85] Ibid., 186 and 205–6. Southgate, op. cit (10), 255–6. Henry Stubbe, *A censure upon certaine passages contained in the History of the Royal Society,* Oxford, 1670, 8, singles out indulgences as a 'primary occasion' of the Reformation.

Thomas White and his fellow-traveller, Kenelm Digby, it should now be said, were not such ivory-tower intellectuals that they believed their new philosophical theology would be sufficient to achieve the ecumenical ends for which they yearned. Both men, and their devoted group of followers who were mostly, like White, secular priests, were actively engaged in various political machinations all bent upon the same objective. The full details of these political manoeuvres are extremely complex and have yet to be fully charted, but the broad outlines are perfectly clear.[86] In 1639 Digby had been prevailed upon by Queen Henrietta Maria to ask English Catholics for funds which would enable King Charles to put down the Scottish rebellion.[87] This marked the beginnings of a close relationship between Digby and the Queen. In 1644 when Henrietta Maria arrived in Paris she made Digby Lord Chancellor of the Court-in-exile, and commissioned him to go to Rome to entreat for Catholic funds. This errand was potentially crucial not only to the plight of the King, now embroiled in the first civil war, but also to the future of English Catholicism. It is not surprising, therefore, that Thomas White chose to accompany his close friend on this mission to Rome.[88]

Needless to say, Digby failed in his endeavour. The Pope's conditions were ludicrously impractical, including, for example, the conversion of Charles.[89] A second and longer sojourn at Rome from 1646 to early in 1648 also produced no results. Digby was profoundly disappointed at his failure to help his King and Queen and there is good evidence to suggest that he was thoroughly disgusted with the Pope, Innocent X.[90] It seems that White too, shared this low opinion of the Papacy and as a result decided to emphasize what previously had been only implicit in his intentions. The main problem confronting English Catholics, and preventing their toleration by English Protestants, was, at its simplest, the belief that no-one could be the servant of two masters: King and Pope. White and his followers, therefore, began to promote a schism between English Catholics and the Papacy. By this time the Blackloists constituted a powerful faction in the Chapter of secular clergy.

Many of the details of their schismatical plotting have been preserved in an account evocatively entitled *Blacklo's cabal*.[91] In September 1647 the

[86] The most detailed sources so far are Bradley, op. cit. (10); Southgate, op. cit. (10); Lewis, op. cit. (10); and Petersson, op. cit. (12).

[87] Petersson, op. cit. (12), 151–2. See also Henrietta Maria, *A coppy of 1. The Letter sent by the Queenes majestie concerning the collection of the recusants mony for the Scottish warre, Apr. 17 1639. 2. The letter sent by Sir K. Digby and Mr. Montague concerning the contribution . . . etc.*, London 1641.

[88] Petersson, op. cit. (12), 215. See also V. Gabrieli, 'La missione di Sir Kenelm Digby alla corte di Innocenzo X (1645–1648)', in Mario Praz (ed.) *English Miscellany*, vol. 5, Rome, 1954, 247–88.

[89] Petersson, op. cit. (12), 216.

[90] Ibid., 218–22. See also Digby's letter to Henry Holden, 18 Nov. 1647, reprinted in R. Pugh, *Blacklo's cabal* (see following note), p. 67.

[91] R. Pugh, *Blacklo's cabal discovered in severall of their letters clearly expressing designs inhumane against*

Blackloists drew up a list of 'Instructions' in which they detailed the concessions which Catholics should make in order to gain toleration of worship. These included taking an oath of loyalty to the English Government, denying fealty to the Pope, the establishment of six or eight Bishops who were completely autonomous and concerned exclusively with spiritual matters, and finally, the Blackloists recommended the expulsion of the Jesuits, who were seen as the biggest obstacle to ecumenism.[92]

By this time, of course, the King's cause was in ruins and the Government which had to be persuaded to grant freedom of worship was that of the rebels. It is an indication of the single-mindedness of the Blackloists that in 1649 shortly after the execution of the King, Digby returned briefly to England and once again negotiated for toleration for his fellow communicants. This time, he was dealing, in secret, with Oliver Cromwell, but like his other efforts it came to nothing. As well as a denial of the temporal power of the Pope, Cromwell wanted English Catholics to provide an army of ten thousand men for the service of the Commonwealth. Neither Digby nor any other Catholic could guarantee that.[93] Even so, Digby did not despair and, as is well known, when Cromwell became Lord Protector in 1653, he returned to England and recommenced his efforts. And so, surprising though it may seem, this man who was once so ardent in his support of the King, now became something of a favourite in the eyes of the Lord Protector.[94]

Meanwhile, the philosophical work of Digby and White was beginning to win the approval of leading Protestant thinkers. When John Webster recommended the reform of the university curriculum in his *Academiarum examen* of 1653 he took for granted the importance of Digby and White:

> What shall I say of the Atomical learning revived by that noble and indefatigable person Renatus des Cartes, and since illustrated and improved by Magnenus Regius, White, Digby, Phocyllides, Holwarda and divers others.[95]

regulars, unjust against the laity, scismatical against the Pope, cruel against orthodox clergymen, and owning the nullity of the chapter, their opposition to the episcopal authority, Douay?, 1680.

[92] Ibid., 33–34 and 36–40. For an account of the Blackloist opposition to the Jesuits see R. I. Bradley, 'Blacklo and the counter-Reformation: an inquiry into the strange death of Catholic England', in C. H. Carter (ed.), *From the renaissance to the counter-reformation: essays in honour of Garret Mattingly,* London, 1966, 348–70. It is evident from Gregorio Panzani, op. cit. (28), that the Jesuits were the major stumbling block to any ecumenical movements. For surprising indications as to just how far the Blackloists were willing to go in their opposition to the Jesuits see M. V. Hay, *The Jesuits and the Popish plot,* London, 1934.

[93] Petersson, op. cit. (12), 55.

[94] Ibid., 251. See also the articles on Digby in *Biographia britannica,* and *DNB.* Digby's close association with Cromwell is perhaps not so surprising as it may seem at a superficial glance. Cromwell seems to have been more inclined to grant freedom of worship than most of his contemporaries. See R. S. Paul, *The Lord Protector: religion and politics in the life of Oliver Cromwell,* London, 1955, 327–8 and 331–3; and Dockery, op. cit. (27), 124–6.

[95] John Webster, *Academiarum examen, or the examination of academies,* London, 1653, p. 53. This is now conveniently reprinted in A. G. Debus (ed.) *Science and education in the seventeenth century: the Webster-Ward debate,* London, 1970.

It would be wrong to assume that Webster was so radical as to be unrepresentative. It has been pointed out that the works of Digby and White frequently appeared alongside those of Descartes, Gassendi and Mersenne on the shelves of Oxford College libraries just after the civil wars.[96] And Thomas Barlow of Queen's College included their works in his recommendations for *A library for younger schollers.*[97] When Webster's book was answered by the more conservative Seth Ward and John Wilkins, the mechanical philosophy of our two Catholics was spared from the vilification heaped upon that of Thomas Hobbes.[98] Indeed, Ward accused Hobbes of plagiarising from 'Des Cartes, Gassendi, S. K. Digby, and others.[99] Two years later Ward, once again, praised 'illustrious' Digby and White in the same breath as Gassendi, and described the two Englishmen as *praeclarissimi.*[100] John Wallis, the Oxford mathematician dedicated his *Commercium epistolicum* to Digby and included in it a letter fulsome in its praise of Digby and declaring that he, Wilkins and Ward were all 'deeply indebted to M. White'.[101] Ralph Bathurst, Fellow of Trinity, has also been recorded giving high praise to Digby and White.[102]

Gratifying though these praises must have been, it is clear that they did little to serve the Blackloists' great purpose. Evidently something more had to be done and in 1655 White made his boldest attempt to ingratiate the Blackloist cause to Oliver Cromwell. In that year he published a political treatise entitled *The grounds of obedience and government.*[103] This work purported to give a strictly rationalist account of the circumstances under which a government should be obeyed and when it should not. White argued that although people prefer to make their own choices about how to act, lack of time and expertise, and the complexities caused by living in large communities made it expedient to entrust certain decisions to others. This delegation of responsibility is voluntary and derives from self-

[96] Charles Webster, op. cit. (3), p. 134.

[97] A. DeJordy and H. F. Fletcher (eds.) *'A Library for younger schollers', compiled by an English scholar-priest about 1655 (Illinois studies in language and literature,* vol. 48), Urbana, 1961, p. 4; cited from Webster, op cit. (3), p. 134.

[98] Seth Ward and John Wilkins, *Vindiciae academiarum,* Oxford, 1654. This included an *Appendix concerning what M. Hobbes and M. Dell have published on this argument.* See Debus, op. cit. (95). Hobbes was attacked because of what was thought to be his atheism. The mechanical philosophy of Digby and White escaped these charges.

[99] Ward, *Appendix,* 53.

[100] Seth Ward, *In Thomae Hobbii philosophicam exercitatio epistolica,* Oxford, 1656, 93.

[101] John Wallis (ed.), *Commercium epistolicum de quaestionibus quibusdam mathematicis nuper habitum . . .* Oxford, 1658, letter to Digby, 3 Sept. 1657, pp. 12–15. I owe this and the references to Ward and Wilkins to Lewis, op. cit. (10), 217–20.

[102] Thomas Warton, *The life and literary remains of Ralph Bathurst, M.D.,* London, 1761, letter to John Sergeant, 4 Nov. 1656, pp. 165–7. See Jones, op. cit. (10), 386.

[103] Thomas White, *The grounds of obedience and government,* London, 1655, reprinted Farnborough, Hants, 1968. Previously this work has been dated as 1649, see H. W. Jones, op. cit. (10), 381–8. However, I have accepted the later date of 1655 which is proposed by G. Anstruther, op. cit. (10), ii, 351; and clearly established in B. C. Southgate, 'Thomas White's *Grounds of obedience and government:* a note on the dating of the first edition', *Notes and queries,* 1981, *28,* 208–9. For a discussion of this work and indications of its relationship to White's theology see J. A. W. Gunn, *Politics and the public interest in the seventeenth century,* London, 1969, 87–97.

interest. Men are best governed when they are themselves convinced of the rightness of the course of action outlined for them. So the governors should aim at what is best for the people.[104] Under these conditions obedience is justified and desirable, but the governors must not exceed or abuse their power. White defines government, then, as 'a power or right of directing the common affaires of a multitude, by a voluntary submission of the communities wills to the will of the Governours'.[105] If, however, the governors should cease to serve the best interests of the people then a revolution to depose that government and replace it with another is justified. Furthermore, White even declares that it is reasonable to kill the dispossessed governor if he is threatening a revenge which is not consonant to the public good.[106] For, 'when ever . . . the common good is clearly on the possessor's side then the dispossessed hath no claime'.[107] Inspired largely by the rationalist politics of Hobbes, the motivation behind this work is perfectly obvious.

Clearly, White was engaged in a blatant attempt to demonstrate to the Lord Protector that Blackloist Catholicism could be counted upon to back him if he granted toleration to this new reformed Church. It hardly needs saying that it all came to nothing. The mistrust of Catholics was evidently so deep-seated that no matter how far White travelled away from the doctrinal norms of the church of his priesthood, he still could not win toleration. What is more, making his political intentions as obvious as he did in *The grounds of obedience*, he had very definitely put all his eggs in one basket. When, in 1660, King Charles II ascended his father's throne, White's political views were, to say the least, extremely embarrassing. He had expressed himself so clearly and so forcefully that it would have availed him nothing to rescind his views on government. Digby, in spite of all, was still Chancellor and friend to Henrietta Maria but even he was compromised now by his association with White, and never really gained the new King's favours.[108] The Blackloists, as a movement, continued their campaigning but White himself retreated increasingly into the background.[109]

Conclusions and implications

We can now begin to draw some conclusions from all of this. The development and coherence of the philosophical and theological thought of Digby and White indicates that the motive force behind all their work, even their mechanist natural philosophy, was to re-establish the one true

[104] White, *Grounds*, 9. [105] Ibid., 47.
[106] Ibid., 147. [107] Ibid., 152.
[108] Petersson, op. cit. (12), 288–92.
[109] White continued to write and publish after the Restoration but he did not play an active role in the Chapter after that date, possibly because of the increasing condemnation of his work by the Holy Office (his works were censured in 1655, 1657, 1661, and 1663), and no longer seems to have involved himself directly in any of the political endeavours of the Chapter. Lewis, op. cit. (10), 230; Southgate, op. cit. (10), 10; and Anstruther, op. cit. (10), ii, 351.

faith which would be capable of re-uniting all believers. The fact that their natural philosophy was intended to provide a foundation for the establishment of their new theology did not go unrecognised by their contemporaries. One of their most eloquent critics drew up an analysis of the development of their work:

> Now as the order in which this new fabrick of Purgatory, and indeed the whole new system of phylosophy and Divinity was made publick; it was (as I take it) this: after the *Book of the immortality of the soul,* fathered on Sir Kenelme Digby; Master White appeared himself on the stage, under the name of Thomas the Englishman . . . where, in a moderate volume entituled Peripatetick Institutions . . . He discovered the great mine of this Phylosophy; here the suttleties of Logick, the secrets of nature, the hidden properties of bodies, both heaven and earthe are layed open . . .[110]

Similarly, another observer declared that Digby

> was easily wrought upon to help to bolster up and spread the Atomical philosophy, which Blacklo persuaded him would shortly prevail in the Christian world . . . which Honour they said he should have of founding this new doctrine, & divinity it self, which was to be new modeled also according to these unheard of principles . . .[111]

It may be that the modern reader will simply suppose that White and Digby were out and out opportunists, willing to sacrifice any Catholic doctrine in order to gain toleration for their fellow communicants. If this were so, however, we would have to ask ourselves why they should have bothered to retain the name of Catholic. If they were willing to sacrifice so much of the doctrine, why not sacrifice the name also? In fact, the only way to do justice to the Blackloist programme is to see it as a major development within a particular tradition of counter-Reformation theology. There were many Catholic thinkers who believed that the counter-Reformation should not be simply a negative reaction to the Protestant Reformation, seeking merely to restore the pre-Lutheran supremacy of the Roman Church. Such men envisaged the counter-Reformation as a new reformation of the Apostolic Church from within. A successful reform would rediscover and re-establish the *true* faith, which, by definition, would be unique and would once again be accepted by all Christians. If we see the Blackloists in this light it is possible to make sense of their activities.[112] They saw no contradiction in arguing for the truth of certain typically Protestant beliefs while still insisting upon the general supremacy of Roman Catholicism. After all, the Blackloists could only be sure they had recovered the true faith of primitive Christianity because it

[110] S. W. *A vindication of the doctrine contained in Pope Benedict XII, his bull and in the general councill of Florence under Eugenius IIII concerning the state of departed souls . . . Wherein the progress of Master White's lately minted purgatory is laid open and its grounds examined etc.,* Paris, 1659, 8.

[111] R. Pugh, *Blacklo's cabal,* from 'The epistle to the Catholic reader', unpaginated [p. 6].

[112] R. I. Bradley, op. cit. (92), especially pp. 351–2; Petersson, op. cit. (12), 223; Dockery, op. cit. (27), 124. White's contempt for the Jesuits (see p. 229 above) can be partially explained by their opposite views as to the meaning of counter-reform.

accorded (so they thought) with the three Roman Catholic rules of faith: scripture, tradition, and reason.

It should now be recognised why White and Digby based their natural philosophy upon Aristotelian principles. For them, Aristotelianism was one of the traditions of the Catholic Church. In spite of their close association with original thinkers like Hobbes, Descartes, and Gassendi, they wanted to eschew originality and to pretend that their advanced mechanical philosophy was all to be found in Aristotle's works. So, when Hobbes wrote that 'Aristotelity' was nothing other than 'a handmaid to the Romane Religion', he took it as a sufficient reason to reject Aristotle.[113] For Digby and White, however, it was necessary, on the same assumption, to re-emphasize Aristotle. Just as Catholic theology was always inextricably linked with Aristotelian philosophy, Digby and White would have judged it fitting, and even inevitable, that their reformed Catholicism should be inextricably linked with their reformed Aristotelianism. In both areas there was nothing arbitrary about their reforms. Their natural philosophy, as much as their theology, was guided by tradition as well as by reason.

Further evidence that White believed he had rediscovered the true faith, to which all Christians should subscribe, is afforded by his determined opposition to the scepticism of the age. The most clear example of this can be seen in his debate with the Anglican Joseph Glanvill over the *Vanity of dogmatising*.[114] For White, scepticism was the 'Mother of Infinite Errors and all Heresies, and that very seducing Philosophy and vain fallacy which the Saints warned by the Apostles, have taught us to beware of'.[115] Throughout his *Answer to the vanity of dogmatising,* therefore, White asserted that the 'Digbaean method' could solve all the problems of natural philosophy: the causes of motion, gravity, light, colours, and so on.[116] And, more importantly, it could solve all the problems of the nature of the soul.[117] White was confident in his rejection of scepticism because he believed he had found the incontrovertible truth which nowhere deviated from the scriptures, from the early traditions of the Church and its (Aristotelian) philosophy, or from the dictates of the most rigorous logic.[118]

[113] Thomas Hobbes, *Leviathan,* London, 1651, part IV, chapter 46, 370.

[114] Joseph Glanvill, *The vanity of dogmatizing: or, confidence in opinions manifested in a discourse of the shortness and uncertainty of our knowledge,* London, 1661. White replied to this in *Scirri, sive sceptices et scepticorum a jure disputationis exclusio,* London?, 1663. Glanvill defended his own position in *Scepsis scientifica: or confest ignorance the way to science . . . with a reply to the exceptions of the learned Thomas Albius,* London, 1665.

[115] Thomas White, *An exclusion of scepticks from all title to dispute: being an answer to the vanity of dogmatizing,* London, 1665, 1–2.

[116] Ibid., 52.

[117] Ibid., 30–3.

[118] One of the most interesting manifestations of White's belief that his rationalist natural philosophy was fully in agreement with scripture is to be found in his *Theological appendix, of the beginning of the world.* This appendix to his *Institutionum peripateticarum,* London, 1646 (English

The Blackloist counter-Reformation programme was an interesting development in its own right and would repay further study. Already, however, we should be able to see that one or two currently cherished historiographical positions need to be qualified in the light of this Catholic contribution to the early development of mechanism in England.

For example, although it is true to say that the Blackloist programme provides confirmation, from an unexpected quarter, of the importance of eschatological concerns during the period, it cannot be used to confirm Charles Webster's interpretation of how science and eschatology were affiliated.[119] Essentially Webster's argument is that eschatological considerations stimulated scientific development as an important means towards 'social and intellectual progress'. Eschatology, therefore, is married to Baconianism in Webster's historiography and is said to promote 'the technical, utilitarian and ethical dimensions of science'.[120] Now, Webster is only able to argue along these lines because he has concentrated on one rather special aspect of eschatology. Throughout *The Great Instauration* Webster insists that 'Puritans' were expecting an imminent millennium: a thousand year rule of the Saints *on earth* during which time society would be transformed into a utopian paradise. The Puritans, therefore, turn to Baconian utilitarian science, he claims, as a necessary first-step towards the establishment of this utopian society.

In fact, Webster's argument relies for its force on an unfortunate failure to distinguish between two eschatological traditions. The millenarianism of which Webster speaks is only a very specialised version of more general ideas which should be called 'apocalyptic'. Admittedly there were those in seventeenth-century England who looked forward to an imminent paradise on earth, where all social injustice would be eradicated for a thousand years prior to the final Judgement Day. These thinkers were millenarians in the true sense, and it may well be that the major motivation behind such beliefs was, or gave way to, a desire for *political* reform.[121] However, the label 'millenarian' is all too often applied to other thinkers who (being happier with the *status quo* perhaps) recognised the dangers inherent in such political cravings and who preferred to believe that the millennium was long past, or well under way, and even nearing its end.[122] Such thinkers, secure in their self-conscious

translation: London, 1656), seeks to show that the Biblical account in *Genesis* is in accord with the principles of mechanical philosophy as expounded by White. Clearly White pre-empted Thomas Burnet's *Sacred theory of the Earth,* London, 1681–9 (in which Burnet tried to explain the events of *Genesis* in Cartesian terms), by over thirty years. I hope to consider the significance of this work and its possible influence on Burnet in a forthcoming paper.

[119] Webster, op. cit. (3), 1–31, 505–9, 517, 520.

[120] Ibid., 507, 520.

[121] The obvious examples of such thinkers are to be found among the ranks of the Fifth Monarchists. See B. S. Capp, op. cit. (43).

[122] P. Toon (ed.), op. cit. (43), *passim,* especially 8–41 and 104–14. As a further example consider

piety, looked forward to the Day of Judgement itself, not what they would see as the more lowly aim of a political utopia. Their views are more correctly labelled as apocalyptic but amillennial.[123]

Unfortunately we do not have a census detailing which (or even how many) seventeenth-century thinkers were millennialist and which amillennialist. However, it is perfectly clear that the Blackloists were amillennialists. Their concern is only with the state of the human soul after death and after the resurrection of all the dead on the Last Day. There is absolutely no talk of a thousand year rule of the Saints on earth before the Judgement Day. So, it would be short-sighted (not to say blinkered) in the extreme to suppose that the Blackloist mechanical philosophy must have been developed in order to improve man's lot on earth, as a first step to the founding of a utopian society. On the contrary, the Blackloists were concerned with a much more rarefied objective: the establishment of the true faith. They wanted to show that the bodily resurrection, foretold in the scriptures and emphasized in the early traditions of the Church, was in keeping with the third and final rule of faith—reason. Their mechanist physics was developed, then, in order to give philosophical backing to the concept of the resurrection. Their natural philosophy affirmed that the resurrection of the body was not only philosophically sound but philosophically necessary.

Similarly the other leading figure in the early development of the mechanical philosophy in England, Thomas Hobbes, also paid great attention to eschatological concerns. In the final part of *Leviathan* Hobbes discussed the immortality of the soul and the state of affairs after the Last Judgement. There is, however, no discussion of the millennium.[124] Indeed, perhaps it should be added that the prevalence of mortalism during this period tends to suggest that a significant number of thinkers from John Donne and Thomas Browne to Richard Overton, Hobbes and

Henry Hammond, *A premonition concerning the interpretation of the Apocalypse*, in which he declares that the incidents foretold by John all happened very quickly after Christ's resurrection, and even that some of them had already occurred by the time John wrote, *The works*, London, 1684, iii, 855–61, especially 857. For a brief survey of other orthodox views see I. H. Murray, *The Puritan hope*, London, 1971, xvi–xviii, and 39–51.

[123] P. Toon (ed.), op. cit. (43), 17, distinguishes between premillennialism, postmillennialism and amillennialism. Paul Christianson, op. cit. (43), 7, however, warns that these and other distinguishing terms are in danger of causing even greater confusion. I can readily agree with this and I must express my gratitude to Dr. Colin Russell for helping me to make sense of these abstruse matters. Christianson recommends the use of the wider term 'apocalyptic' and reserves the term 'millennial' (or 'millenarian') to refer to the notion of 'a future, collective, imminent transformation of life on Earth'. B. S. Capp is always careful to distinguish between apocalypticism and millenarianism and, essentially, I am in agreement with his position. See, B. S. Capp, '*Godly rule* and English millenarianism' and 'The millennium and eschatology in England', in C. Webster (ed.), op. cit. (1), 386–98, and 427–34. There is one small point of difference between us, however. His characterisation of apocalypticism as 'pessimistic' and millenarianism as 'optimistic' (ibid., 387) is completely subjective and seems to illustrate what could be called 'unbeliever's chauvinism'. Dr Capp seems to be incapable of seeing that a godly man could look forward to the Last Day with great optimism.

[124] Hobbes, op. cit. (113), part III, ch. 38, 238–46. See also Pocock, op. cit. (44), 173–76 (in either edition).

Milton were more concerned with the Last Day than they were with any thousand year preamble.[125]

Providing we bear in mind the distinction between true millenarianism and more general apocalyptic ideas, millenarianism begins to seem less significant than Webster would have us believe. Because Webster believes that most seventeenth-century natural philosophers are simply millenarian, and therefore expecting some sort of political utopia under the rule of the Saints on earth, then it makes perfect sense for him to suppose that they were fired to develop a utilitarian science which then became an indispensable feature in the march of progress. However, if we take the more correct view that most natural philosophers in the period were anticipating the Day of Judgement when all social and political aspirations would be *suspended* once and for all, then Webster's supposition seems far less persuasive.

In the conclusion to *The Great Instauration* Webster claims that he has tried 'to explore the relevance of Puritanism to the investigation of the natural world in terms of contemporary priorities'.[126] It would seem that he has not quite lived up to this ambition. In fact he has exaggerated the role of political utopian motivations (presumably his own priorities) at the expense of more strictly religious motivations. The millenarian-utilitarian-utopian tradition which Webster has charted was undoubtedly significant, but injudicious use of the word 'millenarian' should not be allowed to obscure a separate tradition.[127] That there was a profound interdependence between apocalyptic but amillennial eschatology and the mechanical philosophy is clearly shown by the work of Digby and White. But their work is only the earliest in a tradition later adopted, or rather adapted, by thinkers like Robert Boyle, John Evelyn, Thomas Burnet and others.[128]

[125] On Browne, Hobbes, Overton and Milton see Burns, op. cit. (57); on mortalism in the thought of John Donne see his *Divine poems,* edited by H. Gardner, London, 1952, xliii–xlvi, and 114–7; and John Carey, *John Donne: life, mind and art,* London, 1980, 219–30.

[126] Webster, op. cit. (3), 505.

[127] A number of other writers have allowed themselves to be confused by failing to make the necessary distinction between apocalypticism and millenarianism. For example, J. R. Jacob, *Robert Boyle and the English revolution: a study in social and intellectual change,* New York, 1977, 125, notices that Boyle is too orthodox a thinker to desire the political revolution implied in millenarianism and yet he continues to speak of Boyle's 'millenarianism' throughout the book. Similarly, M. C. Jacob, *The Newtonians and the English revolution, 1689–1720,* Hassocks, 1976, 104 and elsewhere, seems surprised to find 'millenarianism' advocated by many orthodox thinkers. In fact there is no contradiction: the orthodox thinkers she mentions looked forward to the complete destruction of their world followed by the establishment by God of a 'new heaven and a new earth' in which they would be living eternally in heavenly bliss. This is not a political vision but a religious one and it is definitely *not* millenarianism. Burnet, therefore, was not millenarian (contrary to her statement at p. 116), nor was Hobbes (p. 104), and nor was Evelyn (p. 122). Christopher Hill, most recently in *Some intellectual consequences of the English revolution,* London, 1980, 58–61, also sees all shades of apocalyptic thought merely in terms of the 'practical politics' of millenarianism (p. 58) or in terms of a Baconian 'theory of progress' (p. 59).

[128] Robert Boyle, *Some physico-theological considerations about the possibility of the Resurrection, Works,* London, 1772, iv, 191–202; Thomas Burnet, op. cit. (118); John Evelyn left among his manuscript papers, now at Christ Church, Oxford, a tract 'Concerning the millennium' (Evelyn MS 35) in which he insisted that he spoke 'Not as the Millenaries of old' about a thousand year period on this earth, but about an *eternal* period in a 'Renewed Heaven and Earth to come' (f.2). The visions of Burnet and

The Blackloist programme also tends to confirm the belief of many historians that the mechanical philosophy and its associated corpuscularian matter theory were first introduced into the main-stream of English natural philosophy in order to promote a particular ideological purpose. However, the fact that the ideology of Digby and White was neither Puritan, Latitudinarian nor Anglican means that the prevailing historiography has to be qualified yet again. With the addition of Catholicism to these other three religious ideologies it now seems even more obvious than it did before that the mechanical philosophy could be (and was) used to provide a philosophical foundation for almost any ideological standpoint.

It follows that the recent attempt to argue exclusively for 'the Anglican origins of modern science' by establishing 'the relation of matter theory to ideology' in the work of Robert Boyle during the Restoration period is totally inadequate.[129] It can now be seen that Boyle merely had to follow the example set by Digby and White. The Blackloists had demonstrated clearly and explicitly how science could be used for ideological purposes long before the Restoration and it was all there for the taking. There can be no doubt that Boyle knew the work of these two Catholics. Digby had ingratiated himself to Samuel Hartlib in the early 1650s amd may well have entertained hopes that the Blackloist theology would hold some appeal for Hartlib's circle, who were also keen to establish a unified Church.[130]

The fact that the earliest overt attempt to use natural philosophy for ideological purposes was perpetrated by counter-reforming Catholics may also add to our understanding of the very real mistrust of the new philosophy shown by many thinkers. Among the factors which historians have already discerned in this mistrust are a reaction against the dogmatism of Descartes and Hobbes which seemed at the time to lead to materialism and atheism; and a reaction against the various kinds of natural philosophy which became associated with various radical and subversive sects in the revolutionary period.[131] No historian could deny, however, that for most seventeenth-century Englishmen Roman Catholicism presented at least as big a threat to their society as atheism or enthusiasm. The machinations of the Blackloists, therefore, might lie behind Boyle's deliberate attempt to dissociate himself from the

Evelyn are similar in important respects to that of Hobbes who might also have been included in this list (see note 124 above).

[129] J. R. Jacob and M. C. Jacob, 'The Anglican origins of modern science: the metaphysical foundations of the Whig constitution', *Isis,* 1980, *71,* 251–67; p. 251.

[130] Letter from Hartlib to Boyle, May 8, 1654, in Boyle, *Works,* vi, 87. On ecumenism in Hartlib's circle see, for example, J. M. Batten, *John Drury: advocate of Christian reunion,* Chicago, 1944. Drury was one of the leading members of the circle.

[131] M. Hunter, *Science and society in Restoration England,* Cambridge, 1981, 28, 154–5, and 162–87; H. G. van Leeuwen, op. cit. (40); and P. M. Rattansi, 'Paracelsus and the Puritan revolution', *Ambix,* 1963, *11,* 24–32.

ideological use of natural philosophy in his *The excellency of theology compared with natural philosophy*. In this work Boyle tries to assert the necessity for revealed religion and to deny the possibility of using natural philosophy to prove the truths of religion. Boyle tries to argue his case by considering an example which is very reminiscent of Digby's claims:

> If we can prove, by some intellectual operations of the rational soul, which matter, however modified, cannot reach, that it is a substance distinct from the human body, there is no reason why the dissolution of the latter should infer the destruction of the former, which is a simple substance, and as real a substance as matter itself, which yet the adversaries (Epicureans) affirm to be indestructible.[132]

Boyle denies the forcefulness of such arguments by pointing out

> that God may have so ordained, that though the soul of man, by the continuance of his ordinary and upholding concourse, may survive the body, yet . . . it shall be annihilated, when it parts with the body, God withdrawing at death that supporting influence which alone kept it from relapsing to its first nothing. Whence it may appear, that not withstanding the physical proofs of the spirituality and separableness of the human soul, we are yet much beholden to divine revelation for assuring us, that its duration shall be endless.[133]

Generally speaking Boyle's discussion is framed as a critique of Cartesianism but on this particular matter he absolves Descartes. It is Digby who is the real perpetrator of such a dogmatic philosophical theology. Descartes himself denied the possibility of proving the immortality of the soul by rational means and spoke only as a probabilist:

> And that you may not doubt of this I will give you for it his own confession as he freely writ it in a private letter to that admirable lady the princess Elizabeth . . . who seems to have desired his opinion on that important question, about which he sends her this answer, *Pour ce qui &c, i.e.* As to the state of the soul after this life my knowledge of it is far inferior to that of monsieur (he means Sir Kenelm) Digby. For, setting aside that which religion teaches us of it, I confess, that by mere natural reason we may indeed make many conjectures to our own advantage, and have fair hopes, but not any assurance.[134]

The ideological use of the new philosophy by the Blackloist Catholic faction may also have prompted, at least in part, Henry Stubbe's attacks on the Royal Society as a crypto-Catholic enclave.[135] In a series of pamphlets published in 1670 Stubbe vilified the Royal Society on a number of grounds. Considering the anti-Catholic feelings which raged throughout the Restoration period it is quite likely that the most damaging

[132] Boyle, *Works*, iv, 12. [133] Ibid., 12–13. [134] Ibid., 13.

[135] Stubbe is a complex character and the motivation that lay behind his pronouncements is often obscure. For a brief account of his attacks on the Royal Society see M. R. G. Spiller, *'Concerning natural experimental philosophie': Meric Casaubon and the Royal Society*, The Hague, 1980, 25–30; and Hunter, op. cit. (131), chapter 6, especially 136–40 and 155.

of Stubbe's accusations was that the Royal Society was bent on introducing 'a Popish implicit faith'.[136] Stubbe's major reason for this belief was spelled out in his *Campanella reviv'd.*[137] He alleged that the encouragement to study natural philosophy was a plot to distract the greatest minds while the Roman Church sought to re-establish itself.[138] It is quite clear that he associated the Blackloists with such plots because he referred explicitly to 'the doctrines of Mr. White, Dr. Holden, Serenus Cressy, and such others as endeavour at present (and that with great show of wit and artifices) to seduce the English to that Apostaticall Church'.[139] Although White was never a fellow of the Royal Society, Sir Kenelm Digby was and Stubbe made a point of denouncing him as 'the Pliny of our age for lying'.[140]

In view of the virulence of anti-Catholic feeling in England during the Interregnum and the Restoration it is hardly surprising that Digby and White failed to persuade their fellow countrymen that they had rediscovered the true faith. Indeed, the Jesuit scholar, R. I. Bradley has forcefully argued that the Blackloists unwittingly eroded what remained of the Catholic Church in England and brought it to its final collapse.[141] The final irony is the fact that no matter how far they compromised themselves in the eyes of their mother church, Digby and White were still regarded fearfully as Catholics by English Protestants. In spite of the willingness of the Blackloists to deny the secular power of the Pope, and in spite of White's explicit denial of Papal infallibility in his *Tabula suffragiales* of 1665,[142] Digby was still said to have 'pursued nothing with so much vehemence as the establishment of Popery in England',[143] and Thomas White was always regarded as 'the great Papist'.[144]

[136] Henry Stubbe, *A censure upon certain passages contained in the History of the Royal Society,* London, 1670, 40.

[137] Henry Stubbe, *Campanella reviv'd: or, an enquiry into the history of the Royal Society,* London, 1670.

[138] See Spiller, op. cit. (135), 29–30.

[139] Henry Stubbe, *A censure,* op. cit. (136), 13.

[140] Henry Stubbe, *The Plus ultra reduced to a non plus,* London, 1670, 161.

[141] R. I. Bradley, op. cit. (10) and (92).

[142] Thomas White, *Tabula suffragiales determinandis fidei ab ecclesia catholica fixae . . .,* London, 1655. White's denial of Papal infallibility was by no means as unprecedented as many of his other beliefs. In fact it was a much disputed opinion within the Roman Catholic Church at that time.

[143] *Biographia britannica* article, 194 (in the footnotes).

[144] Richard Baxter, *Reliquae Baxterianae: or, Mr. Richard Baxter's narrative of the most memorable passages of his life and times,* London, 1696, 118. For an interesting account of another group of heterodox Catholics who shared some of their theological doctrines with the Puritans but who were also dismissed as straightforward Catholics by English Protestants see: Robin Briggs, 'The Catholic Puritans: Jansenists and Rigorists in France', in D. Pennington and K. Thomas (eds.), *Puritans and revolutionaries: essays in seventeenth-century history presented to Christopher Hill,* Oxford, 1978, 333–54.

III

OCCULT QUALITIES AND THE EXPERIMENTAL PHILOSOPHY: ACTIVE PRINCIPLES IN PRE-NEWTONIAN MATTER THEORY

> *Occult* ... 3c. Applied in early science or natural philosophy to physical qualities not manifest to direct observation but discoverable only by experiment, or those whose nature was unknown or unexplained; latent; also *transferred sense* treating of such qualities, experimental. *Obsolete* except *Historical*....
>
> *Oxford English Dictionary*

Right from its inception at the beginning of the seventeenth century the mechanical philosophy was beset with problems. In part, this was an inevitable result of the hubris of the early system builders, Descartes, Gassendi, Digby and Hobbes.[1] Each of these thinkers wanted to substitute their own particular system for the moribund scholasticism in which they were trained, believing that their principles were capable of explaining all natural phenomena in a way that was far clearer and more easily intelligible than any peripatetic accounts. The easy intelligibility was held to result from the fact that the mechanical philosopher "made use of sensible principles such as are matter, Locall Motion, Magnitude, figure, situation &c. so that when hee speakes, it is possible to understand what he means".[2] As is well known, however, the attempt to explain all phenomena in terms of the rearrangement in space or change of motion of invisible corpuscles of matter soon ran into insuperable difficulties. The analogy between familiar, everyday phenomena and the microstructure of matter, which seemed simple and plausible at a superficial glance, broke down very rapidly under close scrutiny.

One of the most fundamental objections to the mechanical philosophy gave rise to a scepticism about the possibility of causal explanations based on 'communication by impulse'. Scepticism on this matter is usually associated with David Hume but it can be traced back much earlier.[3] The scholastic trained contemporary of Descartes would need to know how motion, a quality or mode of individual bodies, could be transferred from one body to another simply because their bodies came into contact. When Henry More considered the Cartesian account of the transferance of motion he felt obliged to interpret it in a quasi-vitalistic way:

> I feel more disposed to believe that motion is not communicated, but that from the impulse of one body another body is so to speak roused into motion, like the mind to a thought on this or that occasion, and that the body does not take as much motion as it needs for movement, being reminded of the matter by the other body....

More reiterated this interpretation some years later in 1655:

> ... he [Descartes] is fabricating some kind of life in that when two bodies meet, he is able to accommodate their motions so that each of them, notified by the other, the one about acceleration of its motion, the other about retardation of its motion, finally agrees on the same course of motion. And it is the same thing for the other laws of transport. For Descartes himself scarcely dares to assert that the motion in one body passes into the other .... But since no motion passes from one body into another, it is manifest that one arouses the other from sleep as it were, and in this way aroused bodies transfer themselves from place to place by their own force; which property of bodies I consider as some shadow and image of life.[4]

Underlying this and many similar objections from a number of leading philosophers was the conviction that a system of natural philosophy based solely on a concept of passive, inert matter endowed only with the 'passive power' of its inertia, could not possibly work.[5] Nevertheless, it is usually assumed that the mechanical philosophy carried on along Cartesian lines, in which explanations were couched in terms of the behaviour of totally passive particles of matter moving either inertially or as a result of the impulse of other particles of passive matter in motion, until the advent of Isaac Newton. As every historian of science 'knows', the essentially unworkable mechanical philosophy was transformed by the genius of Newton who re-introduced 'occult qualities' into natural philosophy. The triumph of the mechanical philosophy was assured once it was accepted that the system of the world was invested with unexplained active principles, forces and powers which could even operate at a distance. The brilliant stroke of incorporating 'occult qualities' into the mechanical philosophy has frequently been cited as further evidence of Newton's supreme genius. For example, in a recent essay Professor Westfall has argued that Newton's concept of attractive and repulsive forces operating between distant particles was "a major revision" or "a major philosophic innovation" which derived from "the conviction that nature cannot be reduced to the arrangement of inert particles of matter". From the late 1670s Newton moved "steadily away from the orthodox mechanical philosophy", we are told, and increasingly relied on secret principles of motion and the belief that "Nature contains foci of activity, agents whose spontaneous working produces results that cannot be accounted for by the mechanical philosophy's only category of explanation: particles of matter in motion".[6]

Efforts to trace the intellectual origins of Newton's concept of active principles have tended to concentrate either on the influence of alchemy or of Cambridge Platonism on his thinking.[7] The evidence in support of these analyses seems unassailable but they do not provide the full picture. What is in danger of being overlooked is the influence of Newton's older English contemporaries in natural philosophy. The historiographical emphasis on the radical revisionism or innovatory nature of Newton's concept of occult active principles and the concentration on non-mechanist sources of influence like alchemy tends to perpetuate the view that English mechanical philosophy before Newton followed the essentially unworkable Cartesian programme in which matter was completely inert and could only act by virtue of its 'force of motion' in collision with other parts of matter. However, once we lay aside all false preconceptions about the equivalence of Cartesianism and mechanicism and about the transcendant genius of Isaac Newton, we can see that English mechanical philosophers before Newton at the very least entertained the possibility of unexplained active principles or 'foci of activity' in matter and very frequently relied upon them quite explicitly.

In the first section of this paper, therefore, I will give a brief survey of some of the ways in which English mechanical philosophers tried to overcome or rather circumvent the perceived philosophical objections to a strict version of the mechanical philosophy (as exemplified by Cartesianism) by introducing unexplained active principles, and occult qualities into matter theory. Much of what I say will be familiar. Indeed, even Westfall, in an earlier paper, has insisted that the mechanical philosophy was "pervaded with conceptions, disguised by mechanistic veneers, that can only be explained as descendants from the Hermetic tradition". By the (perhaps unfortunate) phrase "Hermetic tradition" Westfall meant that philosophy which habitually referred to occult qualities, and he found hints of this way of philosophizing in Descartes, Boyle and John Mayow. Even so, I believe it is worth restating this in a synoptic way in order to restore the balance to a Newtonian historiography which has recently become hard to disentangle from hagiography.[8] I do not intend to attempt a systematic and definitive account of active principles in pre-Newtonian mechanical philosophy. In a sense, to do so would be to write the entire history of seventeenth century English natural philosophy. However, even this brief discussion should suffice to show that Newton's version of the mechanical philosophy was not so radically different from that of his fellow countrymen. Furthermore, we will be able to see that English mechanical philosophers provided Newton not only with many hints towards his own recension of the mechanical philosophy but also with a ready made and crucially important methodology upon which he could draw to defend his natural philosophy from the attacks of Continental Cartesians.

In the second and third sections of the paper I will try to reinforce my claim that the use of active principles in pre-Newtonian matter theory represents a clear and undeniable tradition in English mechanical philosophy and cannot be dismissed as nothing more than a series of minor aberrations by the writers involved. So, in the second section I try to suggest why active principles played such a prominent part in seventeenth century English natural philosophy. Active principles, I contend, could be used to avoid the theological difficulties which were perceived to stem from systems of passive matter moving merely inertially. If God was no longer required to supervise the running of the universal machine, the atheist might argue that God had never been required since the universe may have existed eternally. The demonstration of active principles in matter could be used to forestall such atheist gambits. At the same time, the natural philosophers' insistence that such active principles could be investigated and described by their techniques ensured that the supernaturalism of religious enthusiasts could not assert itself as the truest account of the world. On the contrary, the mechanical philosopher could present his work as the safest way to understand God's creation and, therefore, as an essential ingredient of true religion.

One of the most important aspects of the mechanical philosophers' attempts to claim that their natural philosophy provided a 'true' and objective account of the world, as a number of historians have recently shown, was the emphasis on the experimental method. The experimentalist could claim that he was revealing only 'facts' about the world, which anyone else could confirm by following the same simple procedures. His work was, therefore, largely free from the bias inherent in all efforts to explain the world along preconceived rationalist lines. The defence of the experimental method as the safe way to truth and thereby the safe way to salvation can be regarded as one of the most characteristic features of seventeenth century natural philosophy. In the third and final section of this paper I argue that the claims about the possible existence of occult active principles in matter was an important aspect of the promotion of the experimental method. The occult qualities or principles of matter, so important for understanding the true nature of God's creation, could only be evinced, it was claimed, by experimental procedures. Thus, the professed belief in occult qualities came to be amalgamated with or embedded into other arguments in defence of the experimental method. For example, I argue that the recent claims for the importance of traditional voluntarist theology in the establishment of the experimental method can be reinforced by looking at the links forged by seventeenth century mechanical philosophers between voluntarist theology and the concept of intrinsically active matter.

THE ACTIVITY OF MATTER

Without so much as a hint of braggadocio Descartes announced at the end of *Principia philosophiae* that "there is no phenomenon in nature that is not comprised in what has been explained in this treatise".[9] Not all his readers were convinced. Among those everyday phenomena which immediately presented themselves to these Cartesian doubters were gravity, magnetism, cohesion (why do the particles of bodies stay together?) and 'spring' (why do the particles of air or vapours stay apart?). However, perhaps most fundamental of all was the problem of motion. Those thinkers who accepted Descartes's rules for the transference of motion from one particle to another and his belief that the total amount of motion in the universe was always constant were simply taking it on trust: nothing was really explained. Other thinkers, therefore, preferred to join More and Leibniz in their belief that the transference of motion implied some kind of active or even vital principle in matter.[10] Besides, there were a number of physical phenomena which seemed incompatible with Descartes's claim that there could be no generation of new motions in the universe. Transference of a fixed quantity of motion from one body to another may seem easy to envisage in simple mechanical systems but it is difficult to imagine how it can explain some quite common chemical phenomena. If heat is simply a manifestation of the seething motions of invisible corpuscles then how are we to explain the supposed increase in particulate motions when two cold liquids mix to give an exothermic reaction? The explosive origin of new motions seems hard to deny when gunpowder is ignited, and surely men and animals initiate new movement merely by exercising their will?[11]

Newton's response, as we all know, was to argue that there must be occult active principles in the world to initiate and preserve motions. As he wrote in Query 31 of the *Opticks*:

> The *Vis inertiae* is a passive Principle by which Bodies persist in their Motion or Rest, receive Motion in proportion to the Force impressing it, and resist as much as they are resisted. By this Principle alone there never could have been any Motion in the World. Some other Principle was necessary for putting Bodies into Motion; and now they are in Motion, some other Principle is necessary for conserving the Motion.

Active principles, he insisted, were necessary to "conserve and recruit" motion, "For we meet with very little Motion in the World besides what is owing to these active Principles".[12] It is clear from the full context of these remarks that

Newton was setting up these ideas in opposition to Cartesianism. In so doing, however, he was merely repeating ideas which were common to other versions of the mechanical philosophy.

We can illustrate this by considering what is, as far as we know, the earliest attempt to develop a fully elaborated system of mechanical philosophy by an English thinker. Walter Warner (1570-1642/3) arrived at his mechanical philosophy by blending elements from the neo-Platonism of Medieval and Renaissance naturalists with ideas from his mentor Thomas Harriot and Galileo. He based his system on four metaphysical principles, three passive, one active. This is how he introduced his active principle:

> ... neither time and space at one, nor time and space with matter are sufficient without the position of some fourth thing for the production of all the species, motions, alterations and affects wch are actually apparent in the universe. Some such fourth thing is therefore to be admitted ... as none of these [phenomena] that are in motu (as they all are) can possibly be salved by the solitary existence of matter we must of necessity acknowledg a fourth thing as a cause of motion wch may therefore well be termed vis or power by the quality of his office.

Warner's active principle or "cause of motion" is variously referred to as *vis radiativa* or "vertue radiative" and we are told that "All bodies have in them an efficient power or vertue".[13]

Warner's system was never published although his papers are known to have influenced the thinking of a number of natural philosophers in the first half of the seventeenth century.[14] So, for a discussion of active principles in matter which is known to have been read by Newton perhaps we should turn to Walter Charleton's *Physiologia Epicura-Gassendo-Charltoniana.*[15] Charleton, closely following Gassendi, accepted the Epicurean notion that matter contained an inherent principle of motion. Needless to say, he dissented from the Epicurean belief in the eternity of the atoms and their inherent motions but otherwise the story is the same: "at their creation God invigorated or impregnated" atoms "with an Internal Energy or Faculty Motive, which may be conceived the First Cause of all Natural Actions or Motions". "Atoms are never totally deprived of their essential Faculty, Mobility", he tells us, "and from that foundation is it that all Concretions derive their Virtue Motive". Putting it another way:

> That same MOTIVE VIRTUE, therefore, wherewith every Compound Bodie is naturally endowed, must owe its origine to the innate and co-essential Mobility of its component particles....[16]

The major source for Charleton's ideas in the *Physiologia* was, of course, Pierre Gassendi whose matter theory was strongly influenced by the Paracelsian Petrus Severinus, and managed to reconcile materialism and vitalism.[17] Moreover,

Charleton embraced Gassendi's version of the new philosophy only after making his own eclectic reconciliation of atomism and Helmontianism. In the Preface to his translation of Van Helmont's *Magnetick cure of wounds* Charleton explained the curative efficacy of the "powder of sympathy" in terms of the "imperceptible Emissions steaming in a semi-immaterial thread of Atomes". He justified this claim on the grounds that

> by a general consent of all the secretaries of Nature, and the undeniable testimony of trivial Experience, every mixt Body, of an unctuous composition doth uncessantly vent, or expire a circumferential steam of invisible Atomes.

His image of all unctuous bodies incessantly "steaming" derives from his professed belief, reminiscient of Warner's, that matter is "richly endowed" with a "radial activity". This radial activity is elsewhere referred to as "the secret power of Natural Actives".[18]

Similarly, in his early essay in natural theology, *The darknes of atheism dispelled by the light of nature*, Charleton was clearly repudiating contemporary Cartesianism when he wrote that only "impardonable incircumspection or forgetfulness" could account for the belief that matter is "absolutely idle, and devoid of all motive or active virtue". For Charleton the "native Tendency", "the faculty of self-motion" inherent in atoms, was an essential element in his proof of the existence of God. The strict version of the mechanical philosophy, in which matter was entirely passive (as in the systems of Descartes and Hobbes), was held by Charleton and by others, as we shall see, to provide too much encouragement to atheism.[19]

The same general approach to the ontological problem of motion can be seen even in a minor thinker like the virtuoso Sir Matthew Hale. Hale was essentially a conservative eclectic who wanted to save the best parts of the scholastic system by giving it a mechanistic gloss. In his *Observations touching the principles of natural motions* of 1677 we can see that he too, like Warner, Charleton and Newton, adopted the premise of active principles in matter on the grounds that matter in itself is essentially passive:

> Matter it self simply considered as such, though it be susceptive of Motion (as we daily see) is not the immediate principle of Motion in those subjects that seem to be self-moving ....
>
> Therefore if there be any things in Nature, that have their Principle of Motion in themselves, we must find out if we can, somewhat besides Matter, that is the immediate root or spring of it.[20]

The analysis of a particular body in terms of substance and accidents is inadequate Hale suggested, and immediately he moved to his desired conclusion:

> There seems therefore to be a third kind of Existence or Entity participating in some respect of the nature of both [substance and accident] and yet differing in other respects from both; for indeed it is an Entity among created Beings belonging to Matter, far more noble than either of the two former, and is that which giveth Life, Vigor, Activity and Motion, immediately next under the Lord of Nature to every self-moving Being.

Like Warner, Hale dubbed this entity which belongs to matter *vis*:

> And this entity I call *Vis* or *Virtue activa*, superadded to Matter, and giving immediately those motions to it, that are specifically appropriate to that *Vis* or *Virtus activa*, and without which, Matter would be stupid, dull, unactive, and alwayes at rest in it self unless accidentally moved *ab extrinseco*.

He went on to distinguish a hierarchy of *virtutes activae* and the least noble of these, which are found in the mineral realm, he called "active qualities". Even these "movent principles", however, can be said to give "a kind of Life ... to the Subject wherein they exist". As Newton was to write in a draft query for the *Opticks* (in about 1705), "We cannot say that all nature is not alive".[21]

These three thinkers, then, Warner, Charleton and Hale, provide us with very clear examples of the belief in the necessary existence of active principles in English mechanical philosophy. Warner writing before 1630, Charleton in the 1650s and Hale in the 1670s all believed that the essentially passive nature of matter could not "save the phenomena" of everyday experience (in spite of the fact that they all knew the principle of inertia). Accordingly, they each developed, in their different ways, a concept of active principles which inhere in matter, having been superadded to it by God. The fact that three such different thinkers developed broadly similar concepts of active principles in matter should surely indicate that there may be a clear tradition of active principles in English matter theory at this time. It is the contention of this paper that there was indeed just such a tradition. Of course, the mechanical philosophy in seventeenth century England was not a monolithic entity: each natural philosopher had his own way of developing the basic enterprise of explaining everything in terms of 'matter and motion' and might cherish his own particular solution to this or that problem. Nevertheless, a constant ingredient in English mechanical philosophy was one variety or another of active principle superadded to matter. It would seem likely, therefore, that a number of the ingredients in this rich broth of ideas would provide food for Newton's thought.

Consider, for example, Henry Power's concept of active spirit. Power averred that motion "is as inseparable an attribute to Bodies, as well as Extension is" and insisted, therefore, that "there can be no rest in Nature more than a Vacuity in matter". He admitted that the origin of this motion was "one of the obscurest things in Nature" but his favoured explanation relied on an inherently active material spirit which was "the main (though invisible) Agent in all Natures three Kingdoms Mineral, Vegetal and Animal". This spirit was held to be composed of "subtle particles" and was diffused "throughout the Universe", pervading all other bodies and so producing the motions of inanimate matter and the vital activity in physiological processes.[22] It is important to note that Power's concept of an active spirit is unequivocally materialistic and, as Charles Webster has shown, it does not derive from the speculations of the Cambridge Platonist, Henry More, but is in fact influenced by Paracelsian doctrines and the speculations of his friend, Francis Glisson, on the intrinsically "energetic nature of substance".[23] Power's speculations about an active material spirit appeared in print in his *Experimental philosophy* in 1664, and so predate Newton's own alchemical musings in his "Propositions" (*c.* 1669) about a "mercurial spirit ... dispersed through all places". Similarly, we can compare Power's pervasive active "subtle particles" with the subtle active matter invoked by Newton in his "Vegetation of metals" in the early 1670s. Here Newton suggested that some very subtle and therefore pervasive kind of matter was inherently active and could activate "grosser matter":

> the immediate seat of ... operations is not the whole bulk of the matter but rather an exceeding subtle and unimaginably small portion of matter diffused through the mass, which, if it were separated, there would remain but a dead and inactive earth.[24]

Newton's subtle active matter is linked in his thoughts to the traditional biomedical concept of *flamma vitalis* and more recent theories of the role of fermentation in vital processes:

> This is the subtle spirit. This is Natures universall agent, her secret fire, ye onely ferment & principle of all vegetation. The material soule of all matter....[25]

Here again Newton's private thoughts seem to echo ideas which had already been published by iatromechanists like Thomas Willis, Richard Lower and John Mayow. Willis and Mayow developed corpuscular systems of philosophy which assumed from the outset a hierarchy from active matter down to passive matter; from spirit and sulphur (Willis) or sulphur and mercury (Mayow) to earth. Different combinations of the various kinds of particle, Willis suggested, gave rise to "Ferments" which were responsible for initiating and conserving many vital processes, and which in adverse circumstances could cause many pathological conditions. Neither Willis nor anyone else try to explain the operation of fermentation in strict mechanical terms, the motions of the "ferment" are simply assumed:

> Fermentation is an intestine motion of Particles ... the Elementary Particles being stirred up into motion, either of their own accord or Nature, or occasionally, do wonderfully move themselves, and are moved; do lay hold of, and abvolve one another....[26]

Similarly, "the cause of Fermentation" is simply included by Newton in his list of "active Principles" without which "the Bodies of the Earth, Planets, Comets, Sun and all things in them, would ... become inactive masses".[27]

Newton's indebtedness to such biomedical ideas was very nearly made public in his "Hypothesis of light" sent to be read at the Royal Society in 1675. Some time after sending his speculative "Hypothesis" to Oldenburg he wrote to ask the secretary of the Royal Society to strike out the suggestion that "the forms of nature may be nothing but aether condensed by a fermental principle". Nevertheless, the "Hypothesis" as it appeared in print spoke freely of "the vital aereal spirit, requisite for the conservation of flame and vital motions", and of a "humid active matter". The "Hypothesis of light" is perhaps closest in its overall impression to John Mayow's philosophy of active 'nitro-aerial' particles. Indeed, the similarities are so striking that even Professor Westfall has remarked upon them. For Westfall, however, Mayow's introduction of occult active principles into the mechanical philosophy was a self-delusion because he "did not explore the apparent contradiction of an active principle in the context of a mechanical philosophy of nature". However, it should be clear from what has been said here that this "contradiction" may be apparent to Westfall but it would not have been to an English contemporary of Mayow's. The mechanical philosophy, as it was presented by English natural philosophers, had a strong tradition of active principles.[28]

Robert Boyle who admitted to being "so shy of laying down principles, and sometimes of so much as venturing at explications" was usually non-committal about the theoretical presuppositions underlying his work. Nevertheless, here and there even he acknowledged the need to call upon active principles. A significant example of this is provided in a letter which Boyle wrote to Hartlib sometime before 1660 but which he only chose to publish right at the end of his life in *The general history of the air* (which, in the event, appeared posthumously in 1692). Like Power, Willis and Mayow, Boyle here assumed the existence of "aerious, etherial, luminous" spirits in "all mixed bodies". These spirits are material and "the only principles of energy, power, force and life, in all bodies wherein they are, and the immediate causes through which all alteration comes to the bodies themselves".[29] In his *Suspicions about some hidden qualities in the air* (1674) Boyle talked of "a little vital quintessence" or "some vital substance" in the air. On more than one occasion he speculated that this vital factor, rather than being reducible to the "mechanical affections" of matter, may be "of solar, or astral or some other exotic nature".[30] If Boyle was careful not to use the phrase "occult qualities" in his title to this work he made it clear enough in his preface that they were what concerned him:

> I began some years ago to set down a collection of some new or less heeded observations and experiments relating to the causes and effects of changes in the air, which I referr'd to several heads, ... the last of which was of the occult qualities of the air, supposing there be any such.

While the material gathered by Boyle under some of the other "heads" had to remain unsorted and unpublished, he felt it was "unfit, that anything should perish, that related to so considerable and uncommon a subject, as that of this title".[31]

In his *Cosmicall qualities of things* (1671) Boyle suggested that there may be "a new set of faculties (or powers) and dispositions" which "depend upon some unheeded relations and impressions which these bodies owe to the determinate fabrick of the grand system or world they are part of". It is possible that this new set of faculties may be "saved" to give a strictly mechanistic reading although it seems to me to hint to rather more "exotic natures". Certainly, the appendix to this work provides us with speculations which cannot be reconciled to strict mechanism. In "Cosmicall suspicions" Boyle admitted that some natural phenomena may not be explicable in terms of the "mechanical qualities" of size, shape and motion because they are caused by "particular sorts of corpuscles that have yet no distinct name, which may discover peculiar faculties and ways of working".[32] Elsewhere, he proposed that various "saline substances" may be produced by "some internal thing, that is analogous to a Seminal Principle", and that precious stones may have an "expansive endeavour" or "some Principle of growth" inherent in them.[33]

Even when Boyle confined his explanatory efforts to the familiar principles of matter in motion he did not subscribe to a strictly mechanistic account in which there is no new motion but only redistribution of motion by impact. When Boyle considered the ways in which a very small motion may "tricker" major physical effects we can safely assume that he was casting doubt on the Cartesian notion of simple transference. Boyle's "wonder" at the shaking of the earth and the firm buildings upon it simply at the passing of a cart, for example, made him suspect that this little stimulus caused the earth "to tremble it self".[34] Similarly, Boyle's account of an experiment in *Certain physiological essays* (1661) prompted a Cartesian like Spinoza to object that Boyle seemed "to wish to show ... that the particles of salt carry themselves into the air by their own impulse". Boyle merely replied that he had "made use of the Epicurean principles, which have it that motion is innate in the particles". In so doing Boyle, like Charleton, Willis and all others of a Gassendist persuasion, laid himself open to the charge that he was relying upon an occult quality in matter which outdid "the most credulous Peripatetics". For, as Edward Stillingfleet objected, in his *Origines sacrae* of 1662, to say that "Atoms move *because it is their nature to move*" is

> least of all pardonable in the exploders of substantial forms and occult qualities when the Origine of the whole world is resolved into an occult quality which gives motion to Atoms.[35]

If Boyle ever got 'steamed up' about such charges he might have repeated his words in *The excellency and grounds of the corpuscular or mechanical philosophy* (1674): "I see not, why a philosopher should think it impossible, to make out, ... the mechanical possibility of any corporeal agent, how subtle, or diffused, or active soever it be ...." After all, the *Materia Subtilis* of the Cartesians is, to all intents and purposes, no less active "than the universal spirit of some spagyrists, not to say, the *Anima Mundi* of the Platonists". Accordingly, we frequently see Boyle assuming, for heuristic reasons, that bodies are composed of corpuscles with an inherent vibratory or oscillatory mode. This enabled him to say, for example, that bodies may "owe their elasticity not so much to their structure as their [intestine] motion". The expansion of air upon application of heat, then, takes place because

> heat, that is a kind of motion may make the agitated particles strive to recede further and further from the centres of their motions, and to beat off those that would hinder the freedom of their gyrations, and so very much add to the endeavour of such air to expand itself.

Newton adopted the same idea in his "Hypothesis of light" when he proposed that "the aether is a vibrating medium like air" but "more strongly elastic".[36]

The natural philosopher to make the best use of the concept of incessantly and innately vibrating matter, however, was Robert Hooke. Like Boyle he did not try to explain the vibrations of matter as the result of a series of impacts, he simply made it an *ad hoc* feature of his world system:

> Now that the parts of all bodies, though never so solid do yet vibrate, [Hooke wrote], I think we need go no further for proof, then that all bodies have some degrees of heat in them, and there has not been yet found any thing *perfectly cold*: Nor can I believe indeed that there is any such thing in Nature, as a body whose particles are at rest, or *lazy* and *inactive* in the great *Theatre* of the *World*, it being quite contrary to the grand *Oeconomy* of the Universe.[37]

In his cosmological speculations Hooke used such inherent vibrations of matter (specifically, of the interplanetary aether) to explain light and gravity which he regarded as "the most considerable and most active principles in Nature". The fact that he explained these "active principles" in terms of mechanical vibrations in a universal aether was hardly likely to impress those who demanded a system as rigorously mechanical as Descartes's. In answer to the question, why was the aether incessantly vibrating? Hooke simply reverted to his concept of unexplained vibrating matter:

> Suppose then there is in the Ball of Earth such a Motion, as I, for distinction sake, will call a *Globular Motion*, whereby all the parts thereof have a Vibration towards and fromwards the Center, or of Expansion and Contraction; and this vibrative Motion is very short and very quick, as it is in all very hard and very compact Bodies.[38]

Perhaps we would be closer to Hooke's way of thinking if we spoke of 'pulsations' rather than 'vibrations'. Henry More, writing in 1651, asked the Platonic enthusiast Thomas Vaughan to show him the heart whose systole and diastole produced "the pulse of the world". It would seem that he might have been justified in asking Robert Hooke the same question.[39]

Marie Boas Hall, in her "Establishment of the mechanical philosophy", characterized the new philosophy in terms of the banishment of "sympathies and antipathies, congruities and incongruities, attraction and hostility".[40] And yet we do not have to proceed very far into Hooke's first major work, *Micrographia*, in order to find him explaining a phenomenon by reference to "congruity" and "incongruity". This obvious use of an occult quality in the work of the supreme experimental philosopher has not shaken the prevailing historiography of the mechanical philosophy represented here by Marie Boas Hall. Historians have been quick to point out that Hooke provided a *mechanical* explanation of congruity and incongruity. The fact that this explanation once again relied on the concept of incessantly vibrating matter does not seem to have been registered. Besides, Hooke himself tells us that his experimental demonstration of congruity and incongruity (showing the spontaneous separation of sand from fine gravel when placed on a "nimbly beaten" drumhead) did "not come up to the highest property of Congruity, which is a Cohaesion of the parts of the fluid together, or a kind of attraction and tenacity". Similarly, the highest property of incongruity would, presumably, be a kind of repulsion.[41]

This cannot be dismissed as a trivial failure on Hooke's part to give a satisfactory mechanical explanation of a particularly difficult problem. He went on to develop the concept of incongruity as "a coefficient in the most considerable Operations of Nature". This entered most dramatically into his speculations when he considered the existence of an all pervasive and universally incongruous aether as an explanation of gravity. Hooke asked

> whether the Phaenomena of gravity might not by this means be explained, by supposing the Globe of Earth, Water, and Air to be included with a fluid, heterogeneous to all and each of them, so subtil, as not only to be every where interspersed through the Air (or rather the air through it) but

> to pervade the bodies of Glass, and even the closest Metals, by which means it may endeavour to detrude all earthly bodies as far from it as it can; and partly thereby, and partly by other of its properties may move them towards the Center of the Earth.[42]

This universally incongruous subtle fluid may well have provided a model for Newton's aether speculations. Newton postulated an aether which was "much rarer" within dense bodies than in seemingly empty space. The resulting density gradients in the universe could then be invoked to explain gravitational attraction: "every Body endeavouring to go from the denser parts of the Medium towards the rarer." In earlier aether speculations Newton had even written of "a certain secret principle in nature, by which liquors are sociable to some things, and unsociable to others".[43]

Hooke's universally incongruous aether with its occult power of detrusion or repulsion may seem, at first glance, to be no less remarkable a flight of fancy than Newton's own "non-mechanical" aether.[44] However, we do not have to search too far for a likely source of inspiration. Robert Hooke had a very clear and very familiar exemplar of a fluid whose parts seemed to detrude one another to a remarkable extent. Investigation of this fluid and its capacity for expansion was to engage the intellects of some of the leading mechanical philosophers in England: Hooke himself, Robert Boyle, Henry Power, Sir William Petty and John Mayow.

In 1647 Roberval had shown that a carp's bladder, sealed in a deflated or collapsed state, would expand dramatically, becoming taut as though ready to burst, when placed in a Torricellian vacuum.[45] So began a major effort to understand the mysteries of the 'spring' of the air. Boyle and Hooke were able to manifest many similar examples of this springiness by means of their air-pump. In all cases it seemed that the air was behaving exactly like the material spirits invoked by iatromechanists like Charleton and Willis with their "endeavour to expand themselves, and flye away".[46] The problem for a corpuscularist, like Boyle, was to determine

> how, since the corpuscles of the air are acknowledged to be heavy, and those that remain [after operation of the pump] must be so wonderful thinly dispersed in the cavity of the receiver, they come to be supported and kept, as it were, swimming therein, and do not appear to subside by their own weight, the *Materia subtilis* (though the presence of that should be admitted) not appearing to have gravity wherewith to sustain them; and the vacuum (if that be supposed wherever the aerial particles are not) being too near a-kin to nothing to be able to oppose their descent....[47]

This was essentially a very old problem. Atomist or corpuscularist attempts to explain material properties like weight or density in terms of the packing, or distribution in space of the constituent particles of matter had been rejected by Aristotle on the grounds that it would require what he called "self-determined voids".[48] What Aristotle meant by that is, perhaps, most strikingly illustrated by the contemporary Aristotelian, Sir Kenelm Digby, writing in 1644:

> If such vacuities [between particles] were the cause of rarity, it would follow that fluide bodies being rarer than solid ones, they would be of themselves standing, like nettes or cobwebbes: whereas contrariwise, we see their natures are to runne together, and to fill up every little creek and corner: which effect, following out of the very nature of the things themselves, must needes exclude vacuities out of their nature.[49]

The initial response to such objections was simply to allow that matter *was* structured like "nettes or cobwebbes". Boyle imagined particles of air like tiny springs or like strands of wool which were usually compressed by gravity but could expand when pressure was reduced. Similarly, Newton, even as late as the 1690s, was willing to concede that there must be "a certain wonderful and exceedingly artificial texture of the particles of bodies ... like networks". Elsewhere he wrote of net-like figures, textures and particles.[50] However, as we know from his manuscript fragment entitled *De aere et aethere*, written about 1679, Newton had already conceived of the possibility that "air is composed of the particles of bodies torn away from contact, and repelling each other with a certain large force". Newton's "Suspicions about the cause of this repulsion" were that "God may have created a certain incorporeal nature which seeks to repel bodies" or "that corpuscles may be surrounded by a sphere of most fluid and tenuous matter which admits other bodies into it with difficulty" – what Hooke would call an incongruous fluid.[51]

Professor Westfall has seen these "Suspicions" as "a major revision in the prevailing mechanical philosophy". For Westfall

> The significance of the redirection of his philosophy of nature undertaken about 1679 can scarcely be overstated. What he proposed was an addition to the ontology of nature. Where the orthodox mechanical philosophy of seventeenth-century science insisted that physical reality consists solely of material particles in motion, characterised by size, shape and solidity alone. Newton now added forces of attraction and repulsion, considered as properties of such particles, to the catalogue of nature's ontology.[52]

However, within the context of English mechanical philosophy, Newton was not making such a radical innovation. In fact, the *structural* explanations of the spring of the air, in terms of the extension and compression of long spring-like particles was very quickly dismissed as unsatisfactory by English mechanical philosophers. The expansion of the air due to the uncoiling or straightening out of compressed particles seemed increasingly implausible when experiments with the air-pump showed that the air's "Elasticity is of a vast, and almost unlimited nature, that scarce hath any bounds to it".[53]

One reaction to this, as we have seen with Robert Boyle, was to try to explain the spring of the air in terms of increased vibrations of its particles. However, such increased frequencies and amplitudes of vibration were themselves utterly mysterious since it was recognized that air could and did expand itself without the help of any stimulatory input of other rapidly moving particles, such as particles of fire. As Sir Matthew Hale objected, there was a common notion among natural philosophers

> that if we should suppose any part of the Universe were free from any Body, or had a more subtle Body than that of the common Air, if a portion of the lower Air next the Earth were placed there, it [the air] would upon the account of its own natural Elasticity, expand it self as largely, and it may be much larger than those Artificial Expansions wrought by fire in Aeolipiles, and other Engines....[54]

Such a remarkable capacity for self-expansion envisaged within a corpuscularist frame of reference virtually forced the conclusion that particles of air were able to spread themselves apart in space so that a body of air could be said to contain what Aristotle quite aptly referred to as "self-determined voids". As early as 1652, therefore, Charleton had included the power of repulsion, along with spring ("resilition") in a list of the innate properties of atoms:

> Atoms, in the instant of their creation received immediately from God a faculty of self-motion, and consequently of concurring, crowding, justling, repelling, resilition, exsilition, and reciprocal complectence, concatenation, revinction, &c. according to the respective preordination in the Divine Intellect.[55]

Furthermore when Sir William Petty tried to explain "Springing or Elastique Motions", that "hard Rock in Philosophy ... Which has long lain thwart Us in the way of Our Enquiries", his whole approach took for granted ideas that Newton was later to consider using for the "Conclusio" to his *Principia mathematica*. Newton assumed that there is an "analogy of nature" so that

> Whatever reasoning holds for greater motions, should hold for lesser ones as well. The former depend upon the greater attractive forces of larger bodies, and I suspect that the latter depend upon the lesser forces, as yet unobserved, of insensible particles. For, from the forces of gravity, of magnetism and of electricity it is manifest that there are various kinds of natural forces, and that there may be still more kinds is not to be rashly denied. It is very well known that greater bodies act mutually upon each

> other by those forces, and I do not clearly see why lesser ones should not act on one another by similar forces.[56]

This could almost be seen as a defence of Petty's speculations, published over ten years earlier in 1674, that "every Atome" is "like the Earths Globe or Magnet, wherein are three Points considerable, viz. two in the surface called Poles, & one within the substance, called Center, or rather Byass".[57] Petty went on to suggest that, by virtue of its magnetic nature, an atom "may move about his own Axis, and about other Atoms also, as the Moon does about the Earth; Venus and Mercury about the Sun; and the Satellites Jovis about Jupiter &c." Petty's atoms were modelled upon William Gilbert's conception of the Earth as a huge spherical magnet. Arguing just like Newton from an analogy between large and small, Petty supposed that "the Byas [centre of gravity] of one Atome may have a tendency towards the Byas of another near it", and that

> all Atoms have, like a Magnet, two motions, one of Gravity whereby it tendeth towards the Center of the Earth, and the other Verticity, by which it tendeth towards the Earths-Poles, and whereby Magnets joyn to each other by their Opposite Poles.[58]

It is important to mention here that there is absolutely no question of a Cartesian, mechanistic account of magnetism underlying Petty's theories. He does not believe that magnetic phenomena are brought about by sub-microscopic passive particles shaped like tiny screws and moving merely inertially. For him, the sub-microscopic particles, such that "perhaps a Million do not make up one visible *Corpusculum*", *are themselves magnets*. Each atom, therefore, must exhibit the same 'occult qualities' that we see so strikingly in the familiar magnet, including the occult qualities of attraction and repulsion.[59]

It would seem, then, that the frequent historiographical claims about Newton's radical revision of the mechanical philosophy are somewhat exaggerated. The belief that Newton invigorated the mechanical philosophy by impregnating matter with occult qualities in the form of 'foci of activity', active principles, unexplained faculties and virtues, secret principles of sociability and unsociability and forces of attraction and repulsion, fails to do justice to the fruitful tradition of English mechanical philosophy of which Isaac Newton was a part. Whatever the reactions of Continental natural philosophers to Newton's use of 'occult qualities', not one English mechanical philosopher raised his voice against it. Newton's work did not strike his older contemporaries in England as a dramatic break from their own enterprise. For them the *Principia* was not only a dazzling display of intellectual virtuosity it was also a vindication of their own general approach to the problems of natural philosophy.

WHY ACTIVITY MATTERS

So far we have simply pointed to a few scattered details in the philosophies of a small number of mechanists in seventeenth century England. In order to show that these represent a genuine tradition in mechanical philosophy to which Newton was the heir, we must consider, if again only briefly, the broader context of English natural philosophy. We have already noted that the strict version of the mechanical philosophy, as expounded by Descartes, was represented as implausible on philosophical grounds by a number of thinkers. The general tenor of these arguments was that completely passive matter and a system which did not allow for the generation of new motion in the world could not possibly "save the phenomena". However, we must not forget also the influence of natural theological concerns and the peculiar fears of atheism, irreligion or dangerous heterodoxies in early modern England.

As is well known the mechanical philosophy was pioneered in England by thinkers who saw it as a powerful support for their religious views. The most vigorous early advocate of Cartesianism in England, for example, was the theologian Henry More who believed that the new French system was "the best Engine" against "erroneous fabricks in Relligion" he had ever seen. However, More's evangelical interpretation of Descartes's philosophy was somewhat idiosyncratic. As we have already seen from his response to the Cartesian notion of transference of motion by impact (quoted above), More insisted that Descartes assumed the existence of some kind of active principle in matter which, given the *essential* passivity of matter, testified to the existence of God.[60] More was to bitterly regret his naïvety some years later when he realized that the strict mechanical system of Descartes had been taken up by "atheistical wits" and "mere scoffers at religion". By then, too, there was a home-grown system of mechanical philosophy, rivalling Descartes's in its strictness and eschewing all active principles, which constituted the natural philosophy of the most notorious 'atheist' in England: Thomas Hobbes's *De corpore*.[61]

If More's religious preoccupations and his desire to use Cartesianism in defence of his theology made him slow to see that Descartes's theories could just as easily be used by contemporary 'atheists', other readers were quick to realize the dangers implicit in the Cartesian approach. If the world system was so contrived that once God set it in motion it could continue indefinitely without his further intervention, then it was easy to imagine that God was not necessary at all. Aristotelians and Epicureans alike had argued that the world had always existed and so, when Descartes had seemingly removed the necessity for Aristotle's continually operating "first (or unmoved) mover", it was easy to suppose that the world system had exhibited the kinds of motion and change we daily see throughout eternity.

In order to avoid providing such a hostage to atheism Newton argued that by the passive principle of *vis inertiae* alone "there never could have been any Motion in the World", and so God would be required to provide motion even if the world *had* existed eternally. Boyle made the point clear in a refutation of Lucretius:

> he supposes his eternal atoms to have from eternity been their own movers, whereas it is plain that motion is no way necessary to the essence of matter, which seems to consist in extension .... And accordingly I finde that *Anaxagoras*, though he believed, as Aristotle did after him, that matter was eternal, yet he discern'd that the notion of matter not necessarily including motion, there was a necessity of taking in a *mens*, as he stiles God, to set this sluggish matter a moving.

In other words, the atheist stratagem of proclaiming the world to be eternal can only be successful, Boyle was saying (and speaking for many of his fellow naturalists), if motion is a *logical* concomitant of matter in the same way that extension is. The fact that it is not, since matter can be conceived without contradiction to be at rest, necessarily implies that God gave activity to it.[62]

Of course, Descartes himself tried to forestall any atheistic conclusions from his theories but he used a very different gambit. He engaged God to continually maintain matter in motion from moment to moment — a theological expedient which was emphasized subsequently in Malebranche's occasionalist interpretation of Cartesianism.[63] However, this kind of move was regarded with some suspicion by both theologians and natural philosophers in England. Isaac Barrow, for example, was thinking of Descartes when he wrote: "he speaks unworthily of the Supreme Maker of things who supposed that he created just one homogeneous Matter and extended it blockish and inanimate." What was "unworthy" in this was God's continual labouring with the dross of the world as a mere "carpenter or mechanic repeating and displaying *ad nauseam* his one marionettish feat".[64] For Joseph Glanvill and Henry More this kind of divine immanence was even problematic when considering high level activities like the infusion of souls into unborn foetuses let alone the interactions of base matter. If God was required to animate, by creation of a new soul, every newly conceived foetus in the womb then He would have

> To bear a part amongst Pimps and Bawds, and pocky Whores and Whoremasters, to rise out of his Seat for them, and by a free Act of Creation of a Soul, to set his Seal of Connivance to their Villainies....

Clearly, God the puppetmaster would have to be regarded as a Cosmic accessory before, during and after the fact.[65]

Amongst natural philosophers, on the other hand, appeal to a *Deus ex machina* was generally frowned upon as a betrayal of their tradition and a relinquishment of natural philosophy's claim to reveal God's plan in the Creation. As Boyle pointed out in his examination of *The vulgarly receiv'd notion of nature* (1686), when a philosopher "can say little" in answer to questions

> but that it pleased the Author of the Universe to make them so ... we pretend not to give the particular physical reasons of the things propos'd, but do in effect confess we do not know them.

The alternative, then, was to insist that God created the world in such a way that his intervention was unnecessary. "It much more tends to the Illustration of God's wisdom", Boyle wrote, "to have so fram'd things at first, that there can seldom or never need any extraordinary Interposition of his Power". The wisdom of God, Boyle reiterated, is most clearly demonstrated when the philosopher can show that God

> can make so vast a machine, perform all those many things which he designed it should, by the meer contrivance of brute matter, managed by certain laws of local motion, and upheld by his ordinary and general concourse.[66]

For Boyle, and for English natural philosophers, a continuous recreation of the world from moment to moment would require the extraordinary or absolute power of God. By contrast, the notion of the world as a vast *Automaton* consisting of matter endowed with principles of activity and operating exclusively in accordance with the Laws of Motion required only that God should play a supervisory role. Thus the world is said to be contrived like the Strasbourg Clock rather than

> after the nature of a puppet ... such, that almost every particular motion the Artificer is fain (by drawing sometimes one wire or string, sometimes another) to guide, and oftentimes over-rule the actions of the engine.[67]

Charleton considered it a major philosophical advantage of his system that

> it fitly declares the radical Cause of all Motion, activity or energie in second causes, or natures once removed from the *Primus motor*, God.

Similarly, Sir Matthew Hale believed that

> it far more advanceth the glory of the Divine Wisdom, in that he hath setled such a regular order in things of Nature, that may regularly conduct them to their designed end; then if the Glorious God, or any intelligent Power by him substituted, should by immediate and identical interposition produce every Phaenomena in Nature.[68]

The concept of active principles, therefore, enabled the natural philosopher to account for the physical phenomena without recourse to the perpetual miracle of God's direct intervention. Although Newton, ignoring Christopher Wren's indulgent smile, sometimes felt compelled by his religious obsessions to attribute the cause of gravity directly to God, he always reverted to more naturalistic explanations before too long.[69]

The most vigorously proposed alternative to active principles in matter as an intermediary between God and the world was Henry More's concept of the 'Spirit of Nature'. The reactions of natural philosophers to More's notion show not merely a marked aversion to supernaturalism but also an unmistakable desire to dissociate themselves from More's way of using philosophy to bolster his somewhat idiosyncratic religious opinions.[70] Boyle, Hooke and Hale felt it was necessary to put their refutations into print; but More himself tells us that even before any of these appeared he had been charged:

> That I have introduced an obscure Principle for Ignorance and Sloth to take Sanctuary in, and so to enervate or foreslack the useful endeavours of curious Wits, and hinder the expected progress that may be made in the Mechanick Philosophy; and this, to aggravate the crime, before a competent search be made what the Mechanical Powers of Matter can doe.

More's attempts to rebut this charge in 1662 made no impression on Hooke who made the same point some fifteen years later. The concept of the Spirit of Nature, Hooke wrote, "tends to nothing but the discouraging industry from searching into and finding out the true causes of phenomena". "What should I trouble myself", he went on, "to enquire into that which is never to be understood".[71] Underlying these criticisms was the natural philosophers' belief, which we shall return to in the next section, that active principles, for all their occult nature, were amenable to philosophical investigation while the Spirit of Nature was not. As Sir Matthew Hale put it:

> the substitution of such a vicarious, immaterial, common Principle, ... subordinate to Almighty God, a common Spirit, and different from the particular active principles variously implanted in Physical Bodies, though it be a pleasant supposition, yet it is difficult if not impossible to evince to any tolerable satisfaction.[72]

The critical response of natural philosophers to such religiously inspired supernaturalism is perfectly clear but, it should hardly need saying, their critiques could only be advanced with care. The fear of atheism and irreligion ran high throughout this period and it was all too easy for a philosopher to find his ideas tarred with the brush of atheism. The efforts of English natural philosophers to avoid such imputations and to establish the 'usefulness' of natural philosophy are well known and we need not rehearse them here except in so far as they relate to their use of active principles.[73] However, we must mention one aspect of this tension between religion and science which is germane to our story since it has had historiographical as well as historical consequences.

One of the major reasons behind More's development of the notion of the Spirit of Nature was to avoid the concept of animated, or merely active, matter. His response to Descartes, quoted earlier, in which he seems to suggest his own belief in matter endowed with "some shadow and image of life" was really only loose talk. Even at that early stage (1649-55) More usually wrote in terms of a categorical distinction between matter, which is inactive, and immaterial spirit, which is active. In his earliest writings, however, he was often (as in these early responses to Descartes) lax in his manipulation of these categories. After 1656, however, More tightened up his thinking and henceforth was always insistent on the categorical dichotomy of matter and spirit. One of the reasons for this was his increasing awareness of the dangers of "enthusiastic" beliefs in animate matter. More's *Brief discourse of ... enthusiasm* of 1656 was based on the premise that "Atheism and Enthusiasm, though they seem so extremely opposite one to another, yet in many things do nearly agree".[74] Enthusiastic beliefs in animate matter, therefore, were likely to have dangerous consequences. The actual enthusiastic beliefs which More recorded were

> That Nature is the body of God, nay God the Father, who is also the World, and whatsoever is any way sensible or perceptible.

And

> That everything has *Sense*, *Imagination*, and a *fiducial Knowledge* of God in it, *Metalls*, *Meteors* and *Plants* not excepted.[75]

The significant thing to notice about these beliefs is that matter is said to be not merely active but *perceptive* and in some sense *intelligent*. More sought to oppose these views by insisting on the distinction between dead matter and living spirit but during the course of strengthening his position and hardening his arguments he had to insist more and more upon the *total* passivity of matter. Before long he was not only opposed to ideas of perceptive matter but also to any claims about activity or innate motion within matter.[76]

More was very influential among Anglican theologians and his aggregation of concepts of animated and merely active matter was, to say the least, embarrassing for one or two of the leading mechanical philosophers. Certainly Boyle was always very careful to deny that matter could think or display percipience or appetite. Moreover, so great was his reluctance to offend the sensibilities of Churchmen that, whenever he could without compromising his natural philosophy, he wrote as though matter was entirely passive. Similarly, Isaac Newton, when writing to the theologian Richard Bentley, felt obliged to deny that gravity was inherent in matter ("Pray do not ascribe that notion to me"); but was perfectly willing (when writing in the *Opticks*, for a different audience) to imply that gravity and other active principles were powers or virtues of body.[77]

However, not all natural philosophers were so intimidated by the arguments of More and those theologians who followed him on this issue. Francis Glisson, Regius Professor of Physick at More's own university, even went so far as to take ideas about percipient matter which were current in the medical tradition to their logical conclusion. He developed a complete system of natural philosophy based on *The energetic nature of substance*.[78] Similarly, John Locke, another writer educated in the contemporary medical tradition, felt no compunction to avoid the attribution of the power of thought to matter as a superadded property.[79] Boyle and Newton, by contrast, each with his own reasons for avoiding religious controversy, show clear signs in their writings of speaking for the occasion; sometimes using active principles in matter to argue indirectly for God's existence and sometimes insisting that passive matter can only be moved by a divine agent. The ambivalence of these two most remarkable seventeenth century natural philosophers has had unfortunate historiographical consequences. It has tended to reinforce the simplistic view of seventeenth century matter theory, challenged here, in which matter is held to be passive and inert; and the consequent belief that Newton, for all his seeming hesitation, was the first to introduce active principles into the mechanical philosophy. Moreover, it has even led to what might be seen as a historiographical revival of Henry More's tendentious views in which believers in active matter are, *ipso facto*, held to be subversive radicals while believers in passive matter (a category which, in this revival, is made to embrace virtually every mechanical philosopher in England) are conservative defenders of Church and Crown.[80] In fact, as we have seen, a professed belief in occult qualities and active principles in matter was certainly a legitimate stance for a seventeenth century natural philosopher to take up; and often this kind of matter theory was used to bolster the image of the natural philosopher as natural theologian.

To recapitulate, then, active principles could be claimed to be an essential feature of sound natural philosophy not just for philosophical reasons but also on natural theological grounds. On the one hand they were used to undermine the atheistic interpretation of strict mechanism — that the universe may always have existed with the same amount of motion it now has. As Warner, Charleton, Hale, Boyle and Newton pointed out, passive matter alone could never have given rise to the motions we see in the world, and so God would be required to provide activity even if the world was eternal and uncreated.

On the other hand, active principles provided a series of hypothetical physical entities which fulfilled a useful explanatory role and were susceptible to philosophical investigation. They therefore prevented the heterodox and blasphemous invocation of God as a continual agent immanent in matter and, only slightly less inimical to natural philosophers' concerns, the overrunning of natural philosophy by supernaturalism.

THE MATTER OF ACTIVITY

When Westfall suggested that Newton's triumph in mechanical philosophy was dependent in part on the radical innovation of introducing occult qualities or active principles into mechanist matter theory, he felt the need to say something about Newton's defence of these supposedly outlandish ideas. Newton, after all, "was well aware that he was proposing a major philosophic innovation" and so "he tried to shield himself from expected criticism". This defence included recourse to mathematical realism and an insistence that the truths of natural philosophy could be arrived at phenomenologically in spite of causal nescience. This latter point was made most famously in Query 31 of the *Opticks*; defending the concept of active principles he wrote,

> These Principles I consider, not as occult Qualities, supposed to result from the specifick Forms of Things, but as general Laws of Nature, by which the Things themselves are form'd; their Truth appearing to us by Phaenomena, though their Causes be not yet discover'd. For these are manifest Qualities and their Causes only are occult.[81]

Similarly, I. B. Cohen has recently argued that perhaps the most revolutionary feature of Newton's *Principia* is its elaboration of a new style of mathematical physics. Professor Cohen's analysis and exposition of the 'Newtonian style' is complex and subtle and cannot be summarized here.[82] Suffice it to say, however, that one of the philosophical intentions of the Newtonian style, as Cohen sees it, is to justify a phenomenological position in the face of causal nescience. So, although Newton's critics believed his system was based on "a return to the 'occult qualities' which had supposedly been cast out of natural philosophy", Newton himself replied that

> In this philosophy particular propositions are inferred from the phenomena and afterward rendered general by induction.... And to us it is enough that gravity does really exist and act according to the laws which we have explained, and abundantly serves to account for all the motions of the celestial bodies and of our sea.

This remains sufficient, Newton insisted, even though he has "not been able to discover the cause of those properties of gravity from phenomena, and I frame no hypotheses".[83]

Both Westfall and Cohen seem to imply that this methodological ploy was essentially a new move, developed by Newton to protect his uniquely original philosophy. This is not so. Just as Newton was able to draw on the earlier tradition of English mechanical philosophy for his concept of active principles, so he was able to draw on it for precisely the same kind of phenomenological defence of occult qualities we have just seen. Indeed, this phenomenological approach to the understanding of nature is one of the most characteristic features of seventeenth century English natural philosophy and dovetailed neatly with the other stratagems used to bolster the claims of the new science to establish moral certainties. The frequently proclaimed Baconianism of English natural philosophy, its refusal to deal in mere hypotheses (made famous by various spokesmen for the Royal Society) its mitigated scepticism and its emphasis on experimentalism are all inextricably bound up with, if not derived from, its practitioners' acknowledgment of the impossibility of finding what Joseph Glanvill called "the first springs of natural motions" unless, of course, we simply mean God. A *natural* philosophy (leaving God out of our considerations) which depends on active principles is, essentially, one which must remain aetiologically incomplete. Not all the causal processes underlying physical interactions can be explained, some must remain inscrutable. Nevertheless, the enterprise of natural philosophy need not be abandoned. Though it may be impossible to explain how an occult active principle works it will always be possible (providing it is real) to show that it "does really exist".[84]

So, the existence of "a *Magnetick* way of curing *wounds* by anointing the *weapon*" could be accepted by mechanical philosophers since, as Joseph Glanvill averred, it was "for matter of fact put out of doubt by the Noble Sir K. Digby". Glanvill's subsequent summary of Digby's mechanistic explanation of "this *wonder*" is highly favourable and larded with praise. And yet in the end Glanvill suggests that causal explanations are less important than establishing the fact:

> It is out of my way here to enquire whether the *Anima Mundi* be not a better account, then any *Mechanical* Solutions. The former is more desperate, the latter hath more ingenuity, then solid satisfaction. It is enough for me that *de facto* there is such an entercourse between the Magnetick unguent and the vulnerated body, and I need not be solicitous of the Cause.[85]

Observation and experimental evidence are thus given primacy over causal hypotheses. This attitude even enables Glanvill to entertain the possibility of action at a distance. For example, although a "sympathetic" telegraph involving needles, magnetized by the same magnet, pivoted over "Abecadarian circles"

> May not yet answer the expectation of inquisitive experiment; yet 'tis no despicable item, that by some other such way of magnetick efficiency, it

> may hereafter with success be attempted, when Magnetical History shall be enlarged by riper inspections.[86]

We can see very similar ways of thinking in Boyle's writings. In *The usefulness of experimental naturall philosophy*, for example, Boyle considered how one might explain the "pretended Sympathy" between gold and quicksilver:

> the Naturalist may explicate, by saying, That Gold being the onely Body heavier than Quick-silver of the same bulk, the known Laws of Hydrostaticks makes it necessary that Gold should sink in it, and all lighter Bodies swim on it: But though the cause of this effect be thus plausibly assign'd by deducing it from so known and obvious an affection of Bodies, as Gravity, which every man is apt to think he sufficiently understands; ... yet the Cause and Nature of it [gravity] are as obscure as those of almost any *Phaenomenon* it can be brought to explicate.

Boyle's conclusion to this discussion is simply to acknowledge "an intelligent Author or Disposer of Things", but then this is an avowedly apologetic work.[87] Elsewhere he is content to agree with "Archimedes, Stevinus and others" who "do not take it upon them to assign the true cause of gravity but take it for granted, as a thing universally acknowledged that there is such a quality in the bodies they treat of".[88] And on the equally difficult problem of 'spring' Boyle took a similarly phenomenological position when he insisted that his intention was not "to assign the adequate cause of the spring of the air, but only to manifest, that the air hath a spring, and to relate some of its effects".[89] Speaking of 'spring' in general, he insisted that

> whatever the latent Cause of the Phaenomena may be, the manifest Circumstances of them suffice to show, that bodies, which, as to sense, are in a natural state of Rest, may be in a violent one, as of Tension, and may have, either upon the score of the contexture of the parts among themselves, or upon that of some interfluent subtile matter, or some other Physical Agent, a strong endeavour to fly off or recede from one another....[90]

Spring was, in other words, what Newton would call a manifest quality of which the cause was occult.

In 1716 Leibniz accused Newton of introducing "a chimerical thing, a scholastic occult quality" into his natural philosophy but Boyle was anticipating and responding to such changes as early as 1671:

> And least you should think that under the name of Cosmicall Qualities I should introduce Chimaeras into Naturall Philosophy I must betimes advertise you, that you will meet with divers Particulars in the following Discourse, fit to show that these Qualities are not meerly fictitious

> Qualities: but such, whose Existence I can manifest, not only by considerations not absurd, but also by real Experiments and Physical Phaenomena.[91]

For Robert Hooke it was permissible to introduce "unheard of Powers, Operations, Effects or Motions" into natural philosophy providing they are all "within our reach and command", so that the experimenter can "daily try, see, and find the regular working of" them. Sir William Petty felt that his "Hypothesis of springing or elastique motions" avoided "Chymerical Speculations" because

> My matter is so simple, as I take notice of nothing in each Atome, but of three such Points as are in the Heavens, the Earth, in Magnets, and in many other Bodies. Nor do I suppose any Motions, but what we see in the greater parts of the Universe, and in the parts of the Earth and Sea.

This in turn ensured that Petty's hypothesis was amenable to experimental investigation:

> all the motions I fancy in my Atoms, may be represented in gross Tangible Bodies, and consequently may be made intelligible and examinable.[92]

Clearly Boyle was not the only natural philosopher who believed it was possible to

> discourse like a naturalist about magnets of celestial and other emanations, that appear not to have been consider'd, not to say, thought of, either by the scholastic, or even the mechanical, philosophers.

To discourse about 'latent qualities' or 'occult qualities' could be useful to "the sagacious and inquisitive", in Boyle's opinion, providing the naturalist confined himself to 'matters of fact' or undeniable, everyday phenomena susceptible to experimental manipulation.[93]

The experimental method, therefore, could be insisted upon by English natural philosophers as essential for the development of the mechanical philosophy and its companion, natural theology. As Keith Hutchison has recently shown, the mechanical philosophers regarded it as one of the supreme advantages of their new philosophies that they could accommodate occult qualities in a heuristically useful way. The scholastic distinction between 'occult' and 'manifest' qualities, Hutchison points out, became inappropriate when all phenomena were explained in corpuscularian terms. A manifest quality for scholastics was one which could be immediately apprehended by the senses. Yet the whole thrust of the new philosophies was to explain all physical change in terms, ultimately, of the insensible motions or powers of insensible particles. All secondary qualities thus became occult.[94] If many causal explanations must, therefore, remain elusive, it was nevertheless possible to establish, experientially, the richness of natural effects. What the experimental method could do, in spite of causal nescience, was establish the 'matters of fact'.

Once we accept Hutchison's analysis it becomes easy to see why Newton could treat occult qualities as epistemologically equivalent to 'metaphysical', 'physical' and even 'mechanical' qualities when composing the General Scholium for the *Principia* in 1713. Newton's efforts to persuade his readers that he did not feign hypotheses were not simply a matter of eschewing all occult qualities. On the contrary, hypotheses were defined as "whatever is not deduced from phenomena", and this could apply equally well whether one talks of "Metaphysical or Physical, whether of occult qualities or Mechanical". The converse of this is that occult qualities are just as acceptable in natural philosophical explanations as metaphysical, physical or mechanical qualities, provided they can be demonstrated or deduced "from phenomena". Gravity, therefore, is an occult quality which is not hypothetical because it can be shown that it "does really exist".[95]

Given the widespread use of such phenomenological arguments in late seventeenth century England it seems reasonable to assume that Newton felt, as he said in another context, he had said enough for people to know his meaning. After 1716, however, he began to think that perhaps he had not. It was at the beginning of 1716 (new style) that Newton learned of Leibniz's dismissal of Newtonian gravity as an occult quality, or to be precise, "une qualité occulte scholastique". In Newton's reactions to this charge, a lot hinged upon Leibniz's use of the qualifier 'scholastic'. His first reply, to the go-between Abbé Conti, made the point forcefully:

> He [Leibniz] changes the signification of the words Miracles & Occult qualities that he may use them in railing at universal gravity ... occult qualities are decried not because their causes are unknown, but because the Schoolmen believed that those things wch were unknown to their Master Aristotel, could never be known.[96]

The point is, of course, that Newton is every bit as opposed to "scholastic occult qualities" as Leibniz is, but for Newton that adjectival phrase has a very different "signification" from unqualified "occult qualities", which he takes (in the English sense) to be unobjectionable. It is presumably for this reason that Newton added an anti-Aristotelian rider to his phenomenological defence of occult qualities in the 1717 edition of the *Opticks*. Having said that active principles are "manifest Qualities, and their Causes only are Occult", Newton wrote:

> And the *Aristotelians* gave the Name of occult Qualities, not to manifest Qualities, but to such Qualities only as they supposed to lie hid in Bodies,

> and to be the unknown Causes of manifest Effects: Such as would be the Causes of Gravity, and of magnetick and electrick Attractions, and of Fermentations, if we should suppose that these Forces or Actions arose from Qualities unknown to us, and uncapable of being discovered and made manifest. Such occult Qualities put a stop to the Improvement of natural Philosophy, and therefore of late Years have been rejected.

Newton then felt able to reiterate his own way of using occult principles in mechanical philosophy:

> To tell us that every Species of Things is endow'd with an occult specific Quality by which it acts and produces manifest Effects, is to tell us nothing: But to derive two or three general Principles of Motion from Phaenomena, and afterwards to tell us how the properties and Actions of all corporeal Things follow from those manifest Principles, would be a very great step in Philosophy, though the Causes of those Principles be not yet discover'd.[97]

Newton's defence of occult qualities, then, is perhaps more sophisticated than that of earlier English mechanists but it is not different in kind. His distinction between occult qualities which can be deduced from phenomena and scholastic occult qualities was foreshadowed in the writings of Charleton, Boyle, and Petty and similar ideas were advocated by Glanvill and Hooke.[98] The fact that occult qualities in the mechanical philosophy have a history before Newton should now be seen to be unsurprising. Indeed, the investigation and demonstration of occult qualities in matter, as we have seen, were major factors in establishing experimentalism as the safest and surest way to truth in natural philosophy. Accordingly, the leading members of the Royal Society can be found trying to promote various experimental research programmes as well as the various Baconian 'natural histories' for which the Society is well known. Christopher Wren, for example, lecturing to the Society in 1662 not only suggested the compilation of a "History of Seasons" but also recommended the chemical investigation of bodily fluids:

> ... in the Body of a Man, if we consider it only mechanically, we may indeed learn the Fabrick and Action of the organical Parts, but without Chymistry, we shall be at a loss to know, what Blood, Spirits and Humours are, from the due temper of which (as of the Spring in the Barrel Wheel) the motions of all the Parts depend.

It is important to note, here, that although Wren uses a mechanical image — "the Spring in the Barrel Wheel" — he is actually saying that there is a chemical "Temper" of matter which is the cause of the matter's motion. Wren believed, as Leibniz and Newton were later to do, that chemistry could enable us to understand more fully what Glanvill called the "prime motive causes" in matter.[99]

In a "List of Subjects for Essays" among his papers, Sir William Petty expressed his intention to write of "Vibrations and Pendulum", fermentations, magnets, "Of Ayre and Vacuum" and "Of the application of Gunpowder to other uses than guns". Each of these were regarded as fruitful areas of experimental investigation of the inherent activities of matter. We have seen, for example, how Hooke believed that all matter may be incessantly vibrating. It comes as no surprise, therefore, to learn that Hooke devoted much of his time to extensive experimental investigations of vibrations in springs, strings and pendulums. His metaphysics of vibrating matter may also lie at the root of his musical interests.[100] No doubt Petty himself hoped that an intensive experimental investigation of magnetism could provide support for his hypothesis of magnetic atoms. The applications of gunpowder which he had in mind were, most probably, to do with its use in chemistry as a source of heat or energy and the use of "nitre" in the "oeconomy of nature", including physiological processes. The rich experimental research programme which focused on the role of nitre and culminated with John Mayow's nitro-aerial particle as "a *primum mobile* of the physical world" was, as Robert Frank has shown, one of the most characteristic aspects of seventeenth century natural philosophy in England.[101] The attempts to investigate experimentally the supposed vibrations of matter, the incongruity of aether or the power of air to spontaneously disseminate itself through a vacuum, and the putative 'vital' ingredient of the air gave rise to the extremely fruitful and well-known investigations using the specially developed air-pumps. The importance of this experimental research programme for establishing the 'matters of fact' about occult qualities in matter cannot be overstressed. As works by Charles Webster, Robert Frank and Simon Schaffer make perfectly clear, the pneumatic experiments of Power, Boyle, Hooke and Mayow were intended not only to demonstrate the existence of spring but also to show the association between that occult quality and *spiritus*, soul and vitality. Mayow, for example, interpreted a loss of volume of air (after burning a candle) as a diminution of the air's elasticity due to the consumption of nitro-aerial particles. Thus, the elasticity of the air was proposed as a measure of its inherent activity and vitalistic power.[102]

Of course the experimental method was regarded by English mechanical philosophers as crucially important for a number of other reasons. Experimentalism was well suited to their mitigated sceptical response to the dogmatic excesses (as they saw them) of Roman Catholicism or sectarian enthusiasm. It was important to deny that explanations of physical phenomena could ever be proved apodictically in order to keep their natural philosophies from being commandeered by particular religious or political groups as an undeniable support for their world views. It was always possible, they therefore insisted, that the machinery behind the clock face ran differently to the way men envisaged it.[103] Accordingly, the most appropriate methodology for the mechanical philosophy seemed to most English natural philosophers to be not the implicitly dogmatic rationalism of Descartes or Hobbes but their own experimental method. Naturally, this promotion of the experimental method was not itself innocent of politico-religious intentions but with the emphasis on 'matters of fact' it could be claimed that it was. The importance of occult active principles in religious apologetics which we considered in the previous section and the insistence on the experimental method as the only way to establish the real existence of such occult principles, should be regarded therefore as one more facet of the ideological and apologetic use of the experimental method which has been analysed by Paul Wood, Steven Shapin, Simon Schaffer and others.[104]

Another important element here is the traditional connection between natural philosophy and voluntarist theology. The voluntaristic belief that God can do anything that is not logically impossible merely by the exercise of his will led to a conviction that whatever order and structure there is in the world is not inherent in the nature of things but is merely contingent upon God's fiat. It follows that any attempt to deduce the system of the world by a process of pure reasoning is misconceived. The only way to discover the system of the world, on this view, is by an empirical investigation of the world as it has been created. As Oakley has pointed out, an empiricist approach to natural philosophy must follow from a belief in the "omnipotence and freedom of God".[105] The opposite view, known as necessitarianism, argues that God is constrained, not only by the laws of logic but also by notions like Justice and Goodness, to create only one possible world — the best. According to the necessitarian the order in the world is immanent within it and can be analysed by a process of pure reasoning.

The use of voluntarist theology in defending a concept of active matter is clearly discernible in writers like Charleton, Boyle, Petty and Glisson. The argument was succinctly stated by that famous underwriter of experimental philosophy, John Locke:

> For I see no contradiction in it, that the first eternal thinking Being should, if he pleased, give to certain Systems of created senseless matter, put together as he thinks fit, some degrees of sense, perception, and thought.[106]

While Locke saw no contradiction in it Glisson built his whole system of philosophy upon the assumption that God had given matter the powers of perception and self-movement. As we might expect, he believed that his theories could be confirmed by experiment and hoped that the Royal Society would undertake the required research programme. Glisson's views, however, drew the anger of theologians like More and Cudworth who perceived them as providing hostages to atheism. Boyle seems to have agreed with the theologians on this matter and he was always careful to deny that matter could perceive and think. It should be noted, however, that Boyle, voluntarist that he was, did not deny that God *could* endow matter with such powers. He merely argued that experience showed that God had not done so. More and Cudworth, on the other hand, rejected the very possibility of perceptive or living matter as a *logical* absurdity. They could only do this, though, because they rejected voluntarist theology and embraced necessitarianism. This, in turn, drew the attention of the theologian, Richard Baxter, who repudiated More's necessitarianism in a spirited defence of Glisson's theory of perceptive matter. "I confess", Baxter wrote, "I am too dull to be sure that God cannot endue matter itself with the formal virtue of Perception: ... Where is the contradiction that makes it impossible?"[107]

Francis Oakley and Steven Shapin have both suggested that voluntarist theology is a key to understanding the social and political uses of early modern science. If we accept this then the fact that voluntarist theology was frequently used to insist that God by his arbitrary will could endow matter with intrinsic active principles of motion should reinforce the argument presented here, that active principles were an important part of the mechanical philosophy even before Newton.[108] Thus, when Newton sought to defend his concept of automotive light particles in his "Hypothesis of light" of 1675, he could immediately invoke this voluntarist tradition. When he wrote, "God, who gave animals self-motion beyond our understanding is, without doubt, able to implant other principles of motion in bodies, which we may understand as little", he was saying nothing which would strike his readers as outlandish. On the contrary, it would have been very hard for a contemporary to consistently deny that God's omnipotence and freedom of will could endow matter with innate active principles.[109]

It should be clear from all this that the use of occult qualities or unexplained active principles in matter in the mechanical philosophy was neither new with Newton nor confined to radical thinkers who were forced out of the natural philosophical mainstream because of their dangerous religious and political views. On the contrary, this kind of matter theory was always a major feature of the mechanical philosophy in England. Furthermore, it was so inextricably linked with other major traditions of both natural theology and philosophy; voluntarism, mitigated scepticism, experimentalism; that its importance for seventeenth century English natural philosophy can hardly be denied.

CONCLUSION

When Bernard le Bovier de Fontenelle inaugurated the Newton industry, with his Eloge of 1727, he wrote as a Cartesian. The cautious, halting and speculative matter theories of earlier English mechanists can have made no impression on him, for he held Newton solely responsible for bringing back the occult quality of "attraction" which had been "banished from Physicks ... for ever" by Descartes. It seems that all subsequent Newton scholars have tacitly accepted the Fontenelle thesis.[110] A more careful reading of Newton's immediate precursors from a less Newton-dominated, or a less Descartes-dominated, perspective gives the lie to this approach. In England the threatening political and religious upheavals militated against the easy acceptance of the strict mechanism associated with Descartes's name, and certainly Cartesianism was never propagated by educational establishments in England the way it was on the Continent.[111] At the very least, English natural philosophers adopted the Gassendist view of matter with inherent automotive power, and many of them, as we have seen, took their speculations further. So, while Continental mechanists felt the need to object to Newton's use of occult qualities, no English mechanical philosopher did. Rather than follow Fontenelle, perhaps we should agree with John Keill who wrote in 1702 that

> Although now-a-days the mechanical philosophy is in great Repute, and in this Age has met with many who cultivate it, yet in most of the writings of the Philosophers, there is scarce anything mechanical to be found besides the Name.[112]

Indeed, the difficulties of providing a consistent account of all physical phenomena in strict mechanist terms, without recourse to active principles, taxed even the ingenuity of Descartes. As Westfall himself has pointed out

> In the discussion of many phenomena, of fire, for example, he [Descartes] summoned the subtle matter onto the stage, like a *deus ex machina*, to play the role of an active principle. Of course, Descartes always implied that its "agitation" resulted from mechanical necessity. In fact, he never analysed in detail how it was caused ... but rather dragged it in by brute force whenever a phenomenon appeared to reveal a level of activity not present before. In such passages his subtle matter can only be seen as a mechanist rendition of an Hermetic active principle.

If we ignore the historiographically *passé* use of the adjective "Hermetic" (he was writing in 1972) then we must surely agree with Westfall.[113] However, while Westfall was correct in insisting that, nevertheless, Descartes "generally carried through the program of replacing the inherent activity of bodies with the passivity of matter dominated by mechanical necessity", he was surely mistaken in believing that "much the same can be said of Robert Boyle".[114] As we have seen, Boyle and the other leading mechanical philosophers in England did not merely modify strict mechanism, momentarily, to fudge over internal philosophical difficulties. Their very different version of the mechanical philosophy with its concept of matter imbued with occult active principles was developed in response to various complex and interrelated philosophical, religious and sociopolitical demands.

Scholarly efforts to trace the origins of Newton's concept of force have so far concentrated on three areas. Perhaps the major effort has been to avoid metaphysical considerations and to see Newton's conception as a development internal to the technicalities of the science of dynamics. But this internalist account has been supplemented by important studies of the influence of the Cambridge Platonists and of alchemy on Newton's thought.[115] We should now add another thread to this Newtonian fabric: the influence of English mechanical philosophy with its active principles and its occult qualities or rather, to adapt Newton's phrase, its "manifest Qualities [whose] Causes only are occult".[116] Newton was able to draw from the work of his older contemporaries not only valuable hints about the inherent activity and powers of matter but also a ready-made methodological defence of these speculations in so far as their effects could be established and measured by experiment.[117] Newton was able to alternate between, or combine, this phenomenological approach and the mathematical realism of thinkers like Kepler and Galileo, depending on what was most appropriate for the problem on hand.

If we persist in seeing Newton's matter theory as a radical revision of the earlier mechanical philosophy, we must surely fail to do justice to the richness and complexity of the theories of his immediate precursors. Only by recognizing that Newton's use of occult qualities and active principles is the culmination of seventeenth century English mechanical philosophy will we fully appreciate the significance of thinkers like Charleton, Power, Boyle, Petty and Hooke, to say nothing of lesser thinkers. Unless we can free ourselves from the biassed interpretation of the mechanical philosophy of a thinker like Fontenelle, we might as well believe that God said, Let Newton be! – And all was overshadowed.

ACKNOWLEDGEMENTS

An earlier version of this paper was read at Dr W. F. Bynum's seminar series in the History of Medicine Unit at University College, London (November 1984) and I would like to thank all those present on that occasion for their help and encouragement. I am also extremely grateful to the following who read and commented upon the earlier version: Peter Dear, Michael Hunter, Keith Hutchison, Ted McGuire, Roy Porter, Simon Schaffer, Steven Shapin and an anonymous referee.

REFERENCES

1. I include Hobbes even though his mature system of natural philosophy, *De corpore* was not published until 1665 because he began to develop his system in the 1630s. The delayed publication of his system is in itself an indication of the difficulties which Hobbes had to surmount before he was satisfied. For a discussion of Hobbes's early efforts in natural philosophy see Jean Jacquot, "Sir Charles Cavendish and his learned friends", *Annals of science*, viii (1952), 13-27, 175-91; and Frithiof Brandt, *Thomas Hobbes' mechanical conception of nature* (Copenhagen and London, 1928).
2. The quotation is from "Mr Petty's letter in answer to Mr. More" (*c.* 1647) which is reprinted in Charles Webster, "Henry More and Descartes: Some new sources", *The British journal for the history of science*, iv (1969), 359-77, p. 367.
3. See especially Hume's *An enquiry concerning human understanding* (1748), Section iv, paragraphs 24, 25 and 26. But for earlier examples see the quotations from Henry More given immediately below and Joseph Glanvill, *The vanity of dogmatizing: Or confidence in opinions* (London, 1661), ch. xx and xxi; *idem, Scepsis scientifica: Or, confest ignorance the way to science* (London, 1665), ch. xxiii. See also R. H. Popkin, "Joseph Glanvill: Precursor of Hume", *Journal of the history of ideas*, xiv (1953), 292-303; and R. A. Watson, *The downfall of Cartesianism, 1673-1712: A study of epistemological issues in late seventeenth-century Cartesianism* (The Hague, 1966), 95-98.
4. Henry More, letter to Descartes (23 July 1649), and "Responsio ad fragmentum cartesii", to Claude Clerselier (July-August 1655), both quoted from Alan Gabbey, "*Philosophia cartesiana triumphata*: Henry More (1646-1671)", in T. M. Lennon, J. M. Nicholas and J. W. Davis (eds), *Problems of Cartesianism* (Kingston and Montreal, 1982), 171-250, 211 and 213; see also 192.
5. The two philosophers who spring most readily to mind are Newton, whom we shall discuss below, but see also Ernan McMullin, *Newton on matter and activity* (Notre Dame and London, 1978); and Leibniz. On Leibniz's rejection of passive matter as an ontological impossibility see Walter Pagel, "The religious and philosophical aspects of Van Helmont's science and medicine", *Supplements to the Bulletin of the history of medicine*, ii (1944), 27-34; C. D. Broad, *Leibniz: An introduction* (Cambridge, 1975), 49-71; and Alan Gabbey, "Force and inertia in seventeenth-century dynamics", *Studies in the history and philosophy of science*, ii (1971), 1-67, pp. 1-15. But, of course, not even Descartes managed to confine his system to a concept of passive matter. See ref. 107 below and Mary B. Hesse, "Action at a distance", in E. McMullin (ed.), *The concept of matter in modern philosophy* (Notre Dame and London, 1963), 119-37, p. 122; Thomas L. Prendergast, "Motion, action and tendency in Descartes' physics", *Journal of the history of philosophy*, xiii (1975), 453-62; and Gary C. Hatfield, "Force (God) in Descartes' physics", *Studies in the history and philosophy of science*, x (1979), 113-40.
6. R. S. Westfall, "Newton and alchemy", in Brian Vickers (ed.), *Occult and scientific mentalities in the Renaissance* (Cambridge, 1984), 315-35, pp. 317, 324, 326, 323, 326.
7. The suggestion that Newton's interest in alchemy stemmed from his efforts to understand the active principles in matter was first made in J. E. McGuire, "Force, active principles and Newton's invisible realm", *Ambix*, xv (1968), 154-208, p. 166. Since then the notion has been taken up and elaborated in P. M. Rattansi, "Newton's alchemical studies", and R. S. Westfall, "Newton and the Hermetic tradition", both in A. G. Debus (ed.), *Science, medicine and society in the Renaissance* (New York, 1972), 167-82 and 183-93 respectively; R. S. Westfall, "The role of alchemy in Newton's career", in M. L. Righini Bonelli and W. R. Shea (eds), *Reason, experiment and mysticism in the Scientific Revolution* (London,

1975), 189-232; *idem*, *op. cit.* (ref. 6); and B. J. T. Dobbs, "Newton's alchemy and his theory of matter", *Isis*, lxxiii (1982), 511-28. On the influence of Cambridge Platonism see J. E. McGuire, "Neoplatonism and active principles: Newton and the *Corpus hermeticum*", in R. S. Westman and J. E. McGuire, *Hermeticism and the Scientific Revolution* (Los Angeles, 1977), 95-142.

8. Westfall, "Newton and the Hermetic tradition" (ref. 7), 185. Newtonian hagiography has been taken to new heights in Westfall's biography *Never at rest* (Cambridge, 1980), in which Westfall expresses his conviction that Newton was "wholly other" and that "with him there is no measure" (p. ix).
9. R. Descartes, *Principia philosophiae*, IV, 199: "Qu'il n'y a aucun phainomene en la nature qui ne soit compris en ce qui a esté expliqué en ce traitté."
10. See above, refs 4 and 5.
11. For an excellent discussion of the latter problem together with other difficulties encountered by mechanical philosophers see Alan Gabbey, "The mechanical philosophy and its problems: Mechanical explanations, impenetrability, and perpetual motion", in J. C. Pitt (ed.), *Change and progress in modern science* (Dordrecht, 1985), 9-84. There is, unfortunately, no survey of critical responses to Descartes among natural philosophers and the material which springs most readily to mind is in Newton's work. Consider, for example, his *Questiones quaedam philosophicae*, now published as J. E. McGuire and Martin Tamny (eds), *Certain philosophical questions: Newton's Trinity notebook* (Cambridge, 1983); and *De gravitatione et aequipondio fluidorum*, translated and discussed in A. R. Hall and M. Boas Hall (eds), *Unpublished scientific papers of Isaac Newton* (Cambridge, 1962), 75-156. But see also the works by Glanvill cited above (ref. 3); and Robert Boyle, *An essay of ... languid and unheeded motions* (1685), chapter 5: "Of the propagable nature of motion"; which includes many examples of new motions and "New experiments about explosion" (1672) which can be found in *The works*, edited by Thomas Birch (6 vols, London, 1772), v, 12-18, and iii, 592-5, respectively.
12. Isaac Newton, *Opticks, based on the fourth edition London, 1730* (New York, 1979), 397, 399.
13. British Library, Additional MS 4394, ff. 389r-v, 129v, 386r, 212r. Warner's work has not received the attention it deserves. For preliminary indications of his importance see Jean Jacquot, "Harriot, Hill, Warner and the new philosophy", in J. W. Shirley (ed.), *Thomas Harriot, Renaissance scientist* (Oxford, 1974), 107-28.
14. His papers were known to Thomas Hobbes, John Wilkins, Ralph Bathurst and others. See Jacquot (refs 1 and 13) and [John Wilkins and Seth Ward,] *Vindiciae academiarum containing some brief animadversions upon Mr Webster's book, stiled, The examination of academies* (Oxford, 1654), 7 and 53.
15. Walter Charleton, *Physiologia Epicuro-Gassendo-Charltoniana: Or a fabrick of science natural upon the hypothesis of atoms* (London, 1654). For the influence of this on Newton see R. S. Westfall, "The foundations of Newton's philosophy of nature", *The British journal for the history of science*, i (1962), 171-82; Henry Guerlac, "Newton et Epicure", *Essays and papers in the history of modern science* (Baltimore, 1977), 82-106; and McGuire and Tamny, *op. cit.* (ref. 11), 26-43.
16. Charleton, *op. cit.* (ref. 15), 126, 269.
17. O. R. Bloch, *La philosophie de Gassendi: Nominalisme, materialisme et metaphysique* (The Hague, 1971), 249, 259, 270; and Charles Webster, *From Paracelsus to Newton: Magic and the making of modern science* (Cambridge, 1982), 69.
18. J. B. van Helmont, *A ternary of paradoxes; the magnetical cure of wounds, the nativity of tartar in wine, the image of God in man, translated, illustrated and ampliated by Walter Charleton* (London, 1650), sig. D4v, D2r, C2r.
19. Walter Charleton, *The darknes of atheism dispelled by the light of nature: A physico-theologicall treatise* (London, 1652), 47, 44, 46. See also the following section of this article. It should be

clear from what has been said here that the claim that Charleton's work shows a dramatic change from an "overwhelming Renaissance-alchemical" position to a "blatantly atomistic" and "modern" approach is misconceived. See N. Rattner Gelbart, "The intellectual development of Walter Charleton", *Ambix*, xviii (1971), 149-68, pp. 149-50. Gelbart was extending here a view proposed by P. M. Rattansi, "Paracelsus and the Puritan Revolution", *Ambix*, xi (1963), 24-32. In fact, these views are casualties of the mistaken historiography I am seeking to correct. Providing we realize that it was perfectly possible to amalgamate atomist heuristic with Paracelsian/Helmontian and similarly alchemical and vitalistic world views (see ref. 17), then we can begin to read Charleton's and other English mechanical philosophers' works aright. For another critique of Gelbart's views from a different perspective see Lotte Mulligan, " 'Reason', 'right reason', and 'revelation' in mid-seventeenth-century England", in Vickers (ed.), *op. cit.* (ref. 6), 375-401, pp. 379-82.

20. [Sir Matthew Hale,] *Observations touching the principles of natural motions; and especially touching rarefaction and condensation: Together with a reply to certain remarks touching the gravitation of fluids* (London, 1677), 3-5. Hale has yet to receive full scholarly attention. In the meantime see Sir Matthew Hale, *The works, moral and religious.... To which are prefixed his life and death by Bishop Burnet, D.D. and an appendix to the life, including the additional notes of Richard Baxter* (London, 1805); and Edmund Heward, *Matthew Hale* (London, 1972).
21. Hale, *Observations* (ref. 20), 8, 9, 13, 10. On Newton's drafts for the Queries see McGuire, "Force, active principles and Newton's invisible realm" (ref. 7), 171. The quotation appears in Cambridge University Library, Add. MS 3970, f. 620r.
22. Henry Power, *Experimental philosophy, in three books: Containing new experiments, microscopical, mercurial, magnetical. With some deductions and probable hypotheses, raised from them in avouchment and illustration of the now famous atomical hypothesis* (London, 1664), sig. b4r, b3v, c2r, p. 61.
23. Charles Webster, "Henry Power's experimental philosophy", *Ambix*, xiv (1967), 150-78. Francis Glisson, *Tractatus de natura substantiae energetica, seu de vita naturae eiusque tribus primis facultatibus I. perceptiva, II. appetitiva, et III. motiva* (London, 1672). Glisson's ideas on "the energetic nature of substance" were first mooted in his *Anatomia hepatis* (London, 1654) and there are clear antecedents for them in the medical tradition and particularly in Van Helmont and Harvey. As we saw earlier with Charleton, writers in the medical tradition were often most free with concepts of active matter. See Owsei Temkin, "The classical roots of Glisson's doctrine of irritation", *Bulletin of the history of medicine*, xxxviii (1964), 297-328; Walter Pagel, "Harvey and Glisson on irritability with a note on Van Helmont", *Bulletin of the history of medicine*, xli (1967), 497-514; *idem*, *New light on William Harvey* (Basel, 1976), 34-36; *idem*, *Joan Baptista van Helmont, reformer of science and medicine* (Cambridge, 1982), 120-3; and John Henry, "Medicine and pneumatology: Henry More, Richard Baxter and Francis Glisson's *Treatise on the energetic nature of substance*", *Medical history*, forthcoming. For more general treatments of the role of active matter in the medical tradition, see W. Pagel, *Paracelsus: An introduction to philosophical medicine in the era of the Renaissance* (2nd edition, Basel, 1982); *idem*, *New light on William Harvey* (Basel, 1976), 21-22, 27-28, 63-66, 74-111; and Owen Hannaway, *The chemists and the word: The didactic origins of chemistry* (Baltimore and London, 1975), 24, 29, 34.
24. "Vegetation of metals" is the working title of Newton scholars for Burndy MS 16 with the *incipit* "Of Natures obvious laws & processes in vegetation"; the quotation is from f.5r. It is discussed in Rattansi's and Westfall's essays on Newton's alchemy cited in ref. 7, and in B. J. T. Dobbs, "Newton manuscripts at the Smithsonian Institution", *Isis*, lxviii (1977), 105-

7; and *idem*, *op. cit.* (ref. 7), 517-19. The quotation from the alchemical "Propositions" is taken from the latter, p. 515. For a full survey of Newton's early alchemical writings see B. J. T. Dobbs, *The foundations of Newton's alchemy, or "The Hunting of the Greene Lyon"* (Cambridge, 1975).

25. Burndy MS 16, f.3r. For a discussion of *flamma vitalis* and fermentation in biomedical thought see Douglas McKie, "Fire and the *flamma vitalis*: Boyle, Hooke and Mayow", in E. A. Underwood (ed.), *Science, medicine and history: Essays on the evolution of scientific thought and medical practice written in honour of Charles Singer* (2 vols, London, 1953), i, 469-88; Everett Mendelsohn, *Heat and life, the development of the theory of animal heat* (Cambridge, 1964), 27-66; A. B. Davis, *Circulation physiology and medical chemistry in England 1650-1680* (Lawrence, Kansas, 1973); and R. G. Frank Jr, *Harvey and the Oxford physiologists: Scientific ideas and social interaction* (Berkeley, 1980).
26. Thomas Willis, "A medico-philosophical discourse of Fermentation; or, Of the intestine motion of particles in every body", in *Dr Willis' Practice of physick, being the whole works* ..., translated by S. Pordage (London, 1684), 9. On Willis see Davis, *op. cit.* (ref. 25), 81-90; Frank, *op. cit.* (ref. 25), 165-9, 221-4, 232-7, 248-50; and T. S. Hall, *Ideas of life and matter* (2 vols, Chicago, 1969), i, 312-25. On Mayow see Frank, *op. cit.* (ref. 25), 224-32, 258-74; and Hall, i, 326-36. On Lower see Frank, 188-205, 208-20; and Hall, i, 337-42.
27. Newton, *op. cit.* (ref. 12), 399-400, see also 401. *Cf.* McGuire, "Force, active principles and Newton's invisible realm" (ref. 7), 171; and Dobbs, *op. cit.* (ref. 7), 515-19.
28. The "Hypothesis of light" is conveniently reprinted in I. B. Cohen (ed.), *Isaac Newton's papers & letters on natural philosophy and related documents* (Cambridge, Mass., 1978), 178-90; for the quotations see p. 181. Newton's letter to Oldenburg, 25 January 1675/6 is reproduced in the same collection, p. 254. On Mayow see ref. 26, and Westfall, "Newton and the Hermetic tradition" (ref. 7), 188-90. See also R. S. Westfall, *Force in Newton's physics* (London and New York, 1971), 367, 389-91.
29. On Boyle's reluctance to commit himself to theories see the "Proemial essay" in *Certain physiological essays, written at different times, and on several occasions* (London, 1661), in *Works* (ref. 11), i, 307. See also L. T. More, *The life and works of the Honourable Robert Boyle* (Oxford, 1944), 231-5; H. G. van Leeuwen, *The problem of certainty in English thought 1630-1690* (The Hague, 1963), 91-106; and R. S. Westfall, *Science and religion in seventeenth-century England* (New Haven, 1958), 167-74. *The general history of the air designed and begun* (London, 1692) appears in *Works*, v, 609-743, p. 641.
30. Boyle, *Works* (ref. 11), iv, 85-104, p. 90. Hall, *op. cit.* (ref. 26), i, 291-2, discusses the Paracelsian nature of these and similar ideas in Boyle's writings.
31. Boyle, *Works* (ref. 11), iv, 79.
32. Boyle, *Works* (ref. 11), iii, 306-16, p. 306; 316-25, p. 316. The concept of 'powers' or 'faculties' is a highly complex philosophical notion and should be treated with some care. For philosophical analyses of the concept in both Boyle and Locke, see F. J. O'Toole, "Qualities and powers in the corpuscular philosophy of Robert Boyle", *Journal of the history of philosophy*, xxii (1974), 295-315; Peter Alexander, *Ideas, qualities and corpuscles: Locke and Boyle on the external world* (Cambridge, 1985), especially pp. 150-67; and R. M. Mattern, "Locke on active power and the obscure idea of active power from bodies", *Studies in the history and philosophy of science*, xi (1980), 39-77. O'Toole argues that for Boyle powers cannot simply be reduced to the primary (mechanical) qualities but are a *function* of those qualities (pp. 310-11); Alexander, similarly, argues that powers may be intrinsic properties of things but only by virture of being entailed by the primary qualities (p. 160). Clearly they both feel that powers can be explained, ultimately, in strictly mechanist terms. I differ from them in this. I feel that they are further victims of the prevailing historiography. Alexander, for example, insists that Boyle was "suspicious" of

"occult qualities" and Newtonian forces. As we have seen, this is not necessarily so and we will try to explain why some occult qualities were acceptable to the mechanical philosophers in the third section of this article. Keith Hutchison has already pointed out the tendency of historians (and philosophers) to misread the evidence on this matter and has argued that seventeenth century natural philosophers only rejected certain kinds of occult quality; see his importantly revisionist paper: Keith Hutchison, "What happened to occult qualities in the Scientific Revolution?", *Isis*, lxxiii (1982), 233-53, pp. 235, 252. Of course, given his probabilistic outlook (see ref. 29 and L. Laudan, "The clock metaphor and probabilism: The impact of Descartes on English methodological thought, 1650-65", *Annals of science*, xxii (1966), 73-104), Boyle no doubt thought it possible that one day all powers of bodies would be explained in strict mechanist terms but, by the same token, he could also entertain the possibility that there may be other "ways of working". Such notions were perfectly acceptable providing they could be investigated experimentally (see third section of this article). For an excellent defence of these ideas by a philosopher of science see R. Harré, "Powers", *British journal for the philosophy of science*, xxi (1970), 81-101.

33. Boyle, *Suspicions about some hidden qualities of the air* (1674), *Works* (ref. 11), iv, 85-103, p. 89, and *Of the great effects of languid and unheeded motions* (1685), *Works* (ref. 11), v, 1-31, pp. 26-27.
34. Boyle, *Languid and unheeded motions*, *Works* (ref. 11), v, 14. For further examples see ch. 5 of this essay, "Of the propagable nature of motion". See also Hall, *op. cit.* (ref. 26), i, 287.
35. Spinoza to Oldenburg, April 1662. *The correspondence of Henry Oldenburg*, edited and translated by A. R. Hall and M. B. Hall (13 vols continuing, Madison etc., 1965-), i, 462 (Latin, p. 453). Oldenburg to Spinoza, 3 April 1663, *ibid.*, ii, 42 (Latin, p. 39). Edward Stillingfleet, *The works* (6 vols, London, 1710), ii, 287. The emphasis is Stillingfleet's own.
36. Boyle, *Excellency and grounds*, *Works* (ref. 11), iv, 72; *General history of the air*, *Works*, v, 615. Cohen (ed.), *op. cit.* (ref. 28), 181, 179. See also Robert Boyle, "An explication of rarefaction", *Works*, i, 179-80; *New experiments physico-mechanical touching the spring of the air*, *Works*, i, 12; *The history of fluidity and firmness*, *Works*, i, 387; *Of absolute rest in bodies: An essay of the intestine motions of the particles of quiescent solids, where the absolute rest of bodies is called in question*, *Works*, i, 443-57; "An historical account of a strangely self-moving liquor", *Works*, v, 71-73. And Hall and Hall (eds), *op. cit.* (ref. 11), 224.
37. Robert Hooke, *Micrographia: Or some physiological descriptions of minute bodies made by magnifying glasses, with observations and inquiries thereupon* (London, 1665), 16.
38. Robert Hooke, *The posthumous works*, ed. by Richard Waller (London, 1705), 166, 184.
39. Henry More, *The second lash of Alazonomastix laid on in mercie upon that stubborn youth Eugenius Philalethes: Or a sober reply to a very uncivill answer to certain observations upon Anthroposophia Theomagica and Anima magica abscondita* (London, 1651), 67. Occasionally, of course, it is possible to find Hooke attempting to give a strict mechanical account of such vibrations. *Cf. De potentia restitutiva*: "This Vibrative motion I do not suppose inherent or inseparable from the Particles of body, but communicated by Impulses given from other bodies in the Universe", in R. T. Gunther (ed.), *Early science in Oxford* (14 vols, Oxford, 1921-45), viii, 340. Presumably Hooke's inconsistency stems from the specific occasion of each utterance. An awareness of the kind of criticism that might be levelled at him from a Cartesian or even a theologian like More might stimulate a greater confidence in strict mechanism than on other occasions.
40. Marie Boas Hall, "The establishment of the mechanical philosophy", *Osiris*, x (1952), 412-541, p. 417.

41. Hooke, *Micrographia* (ref. 37), 12-15.
42. *Ibid.*, 28, 22.
43. These speculations appear in Query 21 of Newton's *Opticks* and in his letter to Robert Boyle of February 1679, which is reprinted in Cohen (ed.), *op. cit.* (ref. 28), 250-3. See also Newton's *De aere et aethere* in Hall and Hall (eds), *op. cit.* (ref. 11), 214-28.
44. On Newton's aether see Hall and Hall (eds), *op. cit.* (ref. 11), 183-213; *idem*, "Newton's theory of matter", *Isis*, li (1960), 131-44; Henry Guerlac, "Newton's optical aether", *Notes and records of the Royal Society*, xxii (1967), 45-57; J. L. Heilbron, *Electricity in the 17th and 18th centuries* (Berkeley, 1979); and R. W. Home, "Newton on electricity and the aether", in Z. Bechler (ed.), *Contemporary Newtonian research* (Dordrecht, 1982), 191-213.
45. See Charles Webster, "The discovery of Boyle's Law, and the concept of elasticity of air in the seventeenth century", *Archive for history of exact sciences*, ii (1965), 441-502.
46. Walter Charleton, *Natural history of nutrition, life and voluntary motion: Containing all the new discoveries of anatomist's, and most probable opinions of physicians, concerning the oeconomie of human nature...* (London, 1659), 13; Willis, *op. cit.* (ref. 26), 1.
47. Boyle, "Of the admirably differing extension of the same quantity of air rarefied and compressed", *Works* (ref. 11), iii, 509. See also, *Works*, i, 11-12, 19-20, 178-82; ii, 503-4; iii, 278-9; v, 28.
48. Aristotle, *Physics*, 216b23-34.
49. Sir Kenelm Digby, *Two treatises. In the one of which the nature of bodies; in the other the nature of mans soule is looked into: In way of discovery of the immortality of reasonable souls* (Paris, 1644), 21.
50. Hall and Hall (eds), *op. cit.* (ref. 11), 317 (Latin, p. 314); see also 306 (303), 341 (328). Newton had read Digby's *Two treatises* early in his career. See Westfall, *op. cit.* (ref. 15), 172; J. E. McGuire, "The origin of Newton's doctrine of essential qualities", *Centaurus*, xii (1968), 233-60, pp. 244-6. Westfall has not made this connection between Newton's net-like matter and anti-atomist arguments but has linked it instead to Newton's searches for the alchemical 'net'. See Westfall, *op. cit.* (ref. 8), 389, 508-9; and *idem*, "Alchemy in Newton's career" (ref. 7), 209-10. See also Dobbs, *Foundations of Newton's alchemy* (ref. 24), 161-3.
51. Hall and Hall (eds), *op. cit.* (ref. 11), 223, (Latin, p. 216).
52. Westfall, *op. cit.* (ref. 6), 317; and *idem*, *op. cit.* (ref. 28), 377.
53. [Sir Matthew Hale,] *Difficiles nugae: Or, observations touching the Torricellian experiment, and the various solutions of the same, especially touching the weight and elasticity of the air. The second edition with some occasional additions* (London, 1675), 181. Sir Kenelm Digby calculated that the proportion of empty space to matter in air must be at least as disparate as 7,600 to 1. This was based on the assumption that gold was completely free of void interstices and was 7,600 times denser than air. Digby, *op. cit.* (ref. 49), 20-21. Digby drew his figures from Marinus Ghetaldus, *Promotus archimedis seu variis corporum generibus gravitate et magnitudine comparatus* (Rome, 1603), 32.
54. Hale, *op. cit.* (ref. 53), 183.
55. Charleton, *op. cit.* (ref. 19), 47.
56. Sir William Petty, *The discourse made before the Royal Society, the 26 November 1674. Concerning the use of duplicate proportion in sundry important particulars: together with a new hypothesis of springing or elastique motions* (London, 1674), 4. Hall and Hall (eds), *op. cit.* (ref. 11), 333 (Latin, p. 321). See also McGuire, "Force, active principles and Newton's invisible realm" (ref. 7), 164-6; and *idem*, "Atoms and the 'analogy of Nature': Newton's third rule of philosophizing", *Studies in history and philosophy of science*, i (1970), 3-58.
57. Petty, *op. cit.* (ref. 56), 125-6.
58. *Ibid.*, 126-30.

59. *Ibid.*, sig. A5r. Petty's matter theory has not received the attention it deserves. For an excellent study of his natural philosophy in general see L. G. Sharp, "Sir William Petty and some aspects of seventeenth-century natural philosophy" (D.Phil. thesis, University of Oxford, 1977).
60. Letter from More to Anne, Lady Conway, 5 July 1662, reprinted in M. H. Nicolson (ed.), *Conway letters: The correspondence of Anne, Viscountess Conway, Henry More, and their friends, 1642-1684* (New Haven, 1930), 204. See above at ref. 4.
61. Letter from More to Boyle, 4 December 1671, reprinted in Boyle, *Works* (ref. 11), vi, 513. This letter has frequently been dated to 1665; for a discussion of the dating of this letter and the best discussion of More's relationship to Descartes see Gabbey, *op. cit.* (ref. 4), 284n. On Hobbes as 'atheist' see S. I. Mintz, *The hunting of Leviathan: Seventeenth-century reactions to the materialism and moral philosophy of Thomas Hobbes* (Cambridge, 1970).
62. Newton, *op. cit.* (ref. 12), 397. Boyle, *Some considerations touching the usefulness of experimental natural philosophy* (1663), *Works* (ref. 11), ii, 42-43; see also *Origin of forms and qualities*, *Works*, iii, 15. Consider also ch. 3 of Sir Matthew Hale, *Magnetismus magnus: Or, metaphysical and divine contemplations on the magnet or loadstone* (London, 1695), 13-29: "The evidence of the existence of the Glorious God, from the supposition of a self-moving principle in Nature"; and Charleton, *op. cit.* (ref. 15), 436.
63. See R. Descartes, *Principia philosophiae*, II, 39; and R. Hooykaas, *Religion and the rise of modern science* (Edinburgh and London, 1973), 19-22. The literature on Malebranche is vast but see, for our purposes, Charles J. McCracken, *Malebranche and British philosophy* (Oxford, 1983).
64. Isaac Barrow, *Cartesiana hypothesis de materia et motu haud satisfacit praecipuis naturae phaenomenis* (1652), reprinted in *Theological works*, ed. by A. Napier (9 vols, Cambridge 1859), ix, 79-104. I quote from the extensive translated extracts in Percy H. Osmond, *Isaac Barrow, his life and times* (London, 1944), 31.
65. [Joseph Glanvill and George Rust,] *Two choice and useful treatises: the one Lux orientalis or an enquiry into the opinion of the eastern sages concerning the praeexistence of souls.... The other a discourse of truth ... with annotations on them both* [by Henry More] (London, 1682), 10, and "Annotations upon Lux orientalis" (separately paginated), 9.
66. Boyle, *A free inquiry into the vulgarly received notion of Nature*, *Works* (ref. 11), v, 158-254, pp. 165, 162; see also *A disquisition about the final causes of natural things* (1688), *Works*, v, 399.
67. Boyle, *A free inquiry*, 163, and also 162-3. An understanding of the distinction between God's absolute and ordinary powers or 'concourses' is crucial here. For reliable guidance which relates directly to Boyle's *Notion of nature* see J. E. McGuire, "Boyle's conception of nature", *Journal of the history of ideas*, xxxiii (1972), 523-42. For further discussions see *idem*, "Force, active principles and Newton's invisible realm" (ref. 7), 187-94; Francis Oakley, "Christian theology and the Newtonian science", *Church history*, xxx (1961), 433-57; and *idem*, *Omnipotence, covenant, & order: An excursion in the history of ideas from Abelard to Leibniz* (Ithaca and London, 1984).
68. Charleton, *op. cit.* (ref. 19), 46; [Sir Matthew Hale,] *An essay touching on the gravitation, or non-gravitation of fluid bodies, and the reasons thereof: The second edition with some occasional additions* (London, 1675), 44.
69. Westfall, *op. cit.* (ref. 8), 509-10. The fact that Newton reverted to naturalistic explanations can be seen in the Queries to the *Opticks* but it is also well brought out in David Kubrin, "Newton and the cyclical cosmos: Providence and the mechanical philosophy", *Journal of the history of ideas*, xxviii (1967), 325-46. The evident reluctance of English natural philosophers to involve God directly in their natural philosophies, as Westfall has perceptively noted, "prepared the ground for the deists of the Enlightenment", Westfall, *op. cit.* (ref. 29), 219. Similarly, in a recent essay Westfall has argued that Newton was a

crypto-deist in spite of himself: *idem*, "Newton's theological manuscripts", in Bechler (ed.), *op. cit.* (ref. 44), 129-43, p. 140.

70. I have discussed this more fully in J. Henry, "Henry More *v.* Robert Boyle: The Spirit of Nature and the nature of Providence", forthcoming.
71. Henry More, *The immortality of the soul* in *Collection of several philosophical writings* (London, 1662), 11; Robert Hooke, *Lampas: Or, descriptions of some mechanical improvements of lamps and waterpoises* (London, 1677), reprinted in R. T. Gunther (ed.), *op. cit.* (ref. 39), viii, 188. Boyle's refutations of More appear in Boyle, *op. cit.* (ref. 66), and "An hydrostatical discourse occasioned by some objections of Dr Henry More ..." (1672), in *Works* (ref. 11), iii, 596-628. Hale rebuts More in his *Observations* (ref. 20).
72. Hale, *Observations* (ref. 20), 28.
73. There is a vast body of seventeenth century literature devoted to the refutation of atheism as though it is a serious presence in contemporary mores. Much of this literature attacks the new philosophy as a major source of encouragement for atheists. This topic is yet to receive definitive treatment but in the meantime see Westfall, *op. cit.* (ref. 29); Michael Hunter, *Science and society in Restoration England* (Cambridge, 1981), 162-87; and John Redwood, *Reason, ridicule and religion: The age of enlightenment in England 1660-1750* (London, 1976).
74. Henry More, *Enthusiasmus triumphatus: Or, a brief discourse of the nature, causes, kinds, and cure of enthusiasm* (London, 1656), quoted from *A collection of several philosophical writings* (London, 1662), 1.
75. *Ibid.*, 30, 31; see also 36.
76. I have argued this more fully in Henry, *op. cit.* (ref. 23).
77. Perceptive or appetitive matter was directly linked to the automotive power of matter by Francis Glisson, *op. cit.* (ref. 23), and was an explicit feature of many biomedical theories, see immediately below. On Boyle's rejection of these ideas see ref. 107 below. It should be noted that Bentley's second Boyle lecture was entitled "Matter and motion cannot think: Or a confutation of atheism from the faculties of the soul". The most convenient study of Newton's belief in active matter is McMullin, *op. cit.* (ref. 5).
78. See ref. 23 above.
79. On the contrary, Locke felt compelled by his voluntaristic theology (a theological viewpoint he shared with virtually every mechanical philosopher in seventeenth century England) to *admit* the possibility of thinking matter. See, for example, John Locke, *An essay concerning human understanding* (London, 1706), IV, iii, 6. I will return to the voluntaristic demand for the acknowledgement of active matter as, at least, a possibility below. For a discussion of thinking matter in Locke's work see M. D. Wilson, "Superadded properties: The limits of mechanism in Locke", *American philosophical quarterly*, xvi (1979), 143-50, which should be supplemented with Edwin McCann, "Lockean mechanism", in A. J. Holland (ed.), *Philosophy, its history and historiography* (Dordrecht, 1985), 209-31; and John Yolton, *Thinking matter: Materialism in eighteenth-century Britain* (Oxford, 1984). The claim that Locke's philosophy is indebted to his medical training in various ways has been made in Patrick Romanell, *John Locke and medicine: A new key to Locke* (New York, 1984); see also Kenneth Dewhurst, *John Locke (1632-1704): Physician and philosopher. A medical biography* (London, 1963).
80. This is essentially the thesis of J. R. Jacob and M. C. Jacob, "The Anglican origins of modern science: The metaphysical foundations of the Whig constitution", *Isis*, lxxi (1980), 251-67; and is implicit in much of their other work. See for example, J. R. Jacob, *Henry Stubbe, radical Protestantism and the early Enlightenment* (Cambridge, 1983); and M. C. Jacob, *The Newtonians and the English Revolution, 1689-1720* (Hassocks, 1976). It should be apparent from what has been said here that this kind of affiliation of matter theory to ideology will

not do. There are broadly similar claims made by a number of other historians. Carolyn Merchant, *The death of nature: Women, ecology and the Scientific Revolution* (San Francisco, 1980), for example, wishes to argue that the magical world view, occultism and a vitalistic belief in active matter can be associated, historically, with naturalistic, ecological and even 'feminine' ideals, while the mechanical philosophy insisted upon dead, passive matter and was exploitive, aggressive, antithetical to the concept of nature and anti-feminist. For similar claims see David Kubrin, "Newton's inside out: Magic, class struggle, and the rise of mechanism in the West", in Harry Woolf (ed.), *The analytic spirit: Essays in the history of science in honor of Henry Guerlac* (Ithaca, 1981), 96-121; and Brian Easlea, *Witch-hunting, magic and the New Philosophy: An introduction to debates of the Scientific Revolution, 1450-1750* (Hassocks, 1980). I believe that each of these writers and their theses have fallen victim to the prevailing mistaken historiography of pre-Newtonian matter theory (and like the victims of the vampire they then serve to propagate the same erroneous historiography).

81. Westfall, *op. cit.* (ref. 6), 324. Newton, *op. cit.* (ref. 12), 401. In using the word 'phenomenological' I mean it only in its literal sense: the study of phenomena as opposed to the study of being. I do not intend it to convey all the philosophical baggage attached to the word by Husserl and his followers. It is important to note also that our natural philosophers were not phenomenalists: they did not believe that only phenomena are real. They believed in a reality underlying phenomena, knowledge of which could frequently be inferred (at least within certain limits laid down by their mitigated scepticism — see ref. 99 below) from experimental investigation.

82. I. B. Cohen, "The *Principia*, universal gravitation, and the 'Newtonian style', in relation to the Newtonian Revolution in science", in Bechler (ed.), *op. cit.* (ref. 44), 21-108; and *idem, The Newtonian Revolution, with illustrations of the transformation of scientific ideas* (Cambridge, 1980). The erudition and historical acumen of Professor Cohen together with the complexity and subtlety of his case go a long way to making these much more than hagiography. Nevertheless, I can't help feeling they would be suitable candidates for analysis along the lines suggested in John A. Schuster, "Methodologies as mythic structures: A preface to the future historiography of method", *Metascience*, i (1984), 15-36.

83. Cohen, "The *Principia* ... and the 'Newtonian style' " (ref. 82), 56, and see also 62; Isaac Newton, *Mathematical principles of natural philosophy and his system of the world, translated by Andrew Motte in 1729 revised by Florian Cajori* (Berkeley, 1960), 546-7. For further examples of Newton's insistence that "we know the properties of things from phenomena", and related ideas (including the interesting remark: "I avoid hypotheses whether mechanical or of occult qualities") see Hall and Hall (eds), *op. cit.* (ref. 11), 360-1 (Latin, 356), which is a draft for the General Scholium.

84. Glanvill, *Vanity of dogmatizing* (ref. 3), 210. Of course, the explanatory lacunae also provide an ever-present guarantee that the mechanical philosophy cannot be commandeered for atheist purposes. My claim here, that the ascription of occult qualities in matter as an explanation of a given phenomenon will lead to further experimental investigation of the matter and the phenomenon, has recently been described as "*what scientists actually do*" in Harré, *op. cit.* (ref. 32), 90.

85. Glanvill, *Vanity of dogmatizing* (ref. 3), 207-8. On Digby's "powder of sympathy" see B. J. T. Dobbs, "Studies in the natural philosophy of Sir Kenelm Digby", *Ambix*, xviii (1971), 1-25.

86. Glanvill, *Vanity of dogmatizing* (ref. 3), 202-4, p. 204.

87. Boyle, *Works* (ref. 11), ii, 36-37.

88. Boyle, "Hydrostatical discourse", *Works* (ref. 11), iii, 601; *cf.* 627.

89. Boyle, *New experiments physico-mechanical*, *Works*, i, 12.
90. Boyle, *Languid and unheeded motion*, *Works*, v, 26.
91. H. G. Alexander (ed.), *The Leibniz-Clarke correspondence* (Manchester, 1956), 94; Boyle, *Cosmicall qualities*, *Works*, iii, 307.
92. Hooke, *op. cit.* (ref. 39), 179. See also J. A. Bennett, "Hooke and Wren and the system of the world: Some points towards an historical account", *The British journal for the history of science*, viii (1975), 32-61, pp. 44-47; and *idem*, "Robert Hooke as mechanic and natural philosopher", *Notes and records of the Royal Society of London*, xxxv (1980), 33-48, p. 43. Petty, *op. cit.* (ref. 56), 132-3.
93. Boyle, *Hidden qualities of the air*, *Works* (ref. 11), iv, 96, 88, 79.
94. Hutchison, *op. cit.* (ref. 32). In making his case Hutchison also shows how historians have misinterpreted the many attacks on occult qualities which were prevalent at this time. George MacDonald Ross, "Occultism and philosophy in the seventeenth century", in Holland (ed.), *op. cit.* (ref. 79), 95-115, makes a broadly similar point when he says that "the new philosophers following the 'way of ideas' were even more dependent than the occultists on hidden or occult entities" (p. 102).
95. I quote from the draft Newton sent to Cotes on 2 March 1712/13 which now forms part of the Newton-Cotes Correspondence (Trinity College Library, R.16.38). I have used the translations provided in I. B. Cohen, *Introduction to Newton's 'Principia'* (Cambridge, 1971), 241. Cohen discusses the various drafts of the scholium, pp. 240-5.
96. Letter from Leibniz to Conti, November/December 1715; draft of letter from Newton to Conti, 26 February 1715/16. These are quoted from and fully discussed in A. Koyré and I. B. Cohen, "Newton and the Leibniz-Clarke correspondence, with notes on Newton, Conti & Des Maizeaux", *Archives internationales d'histoire des sciences*, xv (1962), 63-126, pp. 70 and 74; see also pp. 110 and 113 for drafts which make similar comments.
97. Newton, *op. cit.* (ref. 12), 401-2. The first of these extracts does not appear in the earlier editions of the *Opticks* (1704 and 1706 — before Leibniz's attack on occult qualities). See also Simon Schaffer, "Occultism and reason", in Holland (ed.), *op. cit.* (ref. 79), 117-43, pp. 136-9.
98. I have discussed Boyle, Petty, Glanvill and Hooke above; on Charleton see Hutchison, *op. cit.* (ref. 32), 245, where we are told that "Charleton is attacking not occult qualities but the Aristotelians".
99. Christopher Wren Jr, *Parentalia: Or, memoirs of the family of the Wrens ... but chiefly of Sir Christopher Wren ... in which is contained, besides his works, a great number of original papers and records* ... (London, 1750), 221. See also J. A. Bennett, "A note on theories of respiration and muscular action in England, c. 1660", *Medical history*, xx (1976), 56-59. Glanvill, *Vanity of dogmatizing* (ref. 3), 210. On Newton's use of alchemy or chemistry to investigate active principles see the works cited in ref. 7 above. For Leibniz see G. McD. Ross, "Alchemy and the development of Leibniz's metaphysics", *Studia Leibnitiana supplementa*, xxii (1982), 40-45. On the development of new experimental research traditions at this time see T. S. Kuhn, "Mathematical versus experimental traditions in the development of physical science", *The essential tension: Selected studies in scientific tradition and change* (Chicago, 1977), 31-65, pp. 41-52; and William Eamon, "Arcana disclosed: The advent of printing, the Book of Secrets tradition and the development of experimental science in the sixteenth century", *History of science*, xxii (1984), 111-50, pp. 112-13.
100. Marquis of Lansdowne (ed.), *The Petty papers: Some unpublished writings of Sir William Petty, edited from the Bowood Papers* (2 vols, London, 1927), ii, 207-8; see also ii, 266-8. Louise D. Patterson, "The pendulums of Wren and Hooke", *Osiris*, x (1952), 277-321; and Penelope

Gouk, "The role of acoustics and music theory in the scientific work of Robert Hooke", *Annals of science*, xxxvii (1980), 573-605.

101. Frank, *op. cit.* (ref. 25), 261 and *passim.* Nitre also played a part in theories of muscular contraction developed by Walter Charleton, Thomas Willis and others. See Bennett, *op. cit.* (ref. 99), and T. M. Brown, "The mechanical philosophy and the 'animal oeconomy': A study of the development of English physiology in the 17th and early 18th century" (Ph.D. thesis, Princeton University, 1968).

102. Webster, *op. cit.* (ref. 23); Frank, *op. cit.* (ref. 25); Simon Schaffer, "Godly men and mechanical philosophers: Matter and spirit in English natural philosophy in the 1670s", *Science in context*, forthcoming. On Mayow and the elasticity of the air, see Frank, *ibid.*, 262-3. See also A. G. Debus, "Chemistry and the quest for a material spirit of life in the seventeenth century", in M. Fattori and M. Bianchi (eds), *Spiritus: IV° Colloquia Internationale del Lessico Intellettuale Europeo* (Rome, 1984), 245-63.

103. Consider Boyle, *Usefulness of experimental natural philosophy, Works* (ref. 11), ii, 45; or *Certain physiological essays, Works*, i, 355-6; and the secondary sources cited in ref. 29. The ideological uses of 'mitigated scepticism' were, of course, first chronicled in R. H. Popkin's *A history of scepticism* (1960), now revised as *A history of scepticism from Erasmus to Spinoza* (Berkeley, 1979); but see also Barbara J. Shapiro, *Probability and certainty in seventeenth-century England* (Princeton, 1983). I have given indications of the response to the dogmatic use of the mechanical philosophy to promote counter-reforming Catholic theology in John Henry, "Atomism and eschatology: Catholicism and natural philosophy in the Interregnum", *The British journal for the history of science*, xv (1982), 211-39, especially pp. 237-9. For the response to 'enthusiastic' natural philosophy see P. M. Rattansi, *op. cit.* (ref. 19). Although I am critical of Rattansi's paper in ref. 19 above, I fully accept his claim that various Paracelsian/Helmontian world views were associated by contemporaries with subversive 'enthusiasm' in religion. My point is merely that he is mistaken in seeing this as a factor leading to the rejection of concepts of active matter. Owen Hannaway in his study of an earlier debate about Paracelsian philosophy has quite clearly shown that it is possible for a thinker to reject Paracelsian 'enthusiasm' while still accepting Paracelsian active principles in matter; Hannaway, *op. cit.* (ref. 23), 84-85, 90-91, 99 and *passim.*

104. Paul Wood, "Methodology and apologetics: Thomas Sprat's *History of the Royal Society*", *The British journal for the history of science*, xiii (1980), 1-26; *idem*, "Francis Bacon and the 'experimentall philosophy': A study in seventeenth-century methodology" (M.Phil. thesis, University of London, 1978); Steven Shapin, "Pump and circumstance: Robert Boyle's literary technology", *Social studies of science*, xiv (1984), 481-520; Steven Shapin and Simon Schaffer, *Leviathan and the air-pump: Hobbes, Boyle and the experimental life* (Princeton, 1985); and Peter Dear, "*Totius in verba*: Rhetoric and authority in the early Royal Society", *Isis*, lxxvi (1985), 145-61. Laudan, *op. cit.* (ref. 32) has argued that Descartes too was a mitigated sceptic, but for a corrective to this view see G. A. J. Rogers, "Descartes and the method of English science", *Annals of science*, xxii (1966), 73-104. I would go further than Rogers and argue that Descartes was frequently regarded, particularly by theologians, as a dogmatist. For example see Meric Casaubon, *On learning* (1667), reprinted in M. R. G. Spiller, *"Concerning natural experimental philosophie": Meric Casaubon and the Royal Society* (The Hague, 1980), 195-214, p. 203; and Edward Stillingfleet, *op. cit.* (ref. 35), ii, 101. It is even possible for a modern scholar, unconcerned about the subversive implications of Cartesianism, to claim that Descartes was "firmly in the camp of occultist enthusiasts". See G. MacDonald Ross, *op. cit.* (ref. 94), 113, note 54; also note 56.

105. On the importance of this tradition to the development of science see the works by McGuire

and Oakley cited in ref. 67; Hooykaas, *op. cit.* (ref. 63); Margaret J. Osler, "Descartes and Charleton on Nature and God", *Journal of the history of ideas*, xl (1979), 445-56; Eugene M. Klaaren, *Religious origins of modern science: Belief in creation in seventeenth-century thought* (Grand Rapids, 1977), 29-52. Oakley, "Christian theology and the Newtonian science" (ref. 67), 442. Unfortunately there are exceptions to this rule. Descartes and Hobbes both managed to combine voluntarism with rationalism. This needs to be explained. R. Hooykaas, *op. cit.* (ref. 63) noted merely that Descartes's combination of voluntarism and rationalism was "strange" (p. 41) and "paradoxical" (p. 42) but he made no attempt to account for it. Margaret Osler argues that Descartes is a modified necessitarian (necessitarianism being the theological position opposed to voluntarism) but I find this unconvincing. As far as I know there is no study of Hobbes's voluntarism but see J. Barnouw, "The separation of reason and faith in Bacon and Hobbes and Leibniz's *Theodicy*", *Journal of the history of ideas*, xlii (1981), 607-28, pp. 616-19. These are serious lacunae in our understanding of seventeenth century natural philosophy. Nevertheless, irrespective of Descartes's and Hobbes's attitudes, we can safely say that the majority of voluntarist mechanical philosophers professed themselves to be empiricist, and argued that their view of God and his relationship to the world was only compatible with empiricism.

106. Locke, *op. cit.* (ref. 79), IV, iii, 6.

107. Glisson, *De natura substantiae energetica* (ref. 23), sig. A5r. Let two examples of Boyle's rejection of 'sense and appetite' in matter suffice: *High veneration to God* (1685), *Works* (ref. 11), v, 141; and *Notion of Nature*, *Works*, v, 222. Ralph Cudworth, *The true intellectual system of the universe* (London, 1678), 105; Henry More, *Opera omnia* (2 vols, London, 1679), i, 604-11. More's and Cudworth's attacks on Glisson are discussed in Henry, *op. cit.* (ref. 23). More's necessitarianism and his opposition to Boyle's voluntarism are discussed in Henry, *op. cit.* (ref. 70). On Cudworth's necessitarianism see Oakley, "Christian theology and the Newtonian science" (ref. 67), 441; Richard Baxter, *Of the immortality of Mans Soule, and the nature of it, and other spirits ...* (London, 1682), 28-29.

108. Oakley, *Omnipotence, covenant, & order* (ref. 67), 114; Steven Shapin, " 'Of Gods and Kings': Natural philosophy and politics in the Leibniz-Clarke disputes", *Isis*, lxxii (1981), 187-215, p. 192.

109. Cohen (ed.), *op. cit.* (ref. 28), 185. I have yet to come across a natural philosopher in England who was not a voluntarist.

110. Bernard le Bovier de Fontenelle, *The elogium of Sir Isaac Newton* (London, 1728), 15; reprinted in Cohen (ed.), *op. cit.* (ref. 28), 445-74, p. 457. There is, however, at least one honourable exception among Newton scholars: Simon Schaffer, *op. cit.* (ref. 102).

111. See, for example, E. G. Ruestow, *Physics at seventeenth and eighteenth century Leiden* (The Hague, 1973); or compare L. W. B. Brockliss, "Aristotle, Descartes and the New Science: Natural philosophy at the University of Paris, 1600-1740", *Annals of science*, xxxviii (1981), 33-69; with R. G. Frank Jr, "Science, medicine and the universities of early modern England: Background and sources", *History of science*, xi (1973), 194-211, 239-69. Consider also David Hume's puzzlement in 1748: "I must confess, that there is something in the fate of opinions a little extraordinary. Des Cartes insinuated that doctrine of the universal and sole efficacy of the Deity without insisting on it. Malebranche and other Cartesians made it the foundation of all their philosophy. It had, however, no authority in England. Locke, Clarke and Cudworth, never so much as take notice of it, but suppose all along that matter has a real, though subordinate and derived power. By what means has it become so prevalent among our modern metaphysicians?" Hume, *op. cit.* (ref. 3), VIII, i, 57, in the footnote.

112. John Keill, *Introductio ad veram physicam: Seu lectiones physicae* (Oxford, 1702), sig. b1r. I

have taken the translation from the fourth edition: *An introduction to natural philosophy* (London, 1745), iii.

113. Westfall, "Newton and the Hermetic tradition" (ref. 7), 187. Hermeticism is currently out of vogue largely as a result of Westman and McGuire, *op. cit.* (ref. 7); and C. B. Schmitt, "Reappraisals in Renaissance science", *History of science*, xvi (1978), 200-14. For further surveys which suggest that Descartes was forced to smuggle activity into his system see Prendergast, *op. cit.* (ref. 5); and Hatfield, *op. cit.* (ref. 5).
114. Westfall, "Newton and the Hermetic tradition" (ref. 7), 187.
115. For the latter see refs 6 and 7 above. For the former consider René Dugas, *Mechanics in the seventeenth century* (Paris, 1958); John W. Herivel, *The background to Newton's "Principia"* (Oxford, 1965); developments in dynamics are also the major focus of Westfall, *op. cit.* (ref. 28); and Gabbey, *op. cit.* (ref. 5), and Cohen, *op. cit.* (ref. 82). See also D. T. Whiteside, "The mathematical principles underlying Newton's *Principia mathematica*", *Journal for the history of astronomy*, i (1970), 116-38; and *idem*, "Newton the mathematician", in Bechler (ed.), *op. cit.* (ref. 44), 109-27.
116. Newton, *op. cit.* (ref. 12), 401.
117. Once again the reader is referred to the spirited promotion of just this methodological approach in Harré, *op. cit.* (ref. 32).

IV

MEDICINE AND PNEUMATOLOGY: HENRY MORE, RICHARD BAXTER, AND FRANCIS GLISSON'S *TREATISE ON THE ENERGETIC NATURE OF SUBSTANCE*

The nature of the soul and its relationship to the body has always proved problematical for Christian philosophy. The source of the difficulty can be traced back to the efforts of the early Fathers to reconcile the essentially pagan concept of an immaterial and immortal soul with apostolic teachings about the after-life in which all the emphasis is placed upon the resurrection of the body. The tensions between these two traditions inevitably became strained during the sixteenth century when Protestant reformers insisted on a closer adherence to Scripture. Furthermore, even when leaving the problems of Scriptural hermeneutics aside, the dualistic approach to the question, in which soul (or spirit) and body are held to be categorically different in essence, had to overcome a number of intractable philosophical problems. So, it was not simply coincidence that when the new mechanical philosophy began to be formulated in a systematic way in the seventeenth century, it was couched in vigorously dualistic terms. In fact, three of the earliest fully elaborated systems of mechanical philosophy, those of Descartes, Digby, and Charleton, were explicitly intended to provide a philosophical prop for dualist theology.[1] Moreover, it was because of its usefulness in promoting dualism that Cartesianism was first popularized in England not by a natural philosopher but by the Cambridge Platonist and theologian, Henry More.[2]

[1] The literature regarding Descartes is vast but for a very recent treatment see Richard B. Carter, *Descartes' medical philosophy: the organic solution to the mind-body problem*, Baltimore Md., and London, Johns Hopkins University Press, 1983. Sir Kenelm Digby's intentions are made clear in the title of his new system of philosophy: *Two treatises. In the one of which the nature of bodies; in the other the nature of mans soule is looked into: in way of the discovery of the immortality of reasonable soules*, Paris, G. Blaizot, 1644. But see also John Henry, 'Atomism and eschatology: Catholicism and natural philosophy in the Interregnum', *Br. J. Hist. Sci.*, 1982, **15:** 211–239. Walter Charleton's major work, *Physiologia Epicuro-Gassendo-Charltoniana*, London, T. Newcomb for T. Heath, 1654, is described on the title-page as "the first part" and he announces his intention to write a second part on the nature of the soul at the end of the book, p.479. This never appeared in its intended form, but see W. Charleton, *The immortality of the human soul demonstrated by the light of nature. In two dialogues*, London, W. Wilson for H. Herringman, 1657.

[2] Here again the literature is extensive, but all earlier works are effectively superseded by Alan Gabbey, '*Philosophia Cartesiana triumphata:* Henry More (1646–1671)', in T. M. Lennon, J. M. Nicholas, and J. W. Davis (editors), *Problems of Cartesianism*, Kingston and Montreal, McGill-Queen's University Press, 1982, pp. 171–250.

In spite of this powerful support, however, there were still those who regarded the dichotomy between soul and body, so essential to dualism, as fundamentally inimical to any satisfactory explanation of psychosomatic interaction. Accordingly, monistic solutions to this problem also attracted some formidable advocates. Certainly, the greatest of these were the materialist Thomas Hobbes, the hylotheist Baruch de Spinoza, and the vitalist Gottfried Wilhelm Leibniz. But monistic pneumatology also found allies in a well-established tradition of medical thought. In a series of articles, D. P. Walker has shown the links between medical concepts of "spirit" and materialistic views of the human soul, while Walter Pagel has underlined the monistic pneumatology of two of the leading medical theorists in seventeenth-century England, William Harvey and Francis Glisson.[3]

The crucial nature of the ensuing debate between dualists and monists is revealed by the reaction of that champion of Cartesian dualism, Henry More. As is well known, More lost no time in publishing a detailed attack on Hobbes's *De corpore* (1655) in his *The immortality of the soul* (1659), and in the 1670s, he turned his attention to the new threats, Cartesian atheism and Spinozism.[4] However, what has hardly been noticed before is that More also wrote a refutation of Francis Glisson's *Tractatus de natura substantiae energetica* (1672).[5]

Glisson's *Treatise* deserves recognition as one of the most original systems of philosophy to appear in the second half of the seventeenth century and as one of the most profound attempts to develop a monistic solution to the mind-body problem. One reason for the failure of intellectual historians to recognize its importance stems from the commonly held belief that it had absolutely no influence upon Glisson's contemporaries. The aim of this paper is to make a start towards correcting this erroneous view. Glisson's philosophy, as expressed in the *Treatise*, is rich, complex and

[3] D. P. Walker, 'The astral body in Renaissance medicine', *J. Warburg and Courtauld Institutes*, 1958, **21:** 119–133; *idem*, 'Medical spirits in philosophy and theology from Ficino to Newton', in *Arts du spectacle et histoire des idées. Recueil offert en hommage à Jean Jacquot*, Tours, Centre d'Études Supérieures de la Renaissance, 1984, pp. 287–300; *idem* 'Medical spirits and God and the soul', in M. Fattori and M. Bianchi (editors), *Spiritus. IV° Colloquio internazionale del Lessico Intellettuale Europeo, Rome, 7–9 gennaio, 1983*, Rome, Edizioni dell'Ateneo, 1984, pp. 233–244. The first two of these are now reprinted in D. P. Walker, *Music, spirit and language in the Renaissance*, London, Variorum Reprints, 1985. For related material see also D. P. Walker, *Spiritual and demonic magic from Ficino to Campanella*, London, Warburg Institute, 1958. On monism in Harvey and Glisson see Walter Pagel, 'Harvey and Glisson on irritability with a note on Van Helmont', *Bull. Hist. Med.*, 1967, **41:** 497–514; *idem*, 'The reaction to Aristotle in seventeenth-century biological thought', in E. A. Underwood (editor), *Science, medicine and history: essays on the evolution of scientific thought and medical practice written in honour of Charles Singer*, London, Oxford University Press, 1953, pp. 489–509; and *idem*, *Joan Baptista van Helmont: reformer of science and medicine*, Cambridge University Press, 1982, pp. 120–123.

[4] Henry More, *The immortality of the soul so far forth as it is demonstrable from the light of nature*, London, James Flesher for William Morden, 1659. For More's concern with Cartesian forms of atheism see Gabbey, op. cit., note 2 above; for his reactions to Spinoza see R. L. Colie, *Light and enlightenment: a study of the Cambridge Platonists and the Dutch Arminians*, Cambridge University Press, 1957; *idem*, 'Spinoza in England, 1665–1730', *Proc. Amer. Philos. Soc.*, 1963, **107:** 183–219, especially pp. 183–193; and Sarah Hutton, 'Reason and revelation in the Cambridge Platonists and their reception of Spinoza', in K. Gründer and W. Schmidt-Biggemann (editors), *Spinoza in der Frühzeit seiner religiösen Wirkung*, Heidelberg, Verlag Lambert Schneider, 1984, pp. 181–200.

[5] Francis Glisson, *Tractatus de natura substantiae energetica, seu de vita naturae, eiusque tribus primis facultatibus, I. perceptiva naturalibus, II. appetitiva naturalibus, III. motiva naturalibus & c.*, London, E. Flesher for H. Brome and N. Hooke, 1672.

subtle, and manages to be innovatory while drawing heavily upon scholastic traditions and Renaissance philosophical eclecticism. A full exposition of his philosophy has yet to be attempted and is not within the scope of this essay.[6] Here, we will look at Glisson only as he appeared to Henry More and Richard Baxter, two of the leading contemporary English theologians, who explicity dealt with his theories. While considering the background to and the reasons for their interest in Glisson it will become clear that medical theories frequently had a profound influence on philosophy and theology in the early modern period. Before embarking on this, however, it is important to recognize the intimate connexion between Glisson's new system of natural philosophy and his medical concerns, and to this we will turn first.

FRANCIS GLISSON'S MEDICAL PHILOSOPHY

Glisson's output in print is not extensive, and there is a clear continuity of interests and themes running through his published works. The continuity is most clear between the *Anatomia hepatis* of 1654 and his next medical treatise, the *Tractatus de ventriculo et intestinis* of 1677. However, as Temkin and Pagel have pointed out, there are a number of notions employed by Glisson in his early efforts to understand the nature of rickets which foreshadow his later ideas. Temkin has traced Glisson's concept of irritability, most famously expounded in *De ventriculo*, to his description of the pulse in *De rachitide:* as the arteries thrust blood into the parts of the body there is a certain resistance which "provokes" or "irritates" the heart and arteries into increasing the pressure. A strong pulse will not be found, therefore, where resistance is low, as in the rachitic. Pagel, on the other hand, has pointed to Glisson's early use of the concept of "tension" or "tone" in the nerves to account for both sense and locomotion. By brooding on the notion of irritation and the role of the nerves in this process Glisson

[6] Glisson has been accorded a number of brief treatments, some of them very good, but he has not so far been considered more fully. An excellent starting place is Owsei Temkin's article on Glisson in the *Dictionary of Scientific Biography (DSB)* but consider also Pagel's papers cited in note 3 above and any of the following. Kurt Sprengel, *Versuch einer pragmatischen Geschichte der Arzneikunde*, 5 vols., Halle, J. J. Gebamer, 1803, vol. 5, pp. 105–109. C. Daremberg, *Histoire des sciences médicales, comprenant l'anatomie, la physiologie, la médecine, la chirurgie et les doctrines de pathologie générale*, 2 vols., Paris, J.-B Baillière, 1870, vol. 2, pp. 640–649 and 650–672; C.F.M. de Rémusat, *Histoire de la philosophie en Angleterre depuis Bacon jusqu'à Locke*, 2 vols., Paris, Didier, 1875, vol. 2, pp. 163–168; Owsei Temkin, 'The classical roots of Glisson's doctrine of irritation', *Bull. Hist. Med.*, 1964, **38:** 297–328; R. Milnes Walker, 'Francis Glisson and his capsule', *Ann. R. Coll. Surg. Engl*, 1966, **38:** 71–91; Audrey B. Davis, 'Some implications of the circulation theory for disease theory and treatment in the seventeenth century', *J. Hist. Med.*, 1971, **26:** 28–39; *idem*, 'The circulation of the blood and chemical anatomy', in A. G. Debus (editor), *Science, medicine and society in the Renaissance: essays in honour of Walter Pagel*, 2 vols., New York, Science History Publications, 1972, vol. 2, pp. 25–37; Nikolaus Mani, 'Biomedical thought in Glisson's hepatology and in Wepfer's work on apoplexy', in Lloyd G. Stevenson (editor), *A celebration of medical history*, Baltimore Md., and London, Johns Hopkins University Press, 1982, pp. 37–63. Jeffrey Boss, '"Doctrina de circulatione sanguinis haud immutat antiquam medendi methodum". An unpublished manuscript (1662) by Francis Glisson (1597–1677) on implications of Harvey's physiology', *Physis*, 1978, **20:** 309–336; and *idem*, 'Helmont, Glisson, and the doctrine of the common reservoir in the seventeenth-century revolution in physiology', *Br. J. Hist. Sci.*, 1983, **63:** 261–272. Perhaps it is also worth pointing out in this connection that Glisson himself said that the full details of his system were not worked out. He suggested that the Royal Society might be called upon to perfect and complete it. See Glisson, op. cit., note 5 above, *Epistola dedicatoria* (to the Earl of Shaftesbury), sig. alr. The present author is currently working on Glisson and his significance and hopes to add to this literature in the future; this essay is very much a preliminary sortie.

was eventually led to his mature concept of irritability, in which all fibres in the body are said to be "irritable" even without the mediation of nerves.[7]

Glisson's first prolonged efforts, in *Anatomia hepatis*, to define the nature of "irritability" as a biological phenomenon presupposed that nervous sensibility in the irritated part was a *sine qua non*. "All irritation", he wrote, "indicates the existence of perception", and that, in turn, indicates the presence of nerves.[8] Glisson considered different kinds of "irritation" which could activate particular nerves and so lead to bodily changes or movements. The nerves responded to three kinds of activation: perception or sensation, appetite, and muscular movement. This, as Temkin has shown, was entirely in accord with Galen's teachings about the threefold purpose of the nerves: sensibility in the organs of perception, initiation of movement in the organs of locomotion, and in all other organs "the recognition of distressing things" (equivalent to Glisson's concept of appetite).[9]

Glisson developed these ideas in anatomical lectures delivered to the Royal College of Physicians shortly after the publication of his book on the liver. In his lecture on the brain, Glisson averred that its actions and uses consisted "in the exercise of its noble faculties, perception, appetite and motion".[10] He seems to have had no difficulty in locating the faculties of perception and appetite in the brain and he felt confident, therefore, in locating the motive faculty in the same organ:

> For if it be the braine wch perceives, it must be the braine also wch desires & the same wch prosecutes that desire. For the desire were in vaine, if the braine had noe power to move toward a prosecution of the same, & there would be noe animall motion, if the braine had no power to command it.[11]

However, although Glisson was convinced that the motive faculty is located in the brain, he was all too aware of the difficulty of explaining the operation of this faculty:

> the manner how this power or faculty is put in execution, is soe obscure & harde to be conceived, that the minde reflecting upon it selfe, cannot clearly trace its owne footstepps. It is hard indeed to follow the track of the idaea's [*sic*] from the outward senses to the braine, but much more difficult to trace the edicts of the braine back againe to the outward organs of motion, yet this in nature must be done.[12]

Abandoning all pretence to "a clear account of this", Glisson merely offered "some propositions about it". The first of these is that the brain could only excite the nerves to initiate motion by means of its own motion:

[7] F. Glisson, *Anatomia hepatis, cui praemittuntur quaedam ad rem anatomicam universe spectantia et, ad calcem operis, subjiciuntur nonnulla de lymphae ductibus nuper repertis*, London, O. Pullein, 1654; *idem*, *Tractatus de ventriculo et intestinis. Cui praemittitur allius de partibus continentibus in genere; & in specie, de iis abdominis*, London, E. F. for Henry Brome, 1677; F. Glisson, G. Bates, and A. Regemorter, *De rachitide, sive morbo puerili, qui vulgo the rickets dicitur*, London, G. Du-Gardi for L. Sadler and R. Beaumont, 1650. I have used the English Translation of the latter, *A treatise of the rickets: being a disease common to children*, London, Peter Cole, 1651, see ch. 10, especially pp. 86–88, discussed in Temkin, op. cit., note 6 above, p. 294, and ch. 7, pp. 57–71 and ch. 14, pp. 144–150, discussed in Pagel (1967), op. cit., note 3 above, p. 499.

[8] Glisson, *Anatomia hepatis*, ch. 44, p. 397, and ch. 30, pp. 262–266. See Temkin, op. cit., note 6 above, p. 295; and B. L. Ullman, 'Glisson on gall bladder pain', *J. Hist. Med.*, 1955, **10**:112–113.

[9] Temkin, op. cit., note 6 above, p. 302.

[10] British Library, Sloane MS 3306, f.157.

[11] Ibid., f.163. Glisson discussed perception ff.157–62, ff.162–3, and motion ff.163–6.

[12] Ibid., f.163.

> For it is not conceivable, how the nerves should be excited, or they excite the muscles to worke according to the minde of the braine, if they were by no acte of the braine to give a signall to the nerves & outward organs; that acte must be some kinde of motion in the braine itself. For the braine cannot move the nerves, either by its perception or appetite, because those are both immanent actions & cannot reache any thinge with out their owne subject.[13]

The second proposition was that the brain's "imperiall kinde of mandate" to the nerves "proceedes noe further, then to the grosse commande of such or such an action to be done".

> For a beast which also may make such an animall motion, knows not soe much, as that he hath a braine or nerves or muscles. Wherefore, the braine cannot give a particular command to every nerve or muscle, when it excites a motion. But it is content with the giving such a generall excitation, that such a motion may be made. And upon such excitation, the nerves excite the muscles propper, whereby the motion is affected.[14]

The role of the brain in initiating the movements of bodily parts now begins to look precarious, and its supremacy is hardly supported by Glisson's third and final proposition. Here, Glisson unwittingly attributed to the nerves a mind of their own, as it were, which enabled them to carry out "such a grosse command as that last before mentioned". So, "the nerves serving for every motion, may be sufficiently excited" while "other nerves, upon that excitation will not at all be stirred",

> because those nerves onely, which are propper to the action required, have learned by custome & long practice, that they are to move upon such a signal given, and the other nerves have in like manner bene accustomed to sitt still upon the same signall . . . wherefore, I conclude, that the nerves learne to execute the dictates of the braine, upon the signall given, by custome and often experience.[15]

Only a few folios earlier, Glisson had described how memory and learning were special attributes of the perceptive faculty, that "immanent action" which "cannot reache any thinge with out [its] owne subject".[16] Small wonder, then, that Glisson's lecture on the brain and its three noble faculties breaks off abruptly at this point.

However, Glisson was not the only one to be confused about these matters. Even his mentor, the great William Harvey, had expressed contradictory ideas about the seat of perception in his *De generatione animalium* (1651). One moment Harvey implied that irritability depended on the mediation of the brain:

> that which is plainly devoid of all sensation does not seem to be able to be provoked in any way or to be excited to move or initiate any actions. And we have no other sure sign by which we can distinguish an animate and sentient creature from one that is dead and senseless except by its movement which is provoked by some offensive object and which follows immediately upon the experience of this sensation and so proves the existence of sensation. But it is reasonable to discuss this matter further when we treat of the actions and use of the brain.[17]

[13] Ibid.
[14] Ibid., f.165.
[15] Ibid., ff.165–6.
[16] Ibid., f.163.
[17] William Harvey, *Disputations touching the generation of animals*, translated by Gweneth Whitteridge, Oxford, Blackwell Scientific Publications, 1981, ch. 57, pp. 297–298.

And yet, only a few paragraphs earlier, Harvey wrote about motions independent of the brain and other sensory organs:

> Those motions and actions which the physicians call natural because they go on whether we will or no, and which we can neither moderate, accelerate, retard nor inhibit at our pleasure and are therefore independent of the brain, yet they are not performed without any sensation at all . . .. For whatsoever it is that by the diversity of its own movements strives against these things which provoke and molest it, must needs be endowed with sensation.[18]

Significantly, one of Harvey's examples of organs exhibiting innate sensitivity referred to the stomach and guts:

> The stomach and guts attacked by bad humours, often stir up nausea, and belching, rumbling, vomiting and flux of the bowels, and it is not in our power to arrest these movements or to instigate them, and so I do not know of any sense depending on the brain which should excite those parts to actions of this kind.[19]

When Glisson decided to follow up his research into the anatomy and function of the liver, he chose as his subject the stomach and guts. It was during the course of this work that he developed his concepts of natural perceptions, appetites, and natural motility in all the fibres of the body even where no nerves were present. In developing these ideas, Glisson was able to draw not only upon various hints in Harvey, Galen, and Aristotle but also, as Pagel has shown, upon certain ideas found in the works of Joan Baptista van Helmont. In particular, the concept of nerves "learning" from experience whether to respond or not to a "grosse command" of the brain, which seemed so anomalous in his lecture on the brain to the Royal College, gave place to a general claim that all fibre in the body is capable of "learning" and "knowing" by virtue of its archeus or natural perception.[20]

By his own account, Glisson had completed his work on the stomach and intestines by 1662 but he held it back for fifteen years before publishing it. The reason he gave for this leaves us in no doubt as to the importance he placed upon his *Treatise on the energetic nature of substance* for a full understanding of his anatomical work. His first draft of *De ventriculo* "presupposed in many places (although also proving in many places) a general Natural Perception" which had not been written about anywhere. Consequently, "it seemed right", Glisson wrote, that the draft "should be postponed until I could produce and bring into the light another Treatise, forerunner to this, on the Life of Nature".[21] According to a manuscript fragment, Glisson believed that the delay would amount to no more than two years, but clearly he was over-optimistic. "After the triall of a year or two I founde the worke so difficult", he said, "that it seemed to involve the whole body of natural physiology."[22] In the event, his

[18] Ibid., pp. 296–297.

[19] Ibid., p. 297.

[20] On the influence of Galen and Harvey see Temkin, op. cit., note 6 above; on the influence of Aristotle, see Pagel (1953), note 3 above; and on the influence of Harvey and Van Helmont see Pagel (1967), op. cit., note 3 above especially pp. 500–501 and 505, 508–509.

[21] Glisson (1677), op. cit, note 7 above, sig. A3r.

[22] This fragment is a draft, in English, of the dedicatory epistle to the Earl of Shaftesbury. British Library, Sloane MS 574B, f.159v.

Tractatus . . . de vita naturae appeared in 1672, five years before the final version of *De ventriculo*. It is here that Glisson developed and expounded the notion that substance or matter itself has an "energetic nature" and a "life" which is primarily exemplified in those three "noble faculties" he once reserved for the brain: the perceptive, appetitive, and motive faculties.

Glisson clearly agreed with Harvey's sentiments expressed in *De generatione*, that the only way to distinguish between "an animate and sentient creature" and "one that is dead and senseless" was in terms of perception and the ability to move in accordance with those perceptions.[23] The point of Glisson's *De vita naturae*, however, was simply to show that there was no such thing as "dead and senseless" matter. All matter, organic and inorganic alike, was endowed with life, Glisson insisted, and that life was defined in terms of an intrinsic motility which could be activated by inherent perceptivity and appetite (or aversion) within the matter itself. The difference, for Glisson, between what was usually considered to be non-living and living matter was simply one of organization. Provided matter was organized in a suitable way, the natural perception of matter could be "reduplicated" so that it perceived its own perception—giving rise to the sensitive nature of animals—and the *vita naturae* could be "duplicated" or "triplicated" to produce what we recognize as vegetable or animal life. The death of a plant or animal occurs upon the dissolution of this double or triple alliance.[24]

Glisson's views on the nature of plant and animal life led naturally in his *Treatise* to a categorical rejection of the concept of a separate "substantial soul" in animals. This in itself, however, was not a radical departure from contemporary medical thinking: Cartesians too denied the existence of an animal soul, and Gassendists, like Walter Charleton and Thomas Willis, regarded it as a subtle material principle which dissipated on the death of the animal. Furthermore, Glisson was careful to point out that man was endowed with an immaterial rational soul which, being preternatural, fell outside the purview of his natural philosophy.[25] Nevertheless, in view of the vigorous monistic approach to the concept of life expressed in *De vita naturae*, it is easy to imagine that a contemporary reader might have been reminded of the opening words of Thomas Browne's *Religio medici* and thought them a suitable testament for Glisson himself: "For my Religion, though there be severall circumstances that might persuade the world I have none at all, as the generall scandall of my profession, the naturall course of my studies . . . ."[26] In fact, much more research is needed before we can begin to assess the significance of the medical profession's reputation for irreligion and, in particular, Glisson's own attitude to religion.[27] Nevertheless, we can see from Henry More's reaction to Glisson, examined below, that this leading theologian regarded Glisson's monistic philosophy as dangerously atheistic in tendency. So far, we know little of what Glisson's fellow physicians thought about *De natura substantiae*

[23] Harvey, op. cit., note 17 above.

[24] Glisson, op. cit., note 5 above, especially 'Ad lectorem', section 11, sig. b2v-b3r, and ch. 18, section 5, pp. 234–5.

[25] Ibid., 'Ad lectorem', sections 7 and 8, sig. a4v-b1v.

[26] Thomas Browne, *Religio medici*, London, Andrew Crooke, 1643. p.1.

[27] There are indications of an interest in theology in Glisson's manuscripts: British Library, Sloane MS 3315, f.121–131 and Sloane MS 2251, f.117. A catalogue and index of Glisson's papers is being prepared by Dr Jeffrey Boss and I am grateful to him for these references.

energetica, but it is clear from More's attack upon him, and Richard Baxter's defence, that its importance and relevance was recognized beyond the bounds of the medical profession.

In conclusion, then, we can say that although the *Treatise on the energetic nature of substance* is Glisson's only philosophical work, it cannot therefore be dismissed as a mere curiosity. It was developed as an essential complement to his medical writings and fits perfectly naturally into the rest of his *oeuvre*. Having relied heavily, in the *Anatomia hepatis*, on a concept of irritability to account for the periodic expulsion of large amounts of bile from the gall-bladder into the intestine, Glisson subsequently believed that the dependence of irritability on the presence of nerves created more problems than it solved. His development of hints from Aristotle, Galen, Van Helmont, and Harvey and his own researches on the digestive system led him to formulate his concepts of natural perception, natural appetition, and natural motion intrinsic to all body fibres. And these in turn led him to work out the implications of these ideas for general matter theory. The end result, in the *Treatise* of 1672, is a complete system of natural philosophy dealing not only with organic but also inorganic matter and arguing for a materialistic vitalism which could be used to cut the Gordian knot of problems associated with Cartesian dualism by showing that all matter was intrinsically alive and endowed with faculties of perception, appetition, and self-motion.

Henry More and medical ideas about life and the soul

The major intellectual endeavour of More's career was to combat the atheism which he perceived to be increasingly prevalent among his contemporaries (educated and uneducated alike). His earliest publications, therefore, constitute what has recently been called "an important philosophical programme" by which More hoped to prove the existence of God and the immortality of the soul.[28] In fact, it was the nature of the immortal soul which occupied most of More's attention (beginning with his *Platonicall song of the soul*, 1642, and culminating with his *Immortality of the soul* in 1659) because a refusal to believe in an after-life, with its attendant notions of punishments or reward, was seen by More as likely to have the most drastic consequences on social behaviour. The "practical atheist", the libertine scoffer who had no religious and (*ipso facto*, according to More) no moral scruples, was the greatest threat to the smooth running of society because his behaviour was not constrained by the fear of post-mortem punishments.

> These sort of people [More wrote] are very horribly afraid there should be any Spirit, lest there should be a Devil, and an account after this life; and therefore they are impatient of anything that implies it, that they may with a more full swing, and with all security from an after-reckoning, indulge their own Lusts and Humours in this.[29]

[28] Gabbey, op. cit., note 2 above, p. 222.

[29] The quotation is from 'Dr. H. M. his letter, with the postcript to Mr J. G. . . .' which More included in his edition of Joseph Glanvill's *Saducismus triumphatus or, full and plain evidence concerning witches and apparitions*, London for J. Collins and S. Lownds, 1681, pp. 16–27, at p. 23. The *Platonicall song of the soul* can be consulted in Alexander B. Grosart (editor), *The complete poems of Dr. Henry More (1614–1687)*, [Edinburgh University Press] for private circulation, 1878. *The immortality of the soul*, op. cit., note 4 above is more conveniently consulted in Henry More, *A collection of several philosophical writings*, London, James Flesher for William Morden, 1662, which is now reprinted, New York and London, Garland, 1978.

More elaborated his pneumatology, therefore, in an effort to provide the clearest and most forceful arguments for a personal after-life.[30]

In his attempt to give a comprehensive and definitive account of the nature of the soul, More realized that he could not confine himself merely to a discussion of Christian doctrines. In view of his reputation as a leading member of the Cambridge Platonists it is hardly surprising that he should also consider Platonist and neo-Platonist arguments. However, his consideration of current medical theories might seem, from our point of view, somewhat unexpected. But once we recognize the background to seventeenth-century pneumatology it becomes clear that More simply could not ignore medical theories. As D. P. Walker has shown, the eclectic neo-Platonism of the Renaissance gave rise to a number of confusions or contaminations between notions of *spiritus, pneuma,* and *anima.* Medical theories of "spirit" were especially likely to lead to such confusions, Walker has argued, because it was frequently invoked to bridge the metaphysical gap between body and soul, or matter and mind.[31] By the seventeenth century, there was a tradition, well established in spite of its potentially heretical nature, which tended to identify souls with medical concepts of spirit.[32] This tradition could then be drawn upon by the early mechanical philosphers to explain those aspects of mental life which men share with animals. The Cartesian *res cogitans* was concerned only with abstract reasoning; all other aspects of human behaviour could be explained, analogously to animal behaviour, by the motions of animal spirits through the nerves and into the muscles. As Henry More summed it up:

> the *Glandula Pinealis* is the common Sentient or Percipient of all Objects; and without a Soul, by virtue of the Spirits and Organization of the Body, may doe all those feats that we ordinarily conceive to be performed by Soul and Body joined together. For it being one . . . and so handsomely seated as to communicate with the Spirits as well of the posterior as anteriour Cavities of the Brain; by their help all the motions of the Nerves (as well those that transmit the sense of outwards Objects, as of them that serve for the inward affections of the Body, such as Hunger, Thirst and the like) are easily conveighed into it; and so being variously moved, it does variously determine the course of the Spirits into such and such Muscles, whereby it moves the Body.[33]

Hereinafter I shall refer to this as More, *Immortality*. For a further account of More's attitude to post-mortem punishment see D. P. Walker, *The decline of hell: seventeenth-century discussions of eternal torment*, London, Routledge & Kegan Paul, 1964, pp. 122–134. On the rise of concern about atheism during this period see G. E. Aylmer, 'Unbelief in seventeenth-century England', in D. Pennington and K. Thomas (editors), *Puritans and revolutionaries*, Oxford, Clarendon Press, 1978, pp. 22–46; John Redwood, *Reason, ridicule and religion: the age of enlightenment in England, 1660–1750*, London, Thames & Hudson, 1976; Michael Hunter, *Science and society in Restoration England*, Cambridge University Press, 1981, p. 162–187; *idem*, 'The problem of "atheism" in early modern England', *Trans. R. Hist. Soc.*, 5th Series, 1985, **35:** 135–157.

30 I have given a fuller account of More's pneumatology and the reasons for its extreme importance in his philosophical career in 'A Cambridge Platonist's materialism: Henry More and the concept of soul', *J. Warburg and Courtauld Institutes*, 1986, **49:** 172–195.

31 Walker, 'Medical spirits in philosophy . . .', op. cit., note 3 above, pp. 287–289.

32 Ibid., p. 288; and *idem*, 'Medical spirits and God and the soul', op. cit., note 3 above.

33 More, *Immortality*, p. 80. For fuller discussions of Cartesian mechanist biology see L. C. Rosenfield, *From beast-machine to man-machine: the theme of animal soul in French letters from Descartes to La Mettrie*, New York, Oxford University Press, 1940; P. R. Sloan, 'Descartes, the sceptics, and the rejection of vitalism in seventeeth-century physiology', *Stud. Hist. Phil. Sci.*, 1977, **8:** 1–28; and Carter, op. cit., note 1 above.

As I have argued elsewhere, Walker is perfectly correct to suggest that these heretical materialist concepts of soul contaminate even More's own concept of the soul.[34] However, if we ignore these philosophical inconsistencies then it is true to say that More is totally committed to a dualistic belief in the existence of incorporeal substances. Essentially, these incorporeal substances fall under three headings. There is God himself, of course, and there are two different kinds of "created spirits". There are, on the one hand, all the immaterial souls of men, angels, and demons; and on the other hand, there is a ubiquitous "Spirit of Nature" pervading even deep within the densest material body. Now, More ingeniously used the mechanical philosophy to establish the existence of these immaterial substances. His early admiration for Cartesian mechanism stemmed from his belief that Descartes had shown "the just extent of the Mechanical Powers of Matter, how farre they will reach, and where they fall short." It was the short-fall that was important for More, since it enabled him to bring in the Spirit of Nature which was held to be

> A substance incorporeal, but without Sense and Animadversion, pervading the whole Matter of the Universe, and exercising a Plastical power therein according to the sundry predispositions and occasions in the parts it works upon, raising such Phaenomena in the World, by directing the parts of the Matter and their Motion, *as cannot be resolved into mere Mechanical powers.*[35]

If More's enterprise in natural theology was to succeed, therefore, he had to defend the immateriality of the Spirit of Nature at all costs. If he conceded any corporeality in his hylarchic principle, his whole philosophy was in danger of reduction to materialistic mechanism and became indistinguishable from the atheistic philosophy of "the most able Advocate" of materialism, Thomas Hobbes.[36] The Spirit of Nature, therefore, was used to explain various physical phenomena, such as the descent of bodies due to gravity, cohesion, condensation and rarefaction, magnetism, and various other phenomena which proved to be problematic for the mechanical philosophy. On biological and psychological questions, however, More did not want to rely on a universal spirit of nature since it might lead to another heresy: the denial of individual souls in favour of the Averroistic notion of a universal soul.[37] The Spirit of Nature, then, gives place to individual souls in living creatures:

> God is not the immediate Maker of these Bodies . . . Nor, is it congruous to admit that the Plastick faculty of the Soul of the World [i.e. the Spirit of Nature] is the soul contriver of these fabricks of particular creatures (though I will not deny but she may give some rude preparative strokes towards Efformation;) but that in every particular World, such as Man is especially, his own Soul is the peculiar and most perfective Architect thereof, as the Soul of the Worlds is of it.[38]

[34] See note 30 above; and Walker, 'Medical spirits in philosophy', op. cit., note 3 above, p. 289, 294.

[35] More, *Immortality*, 13 and 193.

[36] Ibid., 43. On the contemporary reaction to Hobbes see S. I. Mintz, *The hunting of Leviathan: seventeenth-century reactions to the materialism and moral philosophy of Thomas Hobbes*, Cambridge University Press, 1970.

[37] On Averroistic notions of the soul see P. O. Kristeller, *Renaissance thought and its sources*, New York, Columbia University Press, 1979, pp. 181–196; and *idem*, 'Paduan Averroism and Alexandrism in the light of recent studies', in *Renaissance thought II: Papers on humanism and the arts*, New York, Harper & Row, 1965, pp. 111–118.

[38] More, *Immortality*, 102.

In the *Scholia* added to his philosophical works for the Latin *Opera omnia* of 1679, More delineated more carefully the spheres of influence of the Spirit of Nature and the soul. Subscribing to the Aristotelian belief that the soul is not infused into an embryo until its fortieth day, he argued that the "preparative strokes toward Efformation" of a foetus "are made by the Soul of the World, or by the Spirit of Nature", because

> in Eggs, where the soul of the mother is not present, these first lineaments are formed before the coming in of the Soul of the Chicken, as being not yet organiz'd; But particular Souls are, according to Aristotle, the Actings of an organical Body. But the *Punctum saliens*, or Life point, discovers not any proper sense, but only Life when it withdraws itself from any hurtful touch: as neither do what we call the Plant-Animals which I do not take to live by any proper Soul, but through the Spirit of Nature, which hath either no Sense at all, or what is very dull.[39]

So, although the Spirit of Nature "assists in the very first beginnings of the Generation of Animals" it is superseded, in the higher animals at least, by the creature's individual soul.[40] However, More's "main design" was not to distinguish between the Spirit of Nature and the soul but rather "to demonstrate that there is a Soul or Incorporeal Substance residing in us, distinct from the Body".[41] Accordingly, he was compelled to discuss critically various medical theories current at the time.

For example, in order to reserve for the immaterial soul (or for the immaterial Spirit of Nature in lower animals) the architectonic or organizing power of living creatures, More had to examine and reject a common alternative view. This was the remarkable idea that "the Sun and the Stars are the most Intellectual Beings in the world, and in them is that Knowledge, Counsel and Wisdome by which all Sublunary things are framed and governed". According to this view, the sun and stars

> by their several impresses and impregnations have filled the whole Earth with vital Motion, raising innumerable sorts of Flowers, Herbs and Trees out of the ground. These have also generated the several kinds of living Creatures. These have filled the Seas with Fishes, the Fields with Beasts, and the Aire with Fowles; the Terrestrial matter being as easily formed into the living shapes of these several Animals by the powerful impress of the Imagination of the Sun and Stars, as the Embryo in the womb is marked by the strong fancy of this Mother that bears him.[42]

Outlandish as this may seem to us, it was an extremely common notion in More's day. Similar arguments can be found in the work of Jean Fernel, Marsilio Ficino, Paracelsus, and even Robert Boyle. One of the ancient sources for this view is Aristotle's *De generatione animalium*: "the spirit which is contained in the foamy body of the semen . . . is analogous to the element of the stars", and it is duly considered,

[39] This is taken from the English translation of the scholia which are included in the so-called "Fourth edition" of *A collection of several philosophical writings*, London, Joseph Downing, 1712, p. 105 (of the *Immortality*, like the 1662 edition the treatises are separately paginated). This is called the fourth edition because the original separate works count as the first, the 1662 *Collection* as the second, and the Latin *Opera omnia* (see note 54 below) as the third. On the time of the infusion of the soul into the embryo see John T. Noonan jun, *Contraception: a history of its treatment by the Catholic Theologians and Canonists*, Cambridge, Mass., Belknap Press, 1965, pp. 89–90. The *locus classicus* is usually taken to be Aristotle, *De hist. animal.*, VII, 3, 583b 2–5.

[40] It seems that More did not confine the concept of the immortal soul exclusively to humankind. On the immortality of animals see *Immortality*, 135–136.

[41] Ibid., 88 [mis-numbered as 78].

[42] Ibid., 52.

therefore, by that latter-day Aristotelian biologist, William Harvey, in his own *De generatione*.[43] Nevertheless, More gave the idea short shrift. Donning his Cartesian hat, he insisted that

> the sun and Stars have no immaterial Being residing in them, but are mere Matter consisting of the subtilest Particles and most vehemently agitated. For then we cannot but be assured that there is nothing in them more Divine than what is seen in other things that shine in the dark, suppose rotten wood, glo-worms or the flame of a rush candle.

And the idea that these heavenly bodies could form a creature like "the most curious Automaton that ever was invented by the wit of man" simply by "a mere Rectilinear impress upon the Aether down to the Earth" was dismissed as untenable.[44]

More dealt similarly with other theories which suggested that the soul or the organizing power of the body was merely a special kind of material substance. Drawing from the full range of current medical theories, he refuted in turn the suggestion that actuation and guidance of the body together with perception and the other functions of the common sense could be performed by the body as a whole, the orifice of the stomach, the heart, the brain, the septum lucidum, the pineal gland, "Regius his small and perfectly-solid particle", the spinal marrow, and the animal spirits.[45] We need not consider each of More's arguments in any detail but it should be noted that the major thrust of his discussion was to deny that matter itself is sufficient to explain perception and sentience. "This absurd Principle", as More called it, gave rise to the belief "that Organization may doe strange feats":

> as if they should imagine, that though neither Silver, nor Steel, nor Iron, nor Lute-strings have any Sense apart, yet being put together in such a manner and formed as will (suppose) make a complete Watch, they may have Sense: that is to say, that a Watch may be a living creature, though the several parts have neither Life nor Sense . . .[46]

More was so confident that he could explode this notion that he actually put it forward as an axiom in order to show that Hobbes's "own acknowledged Principles will necessarily inferre the Existence of [immaterial substances] in the World." The twentieth axiom of *The immortality of the soul* reads:

[43] Aristotle, *De gen. animal.*, 736b–737a; William Harvey, *Exercitationes de generatione animalium*, first published London, 1651, ch. 46. On Fernel see Sir Charles Sherrington, *The endeavour of Jean Fernel*, Cambridge University Press, 1946; on Ficino see Walker (1958) op. cit., note 3 above, especially pp. 3–59; for Paracelsus see Walter Pagel, *Paracelsus: an introduction to philosophical medicine in the era of the Renaissance*, Basle, Karger, 1982, especially pp. 117–125. On Boyle see L. Thorndike, *A history of magic and experimental science*, 8 vols., New York, Columbia University Press, 1923–1958, vol. 8, pp. 186–187. For a general treatment of "light metaphysics", the tradition of which these notions are a part, see A. C. Crombie, *Robert Grosseteste and the origins of experimental science, 1100–1700*, Oxford, Clarendon Press, 1953, pp. 104–134; and David C. Lindberg, *Theories of vision from Al-Kindi to Kepler*, Chicago, University of Chicago Press, 1976, 95–99.

[44] More, *Immortality*, 53, 57.

[45] Ibid., 77–94. As well as displaying a sound knowledge of various traditional therories of anatomical psychology, More also reveals here an awareness of recent theories propounded by J. B. van Helmont, Henricus Regius, and René Descartes. More, like Descartes, seems to have regarded Regius in particular as dangerously materialistic in outlook. See Nicolson (editor), op. cit., note 56 below, p. 389; and Sloan, op. cit., note 33 above, pp. 23–28.

[46] More, *Immortality*, 66, 77.

> Motion or Re-action of one part of the Matter against another, or at least a due continuance thereof, is really one and the same with Sense and Perception, if there be any Sense or Perception in Matter.[47]

Of course, the conditional clause is crucial to More's case and it is just that which he then proceeded to deny. The absurdity of Axiom XX could be shown simply by pointing out that wherever there is motion so must there be sense and perception. In other words a dead carcass, a ringing bell, a bent bow, and "every Jack-in-a-box that School-boyes play with" would have to be "living Animals or Sensitive Creatures". Furthermore, on Hobbes's account, the communication of thoughts from one person to another can only take place by the justling of one interlocutor's animal spirits against those of the other, "as if men should knock heads to communicate to each other their conceits of Wit."[48]

The essential point, then, of More's rejection of this mechanical account of perception is that matter is totally passive:

> It is sufficiently manifest from sense and experience that Matter is a principle purely passive, and no otherwise moved or modified then as some other thing moves and modifies it, but cannot move it self at all. Which is most demonstrable to them that contend for Sense and Perception in it. For if it had any such Perception, it would by virtue of its Self-motion withdraw it self from under the knocks of hammers or fury of the fire; or of its own accord approach to such things as are most agreeable to it and pleasing, and that without the help of Muscles, it being thus immediately endowed with a Self-moving Power.

Matter, however, is "so stupid" that it lacks this Power of self-movement, as Hobbes, Descartes, or any strict mechanist would readily agree.[49]

Unfortunately for More, however, not everyone was a strict mechanist and subscribed to this concept of completely passive matter. As More's close associate, Ralph Cudworth, pointed out in his encyclopaedic survey of atheist philosophies, there were those who believed that matter was inherently active. The hylozoist, as Cudworth dubbed such a thinker, considered that all matter could move itself without muscles and that all matter could perceive and thereby could have a kind of sense without sensory organs or nerves.[50] Cudworth was not slow to realize that hylozoism was much more dangerous to religion than the mechanical materialism (or "Atomick Atheism") of a thinker like Hobbes,

[47] Ibid., 5, 59.

[48] Ibid., 66, 86.

[49] Ibid., 60. I use the term "strict mechanist" to refer to anyone who believed that matter was totally passive and inert. By no means all thinkers known to historians as mechanical philosophers were strict mechanists. I have argued elsewhere that the possibility of active matter was widely entertained among natural philosophers in seventeenth-century England: John Henry, 'Occult qualities and the experiemental philosophy: active principles in pre-Newtonian matter theory' *Hist. Sci.*, 1986, **24:** 335–381.

[50] Cudworth is not so careful to distinguish between perception and sensation as Glisson was. Sensation only occurs in "animate" matter, that is to say matter which is sufficiently organized to enable the simple or natural perception of matter to perceive its own perception. This *perceptio perceptionis* or reduplicated perception constitutes sensation. It seems that highly organized, or animate, matter is able to dispose itself in such a way as to provide itself with special organs of sensation. This requires detailed exposition and analysis, but for now see Glisson, op. cit., note 5 above, pp. 208–216; Pagel (1953), op. cit., note 3 above, pp. 505–507.

> For whereas the Atomick Atheism supposes the Notion or Idea of Body to be nothing but Extended Resisting Bulk, and consequently to include no manner of Life and Cogitation in it; Hylozoism on the contrary makes all Body, as such, and therefore every smallest atom of it, to have Life Essentially belonging to it (Natural Perception and Appetite) though without any Animal Sense or Reflexive Knowledge, as if Life, and Matter or Extended Bulk, were but two Incomplete and Inadequate Conceptions, of one and the same Substance, called Body. By reason of which life (not Animal but only Plastical) . . . it is plain that there is no Necessity at all left, either of any Incorporeal Soul in Men to make them Rational, or of any Deity in the whole Universe to solve the Regularity thereof.[51]

Although Cudworth attached no names to the title hylozoist (apart from the ancient who allegedly founded the movement, Strato of Lampsacus), it is perfectly clear even from this brief extract that the Regius Professor of Hebrew had no one else in mind but the Reguis Professor of Physic at Cambridge, Francis Glisson. It was only Glisson who argued in print for the natural perception and the natural appetite of matter and who regarded life and matter as two *conceptus inadequati* of substance.[52]

So if in 1659 More could afford to dismiss the notion of sensitive and perceptive matter as "A thing so foolish and frivolous, that the mere recital of the opinion may well be thought confutation enough with the sober", after 1672 he was compelled to treat the idea much more seriously. For it was in that year that Glisson's *Tractatus de natura substantiae energetica* appeared.[53]

Henry More and Francis Glisson

Henry More returned to philosophical writing from more theological concerns during the preparation of his collected works, in Latin translations, for the so-called *Opera omnia* of 1679.[54] More took the opportunity of the new edition to add scholia to his earlier philosophical works but he also published a number of short pieces for the first time.[55] One of these new pieces, *Ad. V.C. epistola altera*, was concerned to refute the thinker whom he considered to be the most dangerous atheist since Hobbes, Benedict de Spinoza. We know from a letter of More's to Anne, Lady Conway, that this little tract was composed hurriedly while the Latin collection of his works was

[51] Ralph Cudworth, *The true intellectual system of the universe: the first part; wherein all the reason and philosophy of atheism is confuted; and its impossibility demonstrated*, London, Richard Royston, 1678, p. 105.

[52] The fact that Cudworth was attacking Glisson here has been noticed by Arrigo Pacchi, *Cartesio in Inghilterra da More à Boyle*, Roma-Bari, Editori Laterza, 1973, pp. 150–155. Cudworth's concern with hylozoism is discussed, but with no reference to Glisson, in John W. Yolton, *Thinking matter: materialism in eighteenth-century Britain*, Oxford, Basil Blackwell, 1984, pp. 4–13. See also Colie (1957), op. cit., note 4 above, pp. 117–144.

[53] See note 5 above.

[54] Henry More, *Opera omnia, tum quae latine, tum quae anglice scripta sunt; nunc vero Latinitate donata* . . ., 2 vols., London, J. Macock and R. Norton for J. Martyn and Gualt. Kettilby, 1679. The short title is often taken to include *Opera theologica, anglice quidem primitus scripta, nunc vero per autorem latine reddita*, London, J. Macock for J. Martyn and Gualteri Kettilby, 1675, as the first volume of three, but in my references I exclude this and treat the *Opera omnia* as a two-volume work.

[55] These include a number of pieces on cabbalism (*De usu decem Sephirotharum*, *Catechismus cabbalisticus*, and *Fundamenta philosophiae sive cabbalae aeto-paedo-melissaeae* . . .), a refutation of Jacob Boehme's theology (*Philosophiae teutonicae censura*), and the attack on Spinoza (see following note) which concerns us here. For a brief but important discussion of the *Cabbalae aeto-paedo-melissaeae* see C. A. Staudenbaur, 'Platonism, theosophy and immaterialism: recent views of the Cambridge Platonists', *J. Hist. Ideas*, 1974, **35:** 157–169, which corrects the thoroughly unsound S. Hutin, *Henry More, essai sur les doctrines théosophiques chez les Platoniciens de Cambridge*, Hildesheim, G. Olms, 1966.

actually in press. The letter also reveals that he began writing his refutation of Spinoza's *Tractatus theologico-politicus* even before he had finished reading the work. This is what he said to Anne:

> I come thus late to London by reason of Cuperus his Confutation of *Tractatus Theologicopoliticus* which Monsieur Van Helmont gave me at Ragley from a friend in Holland, in quires, which while it was a binding at Cambridge I fell a reading Theologicopoliticus the better to understand Cuperus his confutation when it came from binding. But I found this Theologicopoliticus such an impious work, that I could not forbeare confuting him while a [*sic*] read him . . . Proposing this Confutation of mine shall make up some part of this Philosophicall volumn.[56]

Spinozism, then, was regarded as a philosophical emergency which had to be treated immediately. The urgency of More's response stems in part from the fact that Spinoza, unlike Hobbes and the Cartesians, seemed to argue that matter is inherently active and "alive". This approach, More could hardly fail to recognize, completely undermined his previous attempts to establish the existence of incorporeal animating spirits, all of which were based on the Hobbist or Cartesian assumption that matter is completely passive and inert. It was during the course of his response to Spinozistic "living" matter that More paused in order to reject Glisson's arguments for the "energetic nature of substance". He took his cue from his fellow anti-Spinozist, Franciscus Cuperus:

> So when Cuperus says that modern atheists nowadays believe in the innate life of matter, not in a mechanical mover, he gives a nod to some such hypothesis as the most distinguished Doctor Francis Glisson attempted to describe, with very accurate and clear arguments, in his Treatise *de Natura Substantiae Energetica*, or, *de Vita Naturae*. Whereas Spinoza contends that substance, in so far as it is substance, exists *a se*, indeed is that which subsists by itself so he [Glisson] holds that substance, *qua* substance, lives by the necessary power of its Nature, that is, it perceives, it desires and it moves itself [*percipere, appetere, seseque movere*]: which faculties; perceptive, appetitive and motive; he holds, accordingly, to be immediately present in matter itself.[57]

There follow eight closely argued folio pages against Glisson's philosophy and what More saw as its atheistic implications. It is clear from these pages that More had given close attention to Glisson's *Tractatus*, and it is easy to see why. Glisson took the concept of inherently active matter further than any of his contemporaries, including Spinoza.

Since Glisson believed that the energetic nature of substance was the internal principle of movements in matter he tried to take evidence for his case "from various motions observed in Nature". Glisson understood the word "motion", in the

[56] Henry More to Anne, Lady Conway, 3 April [1677], in Marjorie H. Nicolson, *Conway letters: the correspondence of Anne, Viscountess Conway, Henry More, and their friends, 1642–1684*, New Haven, Conn., Yale University Press, 1930, p. 429. Franciscus Cuperus [Frans Cuiper] (1629–92) published a refutation of Spinoza under the title, *Arcana atheismi revelata, philosophice et paradoxe refutata examine Tractatus Theologico-Politici*, Rotterdam, Isaac Naeranus, 1676. For a discussion of this work and its author see Colie, op. cit., note 4 above, pp. 74–75, 97–116. More's "confutation" of Spinoza appeared in volume 1 of the two-volume *Opera omnia* (see note 54 above), as *Ad V.C. epistola altera quae brevem Tractatus Theologico-politici confutationem complectitur, paucaque sub finem annexa habet de libri Francisci Cuperi scopo, cui titulus est, Arcana Atheismi revelata*, pp. 565–614. The identity of V.C. still engages scholarly speculation: there is a footnote about it in Gabbey, op. cit., note 2 above, pp. 214–215; and see also Paolo Cristofolini, *Cartesiani e sociniani: studio su Henry More*, Urbino, Argalia Editore, 1974, pp. 139–141. Cristofolini also reprints the latin text of *Ad V.C. epistola altera*, pp. 139–206.

[57] More, *Opera omnia*, **1:** 601. Glisson is mentioned on p. 601 and examined in detail pp. 604–611. Translations from this are mine.

Aristotelian sense, to mean "change" and so a number of his examples of motion may seem unfamiliar to us. Not only did he argue from the motion "of the Heavens and Planets and even the Earth itself around the Sun", the motions caused by *horror vacui* and "the motion of Rarefaction and Condensation" but also from "the motion of solidity" [*motus Antitypiae*], "the motion of coherence" [*motus nexus*], and "the motion of freedom" [*motus libertatis*] (by which he meant the elastic tendency of a compressed body, say, to return to its former state when external pressure is removed).[58] Henry More attempted to refute each of these in turn. The resistance or solidity of matter cannot be used to prove the vitality of matter, he insisted, because "substance qua substance" is sufficient to explain solidity. By which, presumably, he meant that solidity is part of the definition of matter.[59] Elsewhere he called upon the mechanical philosophy in an effort to show that Glisson's vitalism was an unnecessary hypothesis. The coherence of matter can be explained more simply (so More believed!) in terms of Cartesian circular motion, the plenum, and the non-interpenetrability of matter. Similarly, "the *Motus Coelorum*, that is, the Vortexes, does not indicate any perceptive life in the matter of the vortexes, since such a portion of motion is impressed on them mechanically, or much more likely by some immaterial principle, that they revolve freely in these gyrations entirely without self-perception."[60] Needless to say, the immaterial principle alluded to here is More's "Spirit of Nature", and he invoked that in his rejection of all Glisson's other examples of different kinds of motion. "The parts of matter", for example, "cohere and are connected by the one universal bond of the Spirit of Nature which holds together and actuates the whole material world." It is the hylarchic principle, then, which accounts for *fuga vacui* and for *motus libertatis*. When air is caused to expand by heating, for example, it is merely due to increased vibration of the particles and, on cooling, the air returns to its former volume "not by its own power but by virtue of the Universal Hylostatic Spirit."[61]

[58] I am paraphrasing, and quoting from More's account of Glisson's theories. See More, *Opera omnia*, **1:** 605–606. In fact, More was simplifying Glisson's account. Glisson distinguished between perpetual and non-perpetual motions. The perpetual motions are *motus antitypia*, *motus nexus* and *motus corporum coelestium* and they are discussed in the *Tractatus*, note 5 above, pp. 352, 354 and 355 respectively. Non-perpetual motions are more complicated and less clear. At ch. 25 (pp. 356–363) we learn that there are three kinds of non-perpetual motion depending on whether the motive principle is internal, external, or a mixture of both. In order to establish his claims about the innate motility of all matter, Glisson discounted the first of these (which is not in contention) and turned his attention to the last two kinds which he called, in keeping with the common opinion, "inanimate". Inanimate motions are divided into five classes which embrace the other examples of "Species" of motion which More singled out. *Motus libertatus* is discussed at p. 375, *motus condensatione et rarefactione* at p. 376, *fuga vacui* at p. 367, but see also p. 432.

[59] More, *opera omnia*, **1:** 605.

[60] Ibid.

[61] Ibid., p. 606. More used similar arguments, employing the Spirit of Nature to account for phenomena which could be explained perfectly adequately without it, in his *Enchiridion metaphysicum: sive de rebus incorporeis succinta & luculenta dissertatio. Pars prima: de existentia & natura rerum incorporearum in genere*, London, E. Flesher, 1671. Consequently, he had already attracted the attention of the two leading experimental philosophers in England, Robert Boyle and Robert Hooke, both of whom repudiated his ideas in print. See Robert Boyle, *An hydrostatical discourse occasion'd by some objections of Dr. Henry More against some explications of new experiments* . . ., London, for Richard Davis, 1672; and Robert Hooke, *Lampas: or, descriptions of some mechanical improvements of lamps and waterpoises* . . ., London, John Martyn, 1677, reprinted in R. T. Gunther (editor), *Early science in Oxford*, 15 vols., Oxford, [for the author], 1923–1967, vol. 8 (1931), pp. 182–195. In spite of this and another polemic on the same issue with the scientific virtuoso, Sir Matthew Hale, More remained recalcitrant. I have examined More's relationship with

More's "confutation" of Glisson's internal principles of motion, philosophically unsatisfactory though it is, stands representative of all his arguments. He never met Glisson on his own terms but was merely content to assert that everything Glisson used to support his case actually supported the existence of the Spirit of Nature. Glisson's arguments, More somewhat petulantly insisted, "do *not* prove that life emanates from matter itself or that life is *not* communicated to it by an immaterial Principle". The point is, of course, that More was driven by his theological concerns to beg the very questions that really divided him from Glisson. At one highly significant point in his discussion More let the reason for his anxiety show: "This argument [that internal motion is a sufficient principle of life] cannot prove that matter is autozoon [self-living] and endowed with perceptive and appetitive life, unless it is previously supposed that there is no God who can create a Spirit purely Plastic and set it over the matter of the world."[62]

According to More, then, this vitalist way of arguing was even more dangerous than Hobbist mechanism. Glisson actually seemed (by implication) to take providence out of God's hands and to implant that too directly within matter when he asked "Is Nature so far fortuitous that it begins a new operation of its own record without any purpose to itself? or does it rather manage all things wisely and not without admirable providence?"[63] It was speculations like these which led More to denounce Glisson's philosophy in the strongest terms: "Only those who deny God and all incorporeal substance; strive to seek the origin of motion and all life which gleams in the universe in matter itself."[64]

More clearly announced that Glisson's philosophy "should certainly carry some weight with those who deny any substance beyond body. For, then it may be reasonable enough to decide that a kind of life is immediately and primaevally present in matter, whose characteristic properties of life; which are thought to be, and are called 'souls' by the vulgar; are only various modifications of matter . . .."[65]

More agreed, therefore, with Franciscus Cuperus that "modern atheists . . . having forsaken mechanical reasons for the created world, take up [a belief] in the primitive life of matter."[66]

---

these and other natural philosphers in a forthcoming paper entitled 'Henry More *v*. Robert Boyle: the Spirit of Nature and the nature of Providence'.

[62] More, *Opera omnia*, **1:** 607.

[63] Ibid. This is a quotation from Glisson's ch. 17, sect. 3 (as More tells us). It is to be found in Glisson, op. cit., note 5 above, p. 226. The point of Glisson's rhetorical questions was to affirm that matter must have perception and appetite: "I am unable to conceive how it happens that motion arises purely from an internal principle, yet that principle neither desires nor perceives."

[64] More, *Opera omnia*, **1:** 607.

[65] Ibid., p. 608.

[66] Ibid. See also p. 611 where More concluded that the "modern atheists", of whom Cuperus speaks, "who profess a vital principal in the created World, but because of life arising from matter itself, and who do not believe that there is anything in nature beyond this living matter, can scrape together for themselves nothing from the writings of Glisson by which they very greatly support their cause." I should point out that the charge of atheism was an extremely serious one at the time and, consequently, More pointed out that Glisson "not only acknowledges but so often insists upon the existence of the Creator God and other incorporeal substances", p. 607. As we shall see, it is perfectly possible to propound a natural philosophy like that of Glisson and still remain within the traditions of voluntarist theology.

It seems likely, then, that it was More's obsessive fear of atheism which inhibited his philosophical analysis of Glisson's *Tractatus* and prevented him from providing any really cogent counter-arguments. More's response is best characterized as merely a series of desperate reiterations of the role of the Spirit of Nature in the fabric of the World. Consider this final example:

> it is thoroughly manifest that the most learned Glisson has put forward no argument of any import, neither *a priori* nor *a posteriori*, which might prove the Energetic Nature of Substance . . . since those patterns of life which seem to appear in matter could be brought about by some immaterial Principle.[67]

What is worse, More compounded this kind of unsatisfactory procedure with an exercise in question-begging of astonishing impudence. The principle which More invoked (against Glisson) to account for movements, perception, and other "patterns of life" must certainly be *immaterial*, More brazenly insisted, "because it dictates each of the motions of Matter".[68]

In defence of Glisson: Richard Baxter's polemic with More

We would be making a serious mistake if we assumed that More's attack on Glisson as a comforter of atheists represented any kind of consensus. Writing at a time when there was little consensus among theologians in general, More was regarded by a number of his contemporaries as precariously idiosyncratic.[69] But what is significant

[67] More, *Opera omnia*, **1:** 608.

[68] Ibid. More's failure to confront Glisson's arguments and his monotonous reminders of the existence of a "Spirit of Nature", which he claimed to have proved in his earlier writings, clearly testify to the intense fear of atheism that gripped him. But I think the marked intensity of More's commitment to the dualist distinction between matter and a preternatural actuating principle also illustrates an important insight to be found in R. S. Westfall, *Science and religion in seventeenth-century England*, New Haven, Conn., Yale University Press, 1958. Westfall, speaking of seventeenth-century natural theologians, said that "Despite the natural piety of the virtuoso, the scepticism of the Enlightenment was already present in embryo among them" (p. 219). The origins of their doubts, Westfall suggested, lay in the many objections to traditional religion, "which were too close to their own beliefs to be ignored" (p. 145). In developing their natural theologies, therefore, "More than answering hypothetical atheists, they were trying to satsfy their own doubts" (p. 145). More was not a man who suffered from doubts but he was all too aware of how close Glisson's ideas were to his own. Indeed, at one point, More said of Glisson: "If he had put for 'Nature' the 'Spirit of Nature' that would have been enough for us to agree" (*Opera omnia*, **1:** 606). The closeness of Glisson's ideas to those of More and Cudworth have also been noticed by Owsei Temkin (see his article in *DSB*, **5:** 426–427). And yet, no matter how close they seem, the difference between them is absolutely crucial. Perhaps More felt that by expounding Glisson's arguments in order to refute them he would stimulate more interest in them. He certainly felt that he had made this kind of error before. As he wrote to Robert Boyle on 4 December 1671, "I have, from my very first letters to Descartes, till this last book of mine, always expressed my opinion, that this mechanical way would not hold in all phenomena, as I always verily thought: but this would not save us from being accounted amongst the wits, one of their gang; and a perfect Cartesian, . . . and, indeed, no less than an infidel and an atheist", in Robert Boyle, *The works*, 6 vols., London, J. & F. Rivington, L. Davis, *et al.*, 1772, vol. 6, p. 513. In Glisson's case, then, More chose not so much to refute it as to pronounce it to be wrong. For a further discussion of similarity between More's and Glisson's ideas see Henry op. cit., note 30 above.

[69] The fullest discussion of More's unorthodoxy is to be found in Samuel Parker, *A free and impartial censure of the Platonick philosophie*, Oxford, W. Hall for Richard Davis, 1666, especially pp. 79–83. Boyle, op. cit., note 61 above, referred to More's *Principium Hylarchicum* as a "precarious Principle" in theology as well as natural philosophy, as did Sir Matthew Hale, *Observations touching the principles of natural motions; . . . together with a reply to certain remarks touching the gravitation of fluids*, London, W. Godbid for W. Shrowsbury, 1677, p. 282. Moreover, More felt the need to defend his ideas not by recourse to

for our discussion is the fact that just two years after the appearance of More's *Ad. V.C. epistola altera* Glisson's name appeared once again in a theological context; but this time he was cited with approbation. Richard Baxter, in his *Methodus theologiae christianae* (1681), deferred twice to Glisson. Once, in the preface, on the subject of the active principle in nature, and once on the subject of "Virtus activa vitalis".[70] The precise meanings of these references are (at least to this reader) somewhat obscure but, fortunately, Baxter expanded his views on active principles and on Glisson's *De natura substantiae energetica* in a letter to Henry More written in the same year. It was written at More's request because "It seemed good to the worthy Dr. to desire my thoughts of his Description of a Spirit, which he laid down in the first Edition of Mr. Glanvile of Apparitions".[71] The fact that More published a long reply to this private letter (in the second edition of Glanvill's *History of apparitions*) is powerful testimony to More's sensitivity to criticism on this issue and of his protectiveness towards his concept of spirit. This reply, entitled *A letter to a learned psychopyrist*, appeared early in 1682; whereupon, within months, Baxter published his original letter, together with a defence of it in a discourse *Of the nature of spirits*. However, the last word on the matter went to More (again a sign of the importance More attached to his own style of pneumatology), this time writing as the anonymous commentator on Bishop Rust's *Discourse of truth*.[72] This polemic, so far completely overlooked, contains a good deal of interest for the intellectual historian but, for the sake of brevity, we will confine ourselves here to those parts of the dialogue which involve Glisson.[73]

---

philosophical argument but by Scriptural exegesis in his *The apology of Dr. Henry More*, London, J. Flesher for W. Morden, 1664. See also, M. H. Nicolson, 'Christ's College and the latitude-men', *Mod Philology*, 1929, **27:** 35–53.

[70] Richard Baxter, *Methodus theologiae christianae* . . ., London, M. White and T. Snowden, 1681, sig. A5r., p. 177.

[71] Richard Baxter, *Of the immortality of mans soule, and the nature of it, and other spirits. Two discourses: one in a letter to an unknown doubter; the other in a reply to Dr. Henry More's animadversions on a private letter to him; which he published in his second edition of Mr. Joseph. Glanvil's Sadducismus Triumphatus, or history of apparitions*, London, for B. Simmons, 1682, Sig. a4r. See note 29 above for full citation of the first edition of Glanvill's *Saducismus*; the second edition, which includes More's 'Letter to a learned psychopyrist', appeared in the first months of 1682 (new style). More's first letter to Baxter soliciting his opinion on the concept of spirit is extant in the Baxter MSS at Dr Williams's Library, Baxter Letters, III, f.286 (it is dated 25 September 1681).

[72] Joseph Glanvill [and George Rust], *Two choice and useful treatises: the one Lux Orientalis or an enquiry into the opinion of the eastern sages concerning the praeexistence of souls. Being a key to unlock the grand mysteries of providence, in relation to mans sins and misery. The other a discourse of truth, by the late Reverend Dr Rust Lord Bishop of Dromore in Ireland. With annotations on them both*, London, for James Collins and Sam Lowndes, 1682. More's annotations on Rust have a separate title page which reads: *Annotations upon the discourse of truth into which is inserted by way of digression, a brief return to Mr. Baxter's reply, which he calls a placid collation with the learned Dr. Henry More, occasioned by the Doctors answer to a letter of the learned psychopyrist* . . .. The digression is found pp. 180–246. Hereinafter cited as More, *Digression*. The annotations are unsigned but they are almost certainly by More. See Nicolson, op. cit., note 56 above, p. 173; and Walker, op. cit., note 29 above, 126, 131.

[73] The only other discussion of this polemic which I am aware of considers it in the broader context of late seventeenth-century discussions of the nature of spirit: Robert Crocker, 'An intellectual biography of Henry More, 1614–1687', ch. 8 of a thesis to be presented at Oxford University for the DPhil. I am grateful to Mr Crocker for allowing me to read this prior to submission. One of the few historians to embark on a serious study of Richard Baxter in recent years is William M. Lamont but even he failed to notice this polemic. More subsequently accused Baxter of writing against his views on the Apocalypse as revenge against More's "booke of spirits". Lamont points this out but he seems entirely unaware of their earlier polemic. See W. M.

The reason for More's initial concern to refute Baxter's private letter on the nature of spirits is made clear in a letter to Baxter of February 1682. A "psychopyrist", More explained to Baxter, is a thinker who "would make all created Spirits a kinde of fire" and this popular opinion is the only serious threat remaining to "the most true and useful notion of a Spirit", namely More's own. There can be no doubt of the extreme importance he attached to the issue:

> That the world should have a true, easy and useful notion of a Spirit I look upon it to be a thing of exceeding great consequence which made me take this opportunity of settling it by this confutation of the Psychopyrists in my Answer to your letter as I did by the confutation of the Holenmerians and Nullibists in my short Treatise of the True Notion of a Spirit.[74]

Turning to Baxter's letter we can readily agree with More that this author had "closely and compactedly . . . couched therein the . . . strengths of what they have to say for themselves that are Psychopyrists."[75] For example, Baxter wrote that

> Though I run not into the excess of Ludov. Le Grand *de Igne* (nor of Telesius or Patricius)[76] I would *Ignis* were better studied: But this Room will not serve me to say what I think of it. But in brief, He that knoweth that *Ignis* is a Substance, whose Form is the Potentia activa movendi illuminandi, calefaciendi, these as received in a gross Passive Body, being but their Accidents oft, but the Igneous Substance in act operating on them, and conceiveth of Spirits, but as *Ignis eminenter*, that is, of a purer substance; than *Ignis* is, which we best conceive of (next the Formal Virtue) by its similitude, I think knows as much as I can reach of the Substance of Created Spirits. And the Greek Fathers that called Spirits *Fire*, and distinguished *Ignem per formas* into Intellective, Sensitive and Vegetative or Visible Fire, (as it is in *Aere Ignito*) allowing an Incomprehensible Purity of Substance in the higher above the lower (as in Passives Air hath above Water, &c.). I think did speak tolerably, and as informingly as are the notions of Penetrability and Indivisibility; though perhaps these also may be useful.[77]

Lamont, *Richard Baxter and the millennium: Protestant imperialism and the English Revolution*, London, Croom Helm, 1979, p. 45. For a further indication of the extent of More's annoyance about Baxter's critique of his notion of spirits consider the following: "I shall only name one *Disingenuity* more, which was antecedent to them all, and gave occasion both to Mr. Baxters Letter, and to the Doctors Answer thereto, and to this Reply of Mr. Baxter. And that was, That Mr. Baxter in his *Methodus Theologiae* (as he has done also in a little Pamphlet touching Judge Hales) without giving any reasons, which is the worst way of traducing any man or his sentiments, slighted and slurred those two essential Attributes of a Spirit, Penetrability and Indiscerpibility, which for their certain Truth and usefulness the Doctor thought fit to communicate to the World", More *Digression*, p. 202. Baxter's comments on the nature of spirit in his 'Pamphlet touching Judge Hales' (see note 95 below) may be consulted in Sir Matthew Hale, *The works, moral and religious* . . ., London, H. D. Symonds *et al.*, 1805, pp. 95 100. Here also Glisson's theories are invoked (p. 99).

[74] More to Baxter, February 10 1681/2, Dr Williams's Library, Baxter Letters, vol. III, f. 284. I am grateful to the Trustees of Dr Williams's Library for allowing me to consult and quote from the Baxter Letters. More was referring in this letter to his *The easie, true, and genuine notion and consistent explication of the nature of a spirit* . . ., which is an English translation of the last two chapters (27 and 28) of the *Enchiridion metaphysicum* (note 61 above), and which was included in Glanvill, op. cit., note 29 above. "Holenmerians" are those who believe that the soul can be *tota in toto et tota in qualibet parte* while "nullibists" are those, like Descartes, who believe that souls are non-spatial entities and so, in More's terms, "nowhere".

[75] Dr Williams's Library, Baxter Letters, vol. III, f. 284.

[76] I have been unable to trace Lodov. Le Grand and his book *De igne*. For brief accounts of Telesio and Patrizi see P. O. Kristeller, *Eight philosophers of the Italian Renaissance*, Stanford University Press, 1964.

[77] Baxter, *Of the nature of spirits*, op. cit., note 71 above [the second of the *Two discourses*, separately paginated], sig. A5r/v. In his *Letter to a learned psychopyrist*, op. cit., note 71 above, More indicates that he does not know which Greek Fathers Baxter had in mind, p. 234. Neither do I, but for indications of the background to similar ideas see, for example, the articles by Walker cited in note 3 above; and Everett Mendelsohn, *Heat and life, the development of the theory of animal heat*, Cambridge, Mass., Harvard University Press, 1964.

So, Baxter regarded fire as "a Substance permeant and existent in all mixt Bodies on Earth . . . in your Blood it is the prime part of that called the *Spirits*, which are nothing but the Igneous Principle in a pure aerial Vehicle and is the Organ of the Sensitive Faculties of the Soul".[78] Clearly, Baxter would agree with Richard Lower (to take one example from many like-minded physiologists) that "the blood in its course through the arteries is like a flaming fire." Once again we can see the profound influence of medical ideas about spirits and the nature of the soul on a leading theologian.[79]

Among such general indications of a familiarity with current medical thinking in Baxter's discourse on the soul we also find numerous specific references to Glisson. In view of the frequency of these references and Baxter's failure to name any other medical writer (unless we count a passing reference to Daniel Sennert on page 18), it is safe to assume that Glisson was the single most important source for Baxter's views on medical spirits.[80] Indeed, Glisson's opinions are drawn upon right at the beginning of the first letter to More:

> I think you and I are agreed that we cannot conceive of a Spirit *unico conceptu*, but must have two inadequate conceptions of it: of which one is that which Dr. Glisson *De Vita Naturae* calls *conceptus fundamentalis*, and is that which we call *Substantia* . . . the other inadequate conceptus is Formal; and I think you and I are agreed that this is *Virtus Una-tria*, as described by me, *viz*. *Virtus vitalis, vitaliter activa, perceptiva, appetitiva*, as Dr. Glisson speaks.[81]

More's response to this was to argue that from these two inadequate conceptions of spirit "one adequate Conceptus does result" but, thankfully, we do not have to pursue the philosophical niceties of this typical piece of seventeenth-century theology. Suffice it to say that More immediately recognized the danger implicit in Baxter's Glissonian approach, "[which] is not only a mistake but a *mischief*, it implying that the *Virtus Appetitiva* and *Perceptiva*, may be in a substance though Material which betrays much of the succours that Philosophy affords to Religion in the points of the Existence of a God and Immortality of the Soul."[82] The safe way to salvation, More insisted, was through his own recension of Cartesian mechanism. This was a new philosphy which still fulfilled the age-old role of hand-maiden to religion: "This mode of philosophy is the most useful for the best ends, and serves to support the main parts of natural Religion the best; namely the Existence of God, of *Genii* or Angels, and the

[78] Baxter, *Of the nature of spirits*, op. cit., note 71 above, p. 71.

[79] Richard Lower, *Diatribae Thomae Willisii . . . de febribus vindicatio . . .*, London, J. Martyn and J. Allestry, 1665, p. 117; see also pp. 23, 89–107, 110, 114–127. This has been reprinted with a translation as: Kenneth Dewhurst (editor), *Richard Lower's* Vindicatio: *a defence of the experimental method*, Oxford, Sandford Publications, 1983. There would seem to be a lot in common between Willis's pyretology and Baxter's "psychopyrism". For similar ideas see Mendelsohn, op. cit., note 77 above; R. G. Frank, *Harvey and the Oxford physiologists: a study of scientific ideas and social interaction*, Berkeley, University of California Press, 1980; Douglas McKie, 'Fire and the *flamma vitalis*: Boyle, Hooke and Mayow', in Underwood (editor), op. cit., note 3 above, vol. 1, pp. 469–488; Mani, op. cit., note 6 above; A. G. Debus, 'Chemistry and the quest for a material spirit of life in the seventeenth century', in Fattori and Bianchi (editors), op. cit., note 3 above, pp. 245–263.

[80] There is evidence to suggest that Baxter and Glisson were acquainted. See below, notes 95 and 96.

[81] Baxter, *Of the nature of spirits*, op. cit., note 71 above, sig. A2r/v. For "inadequate conceptions" see Glisson, op. cit., note 5 above, sig. a2r/v, and pp. 11–12.

[82] More, *Letter*, op. cit., note 71 above, p. 196, 198; cf. also p. 217.

Immortality of the Soul." To deny this, he continued "deprives us of some of the most considerable aids that Philosophy affords against Atheism and Epicurism."[83]

For his part, Baxter was just as sensitive as More to anything that might threaten sound religion. He had, therefore, added a qualification to his citation of Glisson: "And yet I dare not say that a self-moving principle is proper to a Spirit. Nor do I consent to Campanella *de sensu rerum*, and Dr. Glisson that would make all things alive by an essentiating form in the very Elements."[84] The point of Baxter's qualification, as he later spelled out, was that it is wrong to propose activity or life in matter as a *necessary* or *natural* attribute but that there is no logical contradiction in the concept of living, perceiving matter. As he himself put it;

> I confess I am too dull to be sure that God cannot endue matter itself with the formal virtue of Perception: That you say the Cartesians hold the contrary, and that your writings prove it, certifieth me not . . . That Almighty God cannot make perceptive living Matter, and that by informing it without Mixture [i.e. without adding a separate active substance to it], I cannot prove, or I think you: Where is the Contradiction that makes it impossible?[85]

It is perhaps indicative of Baxter's philosophical sophistication that this same point was made nearly ten years later by John Locke in his *Essay concerning human understanding*.[86] Also worthy of Locke is Baxter's denial of the *essential* dichotomy of body and spirit on the grounds that all we can know are the modifications of substances, not the substances themselves, because we learn everything through sensation or perception.

> For to tell you the truth [Baxter writes], I know nothing at all without the mediation of sense except the immediate sensation itself, and the acts of Intellection, Volition or Nolition, & what the Intellect infereth of the like, by the perception of these . . . Now hence I infer, that I have no sense at all of the difference of a Spirits Substantiality in such modes and accidents from that of Matter, and therefore how can I know it?[87]

The crucial difference between More and Baxter should now be abundantly clear. More, in spite of his many disagreements with Descartes, was always totally committed to the dualist distinction between body and soul (or spirit). Baxter, on the other hand, seems to have had a much clearer sense of the philosophical shortcomings of any dualist system.[88] By insisting so vigorously on the categorical difference between body

[83] Ibid., p. 225.

[84] Baxter, *Of the nature of spirits*, op. cit., note 71 above, sig. A3r.

[85] Ibid., pp. 28–29. Baxter is referring to More, *Letter*, op. cit., note 71 above, p. 198.

[86] John Locke, *An essay concerning humane understanding*, London, Eliz. Holt for Thomas Basset, 1690, bk. 4, ch. 3, sect. 6. For a full discussion of this aspect of Locke's philosophy see Yolton, op. cit., note 52 above, especially pp. 14–28.

[87] Baxter, *Of the nature of spirits*, note 71 above, p. 15.

[88] I am unaware of any extended historical analysis of the fortunes of dualist and monist philosophies but there is a good brief account of the reaction to Cartesian dualism in R. A. Watson, *The downfall of Cartesianism 1673–1712: a study of epistemological issues in late seventeenth-century Cartesianism*, The Hague, Martinus Nijhoff, 1966, especially pp. 105–121, 147. There is much that is germane to the background of seventeenth-century monism in Norman T. Burns, *Christian mortalism from Tyndale to Milton*, Cambridge, Mass., Harvard University Press, 1972. For the end of the century and the story through the eighteenth, see Yolton, op. cit., note 52 above. There was frequent affiliation between monism and

and soul, More found it impossible to give any plausible account of their interaction. Whenever he abandoned rhetoric and seriously tackled the problem, he had to resort to a concept of spirit which was every bit as materialistic as Baxter's "psychopyrism" (as Baxter himself was quick to point out).[89] Baxter, steeped in voluntarist theology, was content to take a monistic solution to the soul-body problem because he believed God, by his will, could endow matter with the attributes of life, perception, and automotion. However, the seventeenth-century reader was always quick to equate monism with atheism (as reactions to Hobbes and Spinoza clearly indicate) and so Baxter's *Of the nature of spirits* is sprinkled with apologetic non-dogmatic expressions of his views.[90] Nevertheless, it remains clear that Baxter is in the tradition of neo-Platonic writers who characterized the system of the World in terms of a hierarchy of forms, descending from the "purely immaterial" (God) through successively less-tenuous material entities like light, aether, fire, and air down to the crass material of earthy substances.[91] In this tradition, fire, being one of the most tenuous and subtle of material substances, is frequently regarded as holding an intermediate position between the sub-lunary world and the higher realm of "pure spirits". By the same token, fire can be equated with the human soul which also holds this intermediary position.[92] It is this neo-Platonic tradition which provides the context for Baxter's rejection of the mechanistic account of fire to which More subscribes. Baxter does not mince his words:

---

medical theories. On this see the works by Walker and Pagel cited in note 3 above, and Walter Pagel, *The smiling spleen: Paracelsianism in storm and stress*, Basle, Karger, 1984, pp. 105–108, 148, 156.

[89] On this see Walker, 'Medical spirits in philosophy', op. cit., note 3 above, pp. 289, 293–296; and Henry, op. cit., note 30 above. See also Baxter, *Of the nature of spirits*, op. cit., note 71 above, sig. A6v, pp. 47–49, 64, 66, 91, and 95–97.

[90] For example, Baxter admitted that when he tried to define the differences between substance and matter he was "at a loss" (sig. A3v), that Glisson's way was merely "as probable" as More's (p. 6) but that he himself preferred to take "the middle way" (p. 7), and still remained content "with an *ignoramus*" on other issues (p. 21). Further evidence for the contemporary reluctance to adopt a monist philosophy is provided by Baxter's statement that he was not an easy convert to the notion of a "Virtue una-trina, vitali, perceptiva, appetitiva" in substance: "Nor did I easily assent to it; nor did Dr. Glisson after 80 years of age, easily procure men to assent to it" (p. 25). As far as I know, Baxter's voluntarist theology has not been noticed before, but for indications of the vigour of this kind of theology in seventeenth-century England, see Francis Oakley, 'Christian theology and the Newtonian science: the rise of the concept of the laws of nature', *Church History*, 1961, **30:** 433–457; J. E. McGuire, 'Boyle's conception of nature', *J. Hist. Ideas*, 1972, **33:** 523–542; and Eugene M. Klaaren, *Religious origins of modern science: belief in Creation in seventeeth-century thought*, Grand Rapids, Mich., W. B. Eerdmans, 1977, pp. 29–52.

[91] On this see Crombie and Lindberg, both cited in note 43 above; as well as A. H. Armstrong, *The architecture of the intelligible universe in the philosophy of Plotinus: an analytical and historical study*, Cambridge University Press, 1940; Kurt Goldammer, 'Lichtsymbolik in philosophischer Weltanschauung, Mystik und Theosophie vom 15. bis zum 17. Jahrhundert', *Studium Generale*, 1960, **13:** 670–682; P. O. Kristeller, *The philosophy of Marsilio Ficino*, New York, Columbia University Press, 1943; and E. E. Maechling, 'Light metaphysics in the natural philosophy of Francesco Patrizi da Cherso', MPhil thesis, University of London, 1977.

[92] This is a prevalent notion and forms the background to the confusion between medical spirits and the soul as described by Walker in the works cited in note 3 above, and suggested in the works by Mendelsohn, Frank, McKie, and Mani cited in note 79 above. Perhaps the clearest example is to be found in the metaphysics of Francesco Patrizi da Cherso, see Maechling, op. cit., note 91 above: and Benjamin Brickman, *An introduction to Francesco Patrizi's 'Nova de universis philosphia'*, New York, Columbia University [PhD thesis], 1941.

> I had as willingly have heard Cartesius tell me any dream else that ever came into his Brain: For this I greatly despise: And wonder not that any man is ignorant of the nature of Spirits, who is so grossly ignorant of the igneous analogical Nature as he was . . . Really when I read how far you have escaped the delusions of Cartesianism, I am sorry that you yet stick in so gross a part of it as this is; when he that knoweth no more than motion in the Nature of Fire, which is the active principle by which mental and sensitive Nature operateth on Man, and Bruits and Vegetables, and all the passive Elements, (if it be not *ipsa forma telluris*) and all the visible actions in this lower World are performed, what can that mans Philosophy be worth? I therefore return your Counsel, study more thoroughly the Nature of Aethereal Fire.[93]

The association of this neo-Platonic tradition with medical theories about the nature of spirits, to which D. P. Walker has alerted us, can now be seen to be fundamental to Richard Baxter's pneumatology. Glisson, however, did not need to single out fire as the active principle in nature since he held all matter to be more or less active. As Glisson was evidently Baxter's major source, it is hardly surprising that we find Baxter equivocating between spirit as fire or as substance endowed by God with perceptive, appetitive, and motive faculties.[94] In spite of this ambiguity one thing is clear; Baxter's theology was strongly influenced by current medical theories.

Conclusions

There is evidence to suggest that Richard Baxter was personally acquainted with Francis Glisson who was physician (as Baxter himself tell us) to his close friend and neighbour, Sir Matthew Hale.[95] This would explain Baxter's familiarity with Glisson's philosophical ideas. But Baxter himself believed that familiarity with Glisson's philosophy was rare. He was even unaware of More's attack on Glisson in the *Epistola altera* since he admonished More:

> I marvel that when you have dealt with so many sorts of Dissenters you meddle not with so subtile a piece as that old Doctor's *de Vita Naturae*: I have talkt with divers high pretenders to Philosophy here of the new strain, and askt them their judgement of Dr. Glissons Book, and I found that none of them understood it, but neglected it as too hard for them, and yet contemned it.[96]

More, naturally enough, was delighted to put his adversary right. In his anonymous commentary on Bishop Rust's *Discourse of truth*, More said of himself

> that the Doctor has not only read that subtile Piece of Doctor Glissons, but understands so thoroughly his Hypothesis, that he has solidly and substantially confuted it. Which he did in a faithful regard to Religion. For that Hypothesis, if it were true, were as safe, if not a safer Refuge

[93] Baxter, *Of the nature of spirits*, op. cit., note 71 above, pp. 57–59. Baxter is returning More's counsel because More had urged Baxter to abandon his neo-Platonic view of fire in favour of the Cartesian account. See More, *Letter to a learned psychopyrist*, op. cit., note 71 above, pp. 226–230.

[94] For instances of this equivocation see Baxter, *Of the nature of Spirits*, op. cit., note 71 above, sig. A3r, and p. 10, but the whole discussion reflects Baxter's uncertainty (as he freely admitted, see note 90 above).

[95] The friendship between Baxter and Hale was documented in Richard Baxter, *Additional notes on the life and death of Sir Matthew Hale*, London, for Richard Janeway, 1682, reprinted in Hale, op. cit., note 73 above. While describing Hale's death, Baxter tells us that "When he found his belly swell, his breath and strength much abate, and his face and flesh decay, he cheerfully received the sentence of death: and though Dr. Glisson, by mere oximel squilliticum, seemed awhile to ease him, yet that also soon failed him . . .", p. 92. (Oximel squilliticum is a composition of vinegar flavoured with squills or sea-onions and honey; it is intended to be both diuretic and expectorant.)

[96] Baxter, *Of the nature of spirits*, op. cit., note 71 above, p. 6.

> for Atheists, then the mere Mechanick Philosophie is: ... But God be thanked Dr. H. More ... has perfectly routed that fond and foul Hypothesis of Dr. Glisson, and I dare say is sorry that so good and old a Knight errant in Theologie and Philosophie as Mr Richard Baxter seems to be, should become benighted, as in a wood, at the Close of his daies, in this most horrid dark Harbour and dismal Receptacle or Rendezvous of wretched Atheists.[97]

The time is long overdue for a further reappraisal of Glisson's influence. Arrigo Pacchi, in his survey *Cartesio in Inghilterra*, has said that contemporaries reacted to Glisson's philosophy with "almost total silence".[98] However, as we have seen, three of the leading theologians in England, Baxter, More, and Cudworth, responded forcefully to his work. Henry Oldenburg gave a careful review of the *De natura substantiae energetica* in his *Philosophical Transactions* and Robert Boyle regarded it as sufficiently threatening to his own natural philosophical enterprise to make it a target for criticism in his *Free inquiry into the vulgarly receiv'd notion of nature*.[99] There are strong indications, also, that Glisson's philosophical ideas influenced Leibniz and Locke, though the evidence has yet to be properly assessed.[100] It is clear, at any rate, that Glisson's views were not dismissed by contemporaries as the irrelevant ramblings of a dotard, but were noticed as an important new system of philosophy.

One final point should be re-emphasized. I have argued that medical theories played an important role in the pneumatology of Henry More and Richard Baxter. More responded in his writings to Harvey, Wharton, and Regius as well as Glisson, and he even adapted J. B. van Helmont's concept of *archeus* for his own use. Baxter drew many of his beliefs about the nature of life and the soul from Jean Fernel, the neo-Platonizing medical tradition of "psychopyrism", to retain More's apt neologism, and the philosophy of Francis Glisson.[101] The importance of such medical concerns for contemporary developments in philosophical theology, therefore, should not be underestimated. I hope I have shown, also, that Glisson's one and only

[97] More, *Digression*, note 72 above, pp. 192–193. More obviously went to some trouble to introduce the puns into this piece, it would be a pity if the reader missed them.

[98] Pacchi, op. cit., note 52 above, p. 151.

[99] *Phil. Trans.*, 1672, **7:** 5076–5077. Robert Boyle, *A free enquiry into the vulgarly receiv'd notion of nature*, London, H. Clark for John Taylor, 1658/6. The major targets for Boyle's attack were latter-day Aristotelians and the Cambridge Platonists. See McGuire, op. cit., note 90 above. What has not been noticed so far is that Boyle was also concerned to argue that "Inanimate bodies can have neither Appetites nor Hatreds" (p. 268). It has been suggested that Boyle was also attacking a specific group of "pagan naturalists" in this work but evidence for this is tendentious, see J. R. Jacob, 'Boyle's atomism and the Restoration assault on pagan naturalism', *Social Studs. Sci*, 1978, **8:** 211–233.

[100] For his influence on Leibniz see Sprengel and Rémusat, both cited in note 6 noted above; and H. Marion, *Franciscus Glissonius quid de natura substantiae seu vita naturae senserit et utrum Leibnitio de natura substantiae cogitanti quidquam contulerit*, Paris, G. Baillière, 1880.

[101] More, *Immortality*, mentions Regius, for example, at 79, 92, 96; Wharton at 93–94; and J. B. van Helmont at 79–80, 171, 174, and 195. Harvey appears in the notes to the 4th ed., note 39 above, pp. 97, 127. More uses Helmont's concept of *archeus* in the notes to his poem, *Psychozoia*, see Grosart (editor), op. cit., note 29 above, p. 139; and in the notes to the 4th ed. of the *Immortality of the soul*, note 39 above, p. 10, and elsewhere throughout his works. Jean Fernel is cited by Baxter in his *Methodus theologiae christianae ...*, note 70 above, p. 176. Baxter also had a copy of Fernel's *Universa medicina*, Paris, 1567, in his personal library. Indeed, there were a significant number of medical books in his library but only one work by Glisson: *Anatomia hepatis*. For the complete list of Baxter's book collection see Geoffrey F. Nuttall, 'A transcript of Richard Baxter's library catalogue; a bibliographical note', *J. Ecclesiastical Hist.*, 1951 and 1952, **2:** 207–21; and **3:** 74–100. There is a concentration of medical works at **3:** 94–95.

"philosophical" book cannot be dismissed as merely a physician's dilettantish excursion out of his own field into philosophy, which adds nothing to my claim that *medical* concepts were important to contemporary philosophy. When Glisson wrote, there was no sharp distinction, as there is today, between philosophy and science, or rather philosophy and natural philosophy. Moreover, it should not be forgotten that Glisson's system of philosophy was specifically and quite explicitly developed to provide the underpinning for his medical ideas on the irritability or natural perception of tissue even where there are no nerves or sensory organs present. Glisson's *Treatise on the energetic nature of substance*, therefore, was a new system of philosophy which grew directly out of the pioneering physiological work pursued by one of Harvey's most able successors. In view of the contemporary recognition of the importance of medical theories to a crucial philosophical debate about the nature of the soul and the validity of dualism, it was inevitable that Glisson's medical philosophy should strike leading thinkers as a work of fundamental importance.

V

The matter of souls: medical theory and theology in seventeenth-century England

Heterodoxy is as old as Christianity itself but irreligion was perceived as something new in the intellectual life of early modern Europe. Atheism was widely seen as a progeny of the crisis of faith engendered by the changes of the renaissance and the reformation. New discoveries in philosophy, as John Donne said, called 'all in doubt', and religious relativism – stemming from an increased awareness of alternative religions in other parts of the world and in the multiplicity of reformed churches – made it easier to deny the absolute truth of any religious claims. Among the remedies which were prescribed for this 'very plague and pest of humane Polities' was a new emphasis on so-called 'natural theology'. It was assumed that even if the atheist remained recalcitrant before Scripture, he could not hold out against 'right reason' and the 'light of nature'. Natural theology was developed, therefore, in order to prove the existence and attributes of God and the immortality of human souls by recourse only to reason and the phenomena of the creation. Proofs of the immortality of the soul were held to be crucially important elements in this natural theology since the fear of post-mortem punishments was regarded as the only guarantee of morality and, therefore, of social and political stability.[1]

In addition to the editors and the members of the conference who offered much helpful advice and criticism, I would like to thank Drs Michael Hunter and Dorothy Watkins for their help in the preparation of this paper.

[1] John Donne, *An Anatomie of the World, the First Anniversary* (London, 1611). Henry More, *An Explanation of the Grand Mystery of Godliness* (London, 1660), book, X, ch. 10., p. 516. See also Nathanael Culverwel, *An Elegant and Learned Discourse of the Light of Nature* (London, 1652). For fuller discussions of the perceived rise of atheism see the following: Paul Hazard, *La Crise de la conscience européene* (Paris, 1935); R.H. Popkin, *The History of Scepticism from Erasmus to Spinoza* (Berkeley, 1979); D. Wooton, *Paolo Sarpi: between Renaissance and Enlightenment* (Cambridge, 1983); D. Wooton, 'Unbelief in Early Modern Europe', *History Workshop*, XX (1985), 82–100; M. Hunter, 'The Problem of Atheism in Early Modern England', *Transactions of the Royal Historical Society*, 5th ser., XXXV (1985), 135–57;

The traditional pneumatology of the Roman Catholic Church and, for that matter, the major reformed churches was, of course, dualistic: body and soul were regarded as categorically distinct entities, one being material, the other immaterial. But dualism was not free from theological difficulties – the Bible being rather equivocal on the issue – and it was always fraught with philosophical difficulties. Nevertheless, it was usually invoked as an essential part of the safe way to salvation. By contrast, all monistic philosophies were generally regarded as subversive and irreligious; even if, like those of Spinoza and Berkeley, they were designed to overcome the difficulties of dualism and to set religion on a firmer foundation.

The most notorious monistic systems of philosophy in seventeenth-century England were the revived ancient philosophy of Epicurus, and the mechanistic philosophy of Thomas Hobbes, both of which were supposed to have many atheist adherents. But even the Cartesian system, which had been formulated in vigorously dualistic terms, was taken up by atheistic 'scoffers' and 'wits' who dismissed Descartes's pronouncements on *res cogitantes* and concentrated solely on a universe of *res extensae*.[2] In view of these perceived threats to sound religion, it is hardly surprising that much of the vast and varied anti-atheist literature which flowed from the presses of seventeenth-century England was directed against the new natural philosophies.

However, there was also a strong element of monistic thought in the medical tradition, and here again this was linked, at least in popular sentiments, to atheism.[3] In part, the physician's reputation for atheism was merely a function of his apparent practice. Whatever a doctor's innermost religious convictions might be, his role was to treat all

R.S. Westfall, *Science and Religion in Seventeenth-Century England* (New Haven, 1958); J. Redwood, *Reason, Ridicule and Religion: the Age of Enlightenment in England, 1660–1750* (London, 1976). On the popular awareness of the cultural relativity of religious belief see Carlo Ginzburg, *The Cheese and the Worms: the Cosmos of a Sixteenth-Century Miller* (London and Henley, 1980), pp. 41–51.

[2] S.I. Mintz, *The Hunting of Leviathan: Seventeenth-Century Reactions to the Materialism and Moral Philosophy of Thomas Hobbes* (Cambridge, 1970). On the atheistic appropriation of Descartes see the letter from Henry More to Robert Boyle, 6 December 1671, in R. Boyle, *Works*, 6 vols. (London, 1772), VI, p. 513; and Alan Gabbey, '*Philosophia Cartesiana Triumphata*: Henry More (1646–1671)', in T.M. Lennon, J.M. Nicholas and J.W. Davis (eds.), *Problems of Cartesianism* (Kingston and Montreal, 1982), pp. 171–250.

[3] See the following, all by Walter Pagel: 'The Reaction to Aristotle in Seventeenth-Century Biological Thought', in E.A. Underwood (ed.), *Science, Medicine and History: Essays on the Evolution of Scientific Thought and Medical Practice Written in Honour of Charles Singer*, 2 vols. (London, 1953), I, pp. 489–509, 'Harvey and Glisson on Irritability with a Note on Van Helmont', *Bulletin of the History of Medicine*, XLI (1967), 497–514; *New Light on William Harvey* (Basel, 1976), pp. 34–6, 52–5; *Joan Baptista Van Helmont: Reformer of Science and Medicine* (Cambridge, 1982), pp. 120–3.

sickness as a purely natural phenomenon. Thus, at a time when many believed that sickness was visited upon mankind by God, the physician was seen everywhere ignoring such religious considerations and treating sickness in an entirely naturalistic way.[4] In the context of medicine, pragmatism looks like materialism. And never more so than in the case of treatment for mental illness. As Timothy Bright wrote in 1586, 'the notable fruit and successe' of medicine in curing madness and melancholy

> hath caused some to iudge more basely of the soule, then agreeth with pietie or nature, & have accompted all maner affection thereof, to be subject to the physicians hand, not considering herein any thing divine, and above the ordinarie events, and natural course of things: but have esteemed the vertues them selves, yea religion, no other thing but as the body hath ben tempered, and on the other side, vice, prophanenesse, & neglect of religion and honestie, to have bene nought else but a fault of humour.[5]

It would seem, then, that Galenic theory, which insisted upon the influence of the temperaments on the human soul, and medical practice, which was successful in treating psychic disorders by somatic means, combined to give medicine an odour of impiety. Moreover, although charges of atheism and irreligion were frequently bandied about in the early modern period as rhetorical devices in censure and polemic, there is clear evidence to suggest that some radically unorthodox thinkers did use current medical theories to bolster their heretical views.[6]

Richard Overton, for example, drew upon traditional medical accounts of the soul to establish his belief that man is 'wholly mortal'. 'Anatomize Man', he wrote, 'take a view of all his *Lineaments* & *Dimensions*, of all his *members* & *faculties*, and consider their state severally', and there will be no need to postulate the existence of a superadded soul.

> So that in veritie [he concluded] Man is but a Creature whose several parts and members are endowed with proper natures or Faculties, each subservient to other, to

[4] Paul H. Kocher, *Science and Religion in Elizabethan England* (New York, 1969), pp. 258–321; R.S. Porter, 'Barely Touching: A Social Perspective on Mind and Body', in G.S. Rousseau (ed.), *The Languages of Psyche: Mind and Body in Enlightenment Thought* (Berkeley, 1990), pp. 45–80.

[5] Timothy Bright, *A Treatise of Melancholie* (London, 1586), sig. *iiir. For a survey of *religious* treatment for psychic disturbance see Michael MacDonald, 'Religion, Social Change and Psychological Healing in England 1600–1800', in W.J. Sheils (ed.), *The Church and Healing* [The Eccelesiastical History Society, *Studies in Church History*, XIX.] (Oxford, 1982), pp. 101–25.

[6] Kocher, *Science and Religion in Elizabethan England*, pp. 239–57, 250–4, 284–305; Owsei Temkin, 'On Galen's Pneumatology', *Gesnerus*, VIII (1951), 180–9, reprinted in Owsei Temkin, *The Double Face of Janus and Other Essays in the History of Medicine* (Baltimore, 1977), pp. 154–61.

make him a living Rational Creature whose degrees or excellences of natural Faculties make him in his kind more excellent than the Beasts.[7]

Similarly, when William Coward had *Second Thoughts concerning Human Soul*, which he published in 1702, he drew upon his medical training. Citing Harvey, Walter Charleton and Daniel Sennert, Coward suggested that 'the Operations of the Soul as it is usually call'd seem extreamly to depend *naturally* upon the Circulation of the Blood'. For Coward, blood was the organizing principle of the body, 'the instrumental cause of Life, or rather of the Operations of *Life*', and so even 'the very Understanding and Apprehension' depended upon nothing more than 'a brisk circulation'.[8]

Lady Anne Conway, author of the remarkable *Principles of the most Ancient and Modern Philosophy* (1690), insisted that the contradictions of dualism could only be avoided by recognizing that 'the Soul is of One Nature and Substance with the Body'. Like Coward, she shared William Harvey's belief that 'the blood would seem to differ nothing from the soul':

Even thus in Man's Body, the Meat and Drink is first changed into Chyle, then into Blood, afterwards into Spirits, which are nothing else but Blood brought to perfection; and these Spirits, whether good or bad, still advance to a greater subtilty or spirituality, and by these Spirits which come from the Blood, we see, hear, smell, taste, feel, and think, yea meditate, love, hate, and do all things whatsoever we do.

The sources of Lady Anne's monistic philosophy are not known with any certainty, but her ideas clearly owe much to the thought of J.B. van Helmont whose son, Francis Mercury, became a close friend of hers in the years preceding the composition of the *Principles*.[9]

During the seventeenth century, then, the reduction of psychic phenomena to somatic states – which Timothy Bright had seen to be implicit in medical practice – became increasingly explicit in medical writings and was taken up by speculative writers with various reformist religious intentions. It is important to note that this medically inspired monism differed significantly from the more familiar monism of mechanist materialists. According to Hobbes and Descartes, matter was completely inert and all vital processes were regarded as mere epiphenomena of the interactions of particles of matter in motion. For William Harvey, Francis Glisson, Henry Power, John

[7] Richard Overton, *Man Wholly Mortal* (London, 1655), pp. 15–16, 8–9; see also pp. 2–3.

[8] William Coward, *Second Thoughts concerning Human Soul* (London, 1704), pp. 87, 92, 96–7, 91.

[9] Anne Conway, *The Principles of the Most Ancient and Modern Philosophy*, ed. Peter Loptson (The Hague, 1982), p. 218. First published in Latin (Amsterdam, 1690), *The Principles* must have been written between 1677 and 1679.

Mayow and others in the medical tradition, matter itself, or some forms of it, was inherently active, self-motive, and endowed with seminal or organizing powers. In this philosophy matter could be held to be not just vital but perceptive, endowed with appetites and aversions and ultimately, capable of thought.[10]

In view of the contemporary equation of monism with atheism we might expect to find such medical theories discussed in the extensive and encyclopaedic anti-atheist literature of seventeenth-century England. For the most part, however, this is not the case. The proverbial atheist, the physician, hardly appears at all in learned attacks on atheism.[11] To a large extent, this can be put down to the fact that the mechanist versions of atheism drew all the fire. For example, the 'Opinion of every Atheist and counterfeit Deist of these times', that matter has 'sensation or thought', was categorized by Richard Bentley as mechanistic: the result of 'such and such determinate Motions [of matter], by the action and reaction of one Particle upon another'. The alternative, vitalistic, view that the 'Faculties of Sensation or Perception' are 'inherent in Matter as such' was simply dismissed as untenable.[12]

Edward Stillingfleet was another who recognized only those theories of thinking matter which depended upon the principle of matter in motion:

> To imagine that the Particles of Matter, as they are in Man, should be capable of Sensation, Memory, Intellection, volition etc. merely because of a different shape, size, & motion from what they have in a piece of Wood, is a riddle that requires a new Configuration of Atoms in us to make us understand.

Similarly, a lesser thinker like Timothy Manlove showed no hesitation in labelling as Hobbist the view that the human soul is 'nothing else but the inflamed glowing Particles of the Blood, called Spirits; which are . . . the Active Principle of Life, Motion, Sense and Understanding in Man, and Beast'.[13]

[10] On Harvey and Glisson see the works by Pagel in note 3 above and J. Henry, 'Medicine and Pneumatology: Henry More, Richard Baxter and Francis Glisson's *Treatise on the Energetic Nature of Substance*', *Medical History*, XXXI (1987), 15–40, esp. 17–22. On Power see Charles Webster, 'Henry Power's Experimental Philosophy', *Ambix*, XIV (1967), 150–78. On Mayow, see R.G. Frank jr., *Harvey and the Oxford Physiologists: Scientific Ideas and Social Interaction* (Berkeley, 1980), pp. 224–32, 258–78.

[11] There were various proverbs about medical atheists, such as: *Ubi tres medici, duo athei*. See Thomas Browne, *Religio Medici* (London, 1643), p. 1; John Donne, 'Why are Courtiers sooner Atheists than Men of Other Conditions?', in John Donne, *Paradoxes and Problems*, ed. with introduction and commentary by H. Peters (Oxford, 1980), pp. 23–4 (first published in 3rd edn, London, 1652). See also Kocher, *Science and Religion in Elizabethan England*, pp. 239.

[12] Richard Bentley, *Matter and Motion Cannot Think* (London, 1692), pp. 14, 15, 17.

[13] Edward Stillingfleet, *Origines Sacrae*, in *Works*, 6 vols. (London, 1710), II, p. 291. Timothy Manlove, *The Immortality of the Soul Asserted and Practically Improved* (London, 1697), p. 5.

The failure of these and other writers to distinguish between the mechanical philosophy and the vitalistic monism of the medical tradition is, if nothing else, clear testimony to the perceived prevalence of Hobbist ideas in late seventeenth-century England. Even so, no matter how overpowering Hobbes's ideas might have been, it remains surprising that, at a time when theologians saw atheists everywhere, there is hardly any critical discussion of medical theories. While Michel Servet's claims, that 'sanguis est ipsissima vita' and that 'the soul as a whole is in the blood, and the soul itself is blood', brought him to the stake in Calvin's Geneva (in 1553), William Harvey's conclusion, in *De Generatione Animalium*, that 'there is no need to search for any spirit distinct from blood', raised not one murmur of dissent from theologians. At a time when the nature of spirit was a crucial issue for right religion, nobody seems to have noticed Harvey's suggestion that the blood, 'furnished with the power of acting above all else' was spirit 'par excellence'. Similarly, although it is clear that orthodox theologians in the 1660s paid scant attention to Biblical statements that 'the blood is the soul', it still seems remarkable that the early experiments on blood transfusions failed to engender any theological response whatsoever.[14]

So, why were the latest medical theories excluded from the anti-atheist literature? Clearly, the answer to this question depends less upon a knowledge of contemporary medicine and more upon a full understanding of the nature of the anti-atheist literature. Michael Hunter has suggested that the focus of this literature was somewhat diffuse because its authors believed atheism to be a rather anti-intellectual movement, inspired more by a scoffing, sarcastic incredulity, and by immorality – so-called 'practical atheism' – than by philosophical critique.[15] Accordingly, if the anti-atheist was to spend any time at all in considering intellectual arguments, he would quite

[14] I quote Servet's *Christianismi Restitutio* (1533), from Walter Pagel, *William Harvey's Biological Ideas: Selected Aspects and Historical Background* (Basel and New York, 1967), pp. 137, 144 (there is further bibliography on Servet at p. 137). William Harvey, *Disputations Touching the Generation of Animals*, trans. G. Whitteridge (Oxford, 1981), pp. 283, 375, see also 240–58. John S. White, 'William Harvey and the Primacy of the Blood', *Annals of Science*, XLIII (1986), 239–55. The similarity between Harvey's views and heterodox ideas is well brought out in Christopher Hill, 'William Harvey and the Idea of Monarchy', in C. Webster (ed.), *The Intellectual Revolution of the Seventeenth Century* (London, 1974), pp. 160–81. Biblical equations between blood and soul are found at Leviticus 17:14; Genesis 9:4; and Deuteronomy 12:23.

[15] Michael Hunter, 'Science and Irreligion: an Early Modern Problem Reconsidered', in R.S. Westman and D.C. Lindberg (eds.), *Reappraisals of the Scientific Revolution* (Cambridge, forthcoming). See also G.E. Aylmer, 'Unbelief in Seventeenth-Century England', in D. Pennington and K. Thomas (eds.), *Puritans and Revolutionaries* (Oxford, 1978), pp. 22–46, and J. Redwood, *Reason, Ridicule and Religion*.

naturally concentrate on those which were most commonly alluded to by the atheists themselves. There can be little doubt, judging from the literature, that the mechanical philosophies of Hobbes and Descartes, or a simplistic version of one of them, were the runaway winners in these stakes. The hallmark of the new mechanical philosophies was their easy intelligibility, and they were all quickly available in English translations. Moreover, the traditions of renaissance scholarship ensured that even controversial works like Epicurus's letters and Lucretius's *De Rerum Natura* were published, studied and translated. The fact that Epicurus's philosophy was expounded in one of the finest examples of Latin poetry ensured that it became suitable reading for gentlemen and it evidently came to enjoy a great vogue in seventeenth-century England.[16]

Medical writings, by contrast, do not seem to have attracted a readership much beyond the ranks of those who had a professional or dilettante interest in medical practice. They were still published almost exclusively in Latin, and they usually took for granted a high level of prior knowledge of medical theory. Even when Francis Glisson stepped out of the purely medical stream in order to present the public with his new system of natural philosophy, he seems to have been constitutionally incapable of changing his habitual way of writing. His *Treatise on the Energetic Nature of Substance* was written in Latin and is incomprehensible without an advanced knowledge of the typical concerns, arguments and tropes of the learned traditions of renaissance humanism. By the time Glisson's book appeared, this kind of philosophizing was already becoming unfamiliar to the new philosophers. As Richard Baxter, who greatly admired the work, wrote:

> I have talkt with divers high pretenders to Philosophy here of the new strain, and askt them their judgment of Dr Glissons Book, and I found that none of them understood it, but neglected it as too hard for them and yet contemned it.[17]

[16] For Epicurean influence see T.F. Mayo, *Epicurus in England (1650–1725)* (Dallas, 1934); G.D. Hadzsits, *Lucretius and his Influence* (London, 1935); W.B. Fleischmann, *Lucretius and English Literature, 1680–1740* (Paris, 1964).

[17] Francis Glisson, *Tractatus de Natura Substantiae Energetica, seu de vita naturae, eiusque tribus primis facultatibus, I. Perceptiva Naturalibus, II. Appetitiva Naturalibus, III. Motiva Naturalibus &c.* (London, 1672). Richard Baxter, *Of the Immortality of Mans Soul, and the Nature of It, and Other Spirits. Two Discourses* (London, 1682), p. 6. There are many indications that mechanical philosophers – particularly the less serious minded of them – were constitutionally unable to understand the older ways of philosophizing. See, for example, Meric Casaubon, 'On Learning' (1667), reprinted (in part) in M.R.G. Spiller, *'Concerning Natural Experimental Philosophie': Meric Casaubon and the Royal Society* (The Hague, 1980), pp. 195–214. See also J.M. Levine, 'Ancients and Moderns Reconsidered', *Eighteenth-Century Studies*, XV (1981), 72–89; and M. Hunter, 'Ancients, Moderns, Philologists and Scientists', *Annals of Science*, XXXIX (1982), 187–92.

But what of those medical writings which were based on the new mechanical philosophy? We might expect them to have had a greater impact upon popular thought. Thomas Willis's *De Anima Brutorum*, for example, looks like the sort of book that might have seemed attractive to those with a penchant for mechanistic atheism. A close look at the contents, however, soon reveals a level of technicality which would have proved very difficult for those outside professional medical circles. For example, Willis's claim that the soul is 'Bipart or Twofold' – being partly material and partly immaterial – is defended on the grounds that it has been

> already, in another place, by a necessary Consequence deduced, from the Life of the Blood, as it were a flame, and from the existency of the Animal Spirits, and as it were lucid or aetherial Hypostasis, asserted and proved.

Those who were not put off by Willis's talk of a 'lucid or aetherial Hypostasis' and wished to follow up his ideas in the other place he mentions would have found themselves having to make sense of Willis's pyretology in his *De Febribus*.[18]

The same point can be made about Harvey's *De Generatione Animalium*. His statement, quoted earlier, that the blood is the supreme examplar of spirit, actually follows upon a medical definition of spirit:

> for us physicians, spirit is that which Hippocrates called *impetum faciens*, namely whatsoever attempts anything by its own endeavour and arouses any motion with agility and vehemence, or initiates any action . . .

Immediately, therefore, the reader feels himself bound up in a discussion which depends upon a mastery of medical presuppositions. Harvey's subsequent elaboration of the claim that blood is spirit requires a detailed grasp of a number of abstruse Aristotelian concepts, including not just the nature of the elements but also the notion of formative powers and virtues, and the operation of teleological principles in nature. Harvey was able to speak glibly in this discussion of the 'celestial nature' of the blood – a nature 'analogous to the element of the stars' – because it was a familiar notion to the university-trained physician. It would, however, have proved a rather difficult notion for

[18] Thomas Willis, *De Anima Brutorum quae Hominis Vitalis ac Sensitiva Est* (Oxford, 1672). I have used the English translation in *Dr Willis's Practice of Physick, being the Whole Works of that Renowned and Famous Physician* (London, 1684), sig. A3v; Thomas Willis, *Diatribae Duae Medico-Philosophicae, quarum prior agit de Fermentatione . . . Altera de Febribus . . .* (London, 1659).

the unitiated, particularly if they had already become familiar with mechanistic ways of philosophizing.[19]

All in all, it seems safe to assume that the 'scoffers' and 'wits', who allegedly propagated atheism and irreligion at Court and in the Coffee House, would have taken none of their ideas from current medical opinion. There were suitable theories of life and soul to be found in Epicurus or Hobbes, which had the advantage of being much easier to grasp. Accordingly, it is easy to see why theologians too excluded the latest developments in medicine from their remit.

It must also be borne in mind that physicians and philosophers were not the only rivals to theologians on the issue of mind–body interaction. Even the most poorly educated person in the land had a common sense of his or her internal, mental life, and many people had a rich experience of the interdependence of mind and body. Indeed, there was a common perception of the link between mental faculties and specific areas of the brain. Richard Overton, for example, wrote that

> experience tells us, if the former Brain-pan be hurt, the Senses are hindered, but the Cogitation remaineth sound. If only the Middle-pan be harmed, the Cogitation is maimed; but the Seat of Sense keeps all the five Senses whole . . . if the Hinder-pan be disordered only, the Memorie alone, and neither Sense nor Cogitation receive harme.

Similarly, Thomas Tryon spoke of the 'common opinion', based on experience, that the different mental faculties have 'distinct Seats, or Cells in the Brain': 'the Phantasie resides in the Formost, the Judgement in the middle, and the Memory in the hindermost Ventricle of the Brain'.[20]

If this kind of common 'experience' of mind–body interdependence created problems for theologians, so did the reformist tradition of founding all one's beliefs on the clear word of scripture. One did not have to be a heresiarch to notice that Christ and the apostles placed all their emphasis not on the immortality of the soul but on the resurrection of the body. Difficulties over this matter troubled even a pious

[19] William Harvey, *Disputations Touching the Generation of Animals*, pp. 374, 378–9. On the role of 'celestial' or 'astral' natures in medical thought see D.P. Walker, 'The Astral Body in Renaissance Medicine', *Journal of the Warburg and Courtauld Institutes*, XXI (1958), 119–33; and D.P. Walker, 'Medical Spirits in Philosophy and Theology from Ficino to Newton', in *Arts du spectacle et histoire des idées. Recueil offert en hommage à Jean Jacquot* (Tours, 1984), pp. 287–300.

[20] Richard Overton, *Man Wholly Mortal*, p. 8. Thomas Tryon, *A Treatise of Dreams and Visions* (London, 1689), pp. 19, 20. On the medieval antecedents of these popular notions see W. Pagel, 'Medieval and Renaissance Contributions to Knowledge of the Brain and its Functions', in F.N.L. Poynter (ed.), *The History of the Brain and its Functions* (Oxford, 1958), pp. 95–114.

Anglican like John Evelyn, as we can see from a letter which Jeremy Taylor sent to him in 1657:

But Sir, that which you check at is the immortality of the soul; that is, its being in the interval before the day of judgement; which you conceive is not agreeable to the apostles' creed, or current of scriptures, assigning (as you suppose) the felicity of Christians to the resurrection.

Taylor is immediately in difficulty because he can hardly deny or play down the importance of the Day of Judgement: 'That the felicity of Christians is not till the day of judgement, I doe believe next to an article of ye creed'. However, he must refute Evelyn's conclusion: 'I cannot allow your consequent that the soul is mortal'. Surprisingly, Taylor seems to take refuge, in this letter, in the heresy of 'soul sleeping' or psychopannychy:

the soul may be immortal, and not beatified till the resurrection. For to be, and to be happy or miserable, are not immediate or necessary consequents to each other. For the soule may be alive and not understand; so our soule, when we are fast asleep, and so Nebuchadnezzar's soule, when he had his lychanthropy. And the Socinians, that say that the soule sleepes, doe not suppose that she is mortal; but for want of her instrument cannot doe any acts of life.[21]

The implication of this is that the soul is completely passive and inert when separated from the body, but this too is an unpleasant doctrine for Taylor to accept. Traditionally, it is the soul which is supposed to activate the body, not the body which activates the soul. As Taylor wrote elsewhere:

it were strange that we should be alive, and live with Christ, and yet do no act of life: the body when it is asleep does many [acts of life]; and if the soul does none, the principle is less active than the instrument.

Taylor's response is simply to invoke the highest faculty of the soul as continuing to be active through this sleep: 'if it [the soul] does any act at all in separation [from the body], it must necessarily be an act or effect of understanding; there is nothing else it can do but this it can do'.[22] This depends, of course, on the assumption that understanding is 'proper to the soul', but, as we have seen, there was a popular notion that it was proper merely to the middle 'Brain-pan'. Richard Overton was doubtless not alone in feeling that if the soul is equated with understanding then

[21] Letter from Taylor to Evelyn, 29 August 1657, reprinted in Taylor, *The Whole Works*, 10 vols. (London, 1856), I, pp. lxxvii. On the heresy of 'soul-sleeping' see N.T. Burns, *Christian Mortalism from Tyndale to Milton* (Cambridge, Mass., 1972); and J. Henry, 'Atomism and Eschatology: Catholicism and Natural Philosophy in the Interregnum', *British Journal for the History of Science*, XV (1982), 211–39.

[22] Taylor, 'Funeral Sermon for Frances, Countess of Carbery', died 9 October 1650, in *The Whole Works*, VIII, p. 440.

all men are born without Souls, and some [die] before they had Souls, as Infants; and some after their Soul is gone, as Mad men that live and perish in their Madness; and some would have *Souls* but by *fits* and *jumps*, as Drunkards, persons with the Falling sickness, &c.[23]

It would be possible to pursue such arguments endlessly through the labyrinthine passages of seventeenth-century theological tomes. Like Taylor, every divine who tried to expound upon the nature of the immortal soul sank into contradiction and incoherence or glossed over the issues so quickly that too many questions remained unanswered. The history of the idea of immortality of the soul, were it ever to be written, would surely reveal the intractability of the philosophical problems it poses. Certainly, the prevalence of Socinian and other monistic philosophies among seventeenth-century sectarian groups testifies to the contemporary awareness of these difficulties, and to the widespread dissatisfaction with the standard theological attempts to defend soul–body dualism. As the Ranter Christmas Carol of 1650 asked of university-trained theologians:

Let them but tell us what a soul is, then
We shall adhere to these mad, brain-sick men.[24]

It should be clear from all this that the authors of the anti-atheist literature had more than enough to contend with in opposing the monistic tendencies of common sense together with the popular materialistic versions of the mechanical philosophy. Small wonder, then, that they did not venture further by engaging with the latest medical theories. It is one thing to try to show the inadequacies of 'common sense' views, but it is altogether another to enter into dispute with physiologists and anatomists armed with a specialist knowledge of the brain and all psychosomatic phenomena.

Besides, it is perfectly clear that, in spite of – or perhaps *because* of – their reputation for impiety, medical writers themselves avoided any direct confrontation with theology. As one theologian, Simon Patrick, pointed out in 1662, it was always possible for an author to entertain 'great Paradoxes' provided he 'had the good manners to make a legg and say, *Omnia Ecclesiae authoritate submittimus*'. This formulaic submission was widely recognized by churchmen and laymen alike as a useful piece of etiquette for avoiding overt clashes of interest, such as

[23] Richard Overton, *Man Wholly Mortal*, pp. 19–20.

[24] *The Arraignement and Tryall of the Ranters* (London, 1650), quoted from Nigel Smith (ed.), *A Collection of Ranter Writings from the Seventeenth Century* (London, 1983), p. 37. For a more detailed example of the intractability of the dualist concept of soul see J. Henry, 'A Cambridge Platonist's Materialism: Henry More and the Concept of Soul', *Journal of the Warburg and Courtauld Institutes*, XLIX (1986), 172–95.

demarcation disputes about who was best qualified to pronounce upon the truth of particular matters.[25]

The pattern of medical writing, accordingly, was to begin with expressions of piety and religious devotion (in some cases reiterated once or twice in the body of the work) before embarking on an analysis of anatomical, physiological or medical notions which was entirely naturalistic and made no reference to theological considerations. We must be wary, however, of assuming that this kind of apologetic minimalism in medical works was merely lip-service. It may seem to the modern reader, for example, that Thomas Willis's careful prefatory distinction, in *De Anima Brutorum*, between animal souls and the rational immortal soul of humans, fails to be sustained in any meaningful way throughout the body of the book. Nevertheless, it would be unsafe to suppose that the dominance of the material, animal soul in his discussion is a sign of covert irreligious sympathies. Willis's deepest beliefs and real intentions in writing about animal souls must always remain inscrutable to the historian, and there is nothing in the evidence provided by his life or his writings which enables us to infer that he was duplicitous. Judging from what evidence there is, it seems most likely that Willis quite simply thought he had successfully excluded the immaterial, immortal soul from his considerations in *De Anima Brutorum*.[26]

It seems safe to conclude, therefore, that seventeenth-century English theologians deemed it unnecessary or inexpedient to include medical theories in their surveys of atheism. However, there were two notable exceptions to this rule – two leading theologians who saw in medicine dangers as real as those presented by mechanistic materialism, and who could not or would not turn a blind eye to it. The Cambridge Platonists, Ralph Cudworth (1617–85) and Henry More (1614–87) made considerable contributions to the anti-atheist literature and, unlike their colleagues, they chose to deal directly and

[25] S[imon] P[atrick], *A Brief Account of the New Sect of Latitude-Men together with some Reflections on the New Philosophy* (London, 1662), p. 23.

[26] Willis, *De Anima Brutorum*, sig. A2v. Paul F. Cranefield, for example, implies that Willis's distinction between animal and rational souls was dictated merely by religious expediency. See, P.F. Cranefield, 'A Seventeenth-Century View of Mental Deficiency and Schizophrenia: Thomas Willis on "Stupidity or Foolishness"', *Bulletin of the History of Medicine*, XXXV (1961), 291–315, esp. 306. The inspiration for Willis's pneumatology came from Pierre Gassendi and one scholar has taken a line similar to Cranefield's in his analysis of the Frenchman. See O.R. Bloch, *La Philosophie de Gassendi: nominalisme, materialisme et metaphysique* (The Hague, 1971). For a corrective see M.J. Osler, 'Baptizing Epicurean Atomism: Pierre Gassendi on the Immortality of the Soul', in M.J. Osler and P.L. Farber (eds.), *Religion, Science and Worldview* (Cambridge, 1985), pp. 163–83.

explicitly with the latest medical opinions. Nevertheless, this should not be taken to invalidate everything that has been said about the lack of theological interest in medical theory. On the contrary, More and Cudworth should rather be seen as exceptions that prove the rule.

The theology shared by the only two Cambridge Platonists who took the trouble to publish their views, was by no means as orthodox as their posthumous reputations might suggest.[27] In particular, their emphasis upon reason ran counter to the more fideistic voluntarist theology which was dominant in England at this time. Voluntarism in theology can be traced back to William of Ockham (*c.* 1300–49) but it received powerful support from Calvinism. The voluntarist assumed that God's most significant attribute was his omnipotent and inscrutable will. Mankind is considered to be utterly incapable of fathoming God's reasons or purposes and must, accordingly, rely upon faith in God's grace. For More and Cudworth, however, this theology had unacceptable moral implications. Cudworth summed it up as the belief

> that there is nothing absolutely, intrinsically, and naturally good and evil, just and unjust, antecedently to any positive command or prohibition of God; but that the arbitrary will and pleasure of God (that is, an omnipotent Being devoid of all essential and natural justice), by its commands and prohibitions, is the first and only rule and measure thereof.

This attitude, according to Cudworth, underwrote the Calvinist doctrine 'that God can justly doom an innocent creature to eternal torment.' It was in order to avoid this theology and its abominable moral corollaries, that More and Cudworth developed their own rationalistic theology in which even God is constrained to obey the dictates of 'eternal and immutable morality'.[28]

[27] The published works of the other Cambridge Platonists, Whichcote, Smith and Culverwel, appeared posthumously. On opposition to the Cambridge Platonists' theology see P[atrick], *A Brief Account of the New Sect of Latitude Men*; Joseph Beaumont, *Some Observations upon the Apologie of Dr Henry More for his Mystery of Godliness* (Cambridge, 1665); Samuel Parker, *A Free and Impartial Censure of the Platonick Philosophie* (Oxford, 1666); F.J. Powicke, *The Cambridge Platonists: A Study* (London and Toronto, 1926); M.H. Nicolson, 'Christ's College and the Latitude-Men', *Modern Philology*, XXVII (1929), 35–53; J. Henry, 'Henry More *versus* Robert Boyle: The Spirit of Nature and the Nature of Providence', in S. Hutton and R. Crocker (eds.), *Of Mysticism and Mechanism: Tercentenary Studies of Henry More (1614–1687)* (The Hague, 1989).

[28] Ralph Cudworth, *A Treatise Concerning Eternal and Immutable Morality* (London, 1731), book I, ch. I. It is perhaps significant that this treatise appeared posthumously. I have used the edition in Ralph Cudworth, *The True Intellectual System of the Universe . . . with a Treatise Concerning Eternal . . . Morality*, to which are added the Notes and Dissertations of Dr J.L. Mosheim, trans. J. Harrison, 3 vols. (London, 1845), III, pp. 517–646, esp. 529, 530. F.J. Powicke, *The Cambridge Platonists: A Study*, pp. 115–16, gives as the reason for Cudworth's failure to complete his publishing ventures, his dismay at the reponse to the first part of his *True Intellectual System*. Cudworth's great work was evidently seen by various contempo-

Entailed in this theological enterprise was the belief that 'things are what they are, not by will but by nature': 'Every thing must by its own nature be what it is, and nothing else.' The rationalist theology of the Cambridge Platonists stood or fell with their ability to demonstrate the essential nature of things. Thus, their dualistic belief in the immaterial nature of the immortal soul could not simply be presented as a matter for faith. Both More and Cudworth felt the need to demonstrate apodictically the essential immateriality of the soul. In a very real sense, then, the writings of More and Cudworth should not simply be seen as further examples of the contemporary anti-atheist literature. In fact, they were also intended to establish, against the prevailing orthodoxy, an alternative theology.[29]

Had the leading Cambridge Platonists been concerned only to refute popular atheism, they too might have felt it permissable to ignore medical notions. Given their wider ambitions, however, it was not possible for them to overlook any theory which might jeopardize their philosophical efforts to establish the essential immaterial nature of the soul. Anything that might blur what they took to be the ontologically necessary distinction between body and spirit, or material and immaterial substance, threatened their attempt to establish the truth of their rational theology and its concomitant, 'eternal and immutable morality'. So, for them, medical theories constituted a set of obstacles which they just could not ignore.

Henry More, the 'Angel of Christ's', was the first to step in where other theologians feared to tread. In his *Antidote against Atheism* of 1653, More considered three medical suggestions for the seat of 'animadversion' and the will: the animal spirits, the brain and the 'Conarion or Pine-kernel'. Whichever we chose, he insisted, must be supposed also to be the seat of memory and reason. But the animal spirits are so 'very thin and liquid' that they could not retain a memory: 'For it is as impossible to conceive Memory competible to such a Subject, as it is how to write Characters in the water or in the wind'.[30]

The brain was judged inadequate on two grounds. Firstly:

raries as Arian, Socinian, deistic, or simply infidel. On the history of the theological belief that God must obey eternal and immutable laws of morality and nature, see A.O. Lovejoy, *The Great Chain of Being: A Study of the History of an Idea* (Cambridge, Mass., 1936), and F. Oakley, *Omnipotence, Covenant & Order: An Excursion in the History of Ideas from Abelard to Leibniz* (Ithaca and London, 1984).

[29] Cudworth, *Eternal . . . Morality*, pp. 531, 522. See, Henry, 'A Cambridge Platonist's Materialism', and Henry, 'Henry More *versus* Robert Boyle'.

[30] Henry More, *An Antidote against Atheism: Or an appeal to the Natural Faculties of the Mind of Man, Whether There Be not a God* (London, 1653), cited from the second edition in Henry More, *A Collection of Several Philosophical Writings* (London, 1662), book 1, ch. 11, § 3, p. 33.

Anatomists tell us, that though the Brain be the instrument of sense, yet it has no sense at all of it self; how then can that that has no sense direct thus spontaneously and arbitrariously the Animal Spirits into any part of the Body? an act that plainly requires determinate sense and perception.

Once again More has linked sense and perception to animadversion and reason, and this brought him to his second point. The substance and structure of 'this laxe pithe or marrow in man's head shows no more capacity for thought, he claimed, than 'a Cake of Sewet or a Bowl of Curds'. Moreover it would be impossible to say in which 'loop or interval' of the brain the various faculties of 'free Phansy' and 'active Reason' reside. He emphasized the unstructured nature of the brain by reference to a case reported by Fontanus 'of a boy at Amsterdam that had nothing but limpid water in his head in stead of Brains; and the Brains generally are easily dissolvable into a watery consistence'.[31]

As for the final suggestion, that the will is enacted by 'that little sprunt piece of the Brain', the conarion, More simply found this ridiculous:

If you heard but the magnificent stories that are told of this little lurking Mushroom, how it does not onely hear and see, but imagines, reasons, commands the whole Fabrick of the body more dexterously then an Indian boy does an Elephant, what an acute Logician, Subtle Geometrician, prudent Statesman, skilful Physician, and profound Philosopher he is, and then afterward by dissection you discover this worker of Miracles to be nothing but a poor silly contemptible Knob . . . would you not sooner laugh at it then go about to confute it?[32]

Six years later, in *The Immortality of the Soul*, More added to these arguments. He now dismissed J.B. van Helmont's theory that the 'Orifice of the Stomack' is the seat of common sense. The 'common sensorium', More asserted, must have the power of moving the body, but

there is no Mechanical reason imaginable to be found in the Body, whereby it will appear possible, that supposing the mouth of the Stomack were the common Percipient of all Objects, it could be able to move the rest of the members of the Body, as we find something in us does.

Van Helmont proposed the pylorus as the seat of the soul because of its extreme sensitivity and behaviour, which he took to be evidence of emotions, deliberations and decision-making. Henry More, however, gave this short shrift: 'The harsh handling of an angry Sore or the treading on a Corn on the Toe, may easily cast some into a swoon, and yet no man will ever imagine the Seat of the Common Sense to be placed in the Foot.' Van Helmont's suggestion that frenzy and madness

[31] *Ibid.*, I, II, 4–5, p. 34; I, II, 7, p. 35. [32] *Ibid.*, I, II, 8–9, p. 35.

may stem from a disturbance of the pylorus is compared by More with the 'furor uterinus'; 'apoplexies and syncopes may well proceed from the wombe but nobody supposes that the wombe is the seat of common sense in women'.[33]

Similarly, More rejected the suggestion, which he attributed to Hobbes but which might equally have been ascribed to Aristotle or Harvey, that the heart is the seat of common sense. Remarkably, given what was known about the blood supply, More repeated the argument that there was no 'Mechanical connexion' between the heart and 'all parts of the body'. Perhaps more interesting, however, is his insistence that the continued motions of the heart would necessarily disturb the tenor of our mental lives:

> it is very unlikely that that part which is so continually employed in that natural Motion of contracting and dilating it self, should be the seat of that Principle which commands Free and Spontaneous progressions: Perceptions also would be horribly disturbed by its squeezing of it self, and then flagging again by vicissitudes. Neither would Objects appear in the same place, or at least our sight not fixt on the same part of the Object, when the Heart is drawn up and when it is let down again.[34]

The 'Spinal Marrow', the 'concourse of all the Nerves of the Body', was also rejected on the grounds that, like the brain, it does not significantly differ from 'the pith of Elder or a mess of Curds'. But if the spinal column's 'dull pasty Matter' is inconvenient for the 'Centre of Perception', so is the 'perfectly-solid, but very small, particle of Matter in the Body', which was invoked by Henricus Regius (Henry le Roy). More objected that a filing of iron or steel may equally well be said to have sense in it before going on to deny that such a small particle could move the body. In order to move the body, it would have to move itself, he claimed, but 'it being more subtile than the point of any needle . . . it must needs passe through the Body and leave it' rather than communicate its motion to the body.[35]

The theory which engaged most of his attention, however, was the Cartesian suggestion that the common sense is confined to the conarion or pineal gland. For his contemporaries this theory, as More

[33] More, *The Immortality of the Soul* (London, 1659). Again, I have used the 2nd edn in *A Collection*, book II, ch. 4, § 3, p. 78; and II, 7, 5, pp. 89–90. For a discussion of Helmont's view see Walter Pagel, 'The Religious and Philosophical Aspects of Van Helmont's Science and Medicine', *Supplements to the Bulletin of the History of Medicine*, II (1944), pp. 37–8.

[34] More, *Immortality of the Soul*, II, 4, 4, p. 78; II, 7, 8, pp. 90–1. Cf. Thomas Hobbes, *De Corpore* (London, 1655), part 4, ch. 25, § 4. Others regarded the constant motion of the heart as a sure sign that it *was* the seat of the soul; see Pagel, 'Medieval and Renaissance Contributions to Knowledge of the Brain and its Functions', p. 104.

[35] More, *Immortality of the Soul*, II, 7, 18, p. 94; II, 4, 8, p. 79. Henricus Regius, *Philosophia Naturalis* (Amsterdam, 1654), book 5, ch. 1.

remarked, bade 'the fairest of anything . . . for the resolution of the Passions and Properties of living Creatures into mere Corporeal motion'. Furthermore, as he knew all too well, there were a number of impious thinkers who had used Descartes' original conception in a thoroughly atheistic way:

The sum of this Abuse [he wrote] must in brief be this, That the Glandula Pinealis is the common Sentient or Percipient of all Objects; and without a Soul, by virtue of the Spirits and Organization of the Body, may doe all those feats that we ordinarily conceive to be performed by Soul and Body joyned together.

Unlike the rather facile arguments against caricatured versions of van Helmont and Regius, some of More's attacks on Cartesian psychology are detailed and trenchant. Consider, for example, his response to Descartes's account of memory, which he accurately summarized:

the transmission of Motion from . . . [an] Object, through the Nerves, into the inward concavities of the Brain, and so on to the Conarion, opens such and such Pores of the Brain, in such and such order or manner, which remain as tracts or footsteps in the presence of these Objects after they are removed. Which tracts, or signatures, consist mainly of this, that the Spirits will have an easier passage through these Pores than other parts of the Brain. And hence arises Memory, when the Spirits be determined, by the inclining of the Conarion, to that part of the Brain where these tracts are found they moving the Conarion as when the Object was present, though not so strongly.

More simply pointed out that this cannot explain our memory of colours which, according to Descartes, are produced only by the different *speeds* of motion of the particles transmitted from an object to the eye–brain system.[36]

More also attacked Descartes where he was already in retreat. Because of his fundamental principle that the amount of motion in the universe was fixed, Descartes could not allow that the soul or its instrument, the pineal gland, could initiate *new* movements of the body (a concept of change based on ceaseless chains of cause and effect entailed the idea that if motion begins somewhere in the universe, a corresponding 'amount' of motion must cease somewhere else). Consequently, he restricted the operation of the pineal gland merely to *redirecting* the constant flow of animal spirits in the body to bring about whatever movements are required. The details of this account required valves in the nerves which, as More pointed out, 'may be onely a mere fancy, these Valvulae in the Nerves not being yet discov-

[36] More, *Immortality of the Soul*, II, 4, 9, p. 79; II, 51, p. 80; II, 5, 7, p. 82–3. More may have had Regius in mind as an atheistic Cartesian; see II, 5, 4–5, p. 82, and II, 7, 13, p. 92. See also P.R. Sloan, 'Descartes, the Sceptics, and the Rejection of Vitalism in Seventeenth-Century Physiology', *Studies in History and Philosophy of Science*, VIII (1977), 1–28, esp. 23–8.

ered by any Anatomist to be part of the Organization of the Body of any Animal'. But More went on to argue that these valves 'would not affect what is aimed at, though they were admitted'. All the conarion can do is direct the animal spirits into one muscle rather than another, 'as Wind in a Bladder'. It follows, according to More, that 'the Tenth part of that force which we ordinarily use to open a mans hand against his will, should whether he would or no easily open it'. 'Mere Mechanical reasons' cannot explain how the spirits in the muscles can be held in place 'in despite of external force'. Instead, we must acknowledge the power of 'the mere *Imperium* of our Soul'.[37]

More conceded nothing whatsoever to any of the medical theories of psychology. He fully intended to reserve all aspects of our mental lives, and much more besides, to the exclusive purview of the immaterial soul. He held the soul responsible not only for higher level mental faculties such as understanding or reason, will, and the self-conscious perception of our 'internal sense', but also for the supposedly 'lower' mental faculties, such as the perception of the external world, the imagination and memory, which doctors had tended to regard as materialistic phenomena. Nevertheless, More still recognized the need to explain psychosomatic interactions and to determine the precise seat of the soul in the body.

His answer to these problems relied upon the animal spirits. Although the animal spirits by themselves were incapable of doing many of the things which anatomists suggested, they were the ideal vehicle for the immaterial soul. More believed that this was 'in a manner the general opinion of all Philosophers' since, for example, those who placed the common sensorium in the heart did so because they believed the left ventricle to be the fountain of these 'pure and subtile Spirits', and even those who held the soul to be corporeal chose 'the freest, subtilest & most active Matter to compound her of'. Descartes and Hermes Trismegistus were clearly in agreement with More on this matter, and even Regius spoke of 'a gale of Spirits' whose circuit in the body accounted for all movements. In view of this, More decided, the bodily site where these spirits accumulate, the 'Acropolis of the Soul', must be, 'the fourth Ventricle' of the brain (the first three being the two 'upper Ventricles' and 'the middle Ventricle').

However, the soul was not confined to this fourth ventricle of the

[37] More, *Immortality of the Soul*, II, 5, 4–6, p. 82. Cf. More, *Appendix to the ... Antidote against Atheism*, in *Collection*, ch. 10, § 6, p. 171. See also Alan Gabbey, 'The Mechanical Philosophy and its Problems: Mechanical Explanations, Impenetrability, and Perpetual Motion', in J.C. Pitt (ed.), *Change and Progress in Modern Science* (Dordrecht, 1985), pp. 9–84, esp. pp. 14–28.

brain. It enacted here its duties as the common sensorium, but it also animated the whole body by spreading itself locally through the nervous system:

it is plain that the main use of the Brain and Nerves is to keep these subtile Spirits from over-speedy dissipation; and that the Brain with its Caverns is but one great round Nerve; as the Nerves with their invisible porosities are but so many smaller productions or slenderer prolongations of the Brain: And so together are but one contained Receptacle or Case of that immediate Instrument of the sensiferous motions of the Soul, the Animal Spirits, wherein also lies her hidden Vehicle of life in this mortal body.[38]

It can now be seen why More was so insistent upon the amorphous nature of that 'Cake of Sewet', the brain. His earliest pronouncements on the unsuitability of the brain for complex functions, in the *Antidote against Atheism*, predated the appearance of Willis's *De Cerebri Anatome* (1664), but even subsequently More made no attempt to take in the latest theories of brain structure. He continued this pretence even when preparing new editions of his philosophical works for the *Opera Omnia* of 1679; the extensive notes and scholia which he added to take account of recent developments offer no second thoughts on the nature of the brain and nerves. Nor did he ever acknowledge the popular view, mentioned above, that different parts of the brain dealt with sense, understanding and memory. It was crucial for More's philosophical theology that all brain functions were performed entirely by the soul. Accordingly, any suggestion that the brain had a complex structure was potentially embarrassing – this was a time when almost everyone agreed that nature did nothing in vain.

Furthermore, although More invoked animal spirits, even going so far as to admit that 'the Soul of her self, without the assistance of the Spirits' is incapable of 'Sensation & other [mental] Operations', his ideas are very different from those of current medical writers. While dualist natural philosophers like Descartes, Gassendi, Willis and Charleton tried to demarcate clearly and specifically what mental functions were carried out by the incorporeal soul and what by animal spirits, More attributed virtually everything to the immaterial soul. Indeed, it is difficult to be sure what role the spirits play in his system beyond providing a material sign that the soul is present. The spirits do not produce muscle contractions which are brought about by the 'mere *Imperium* of our Soul'; they merely assist in the act of sensation where 'the arrival of motion from the Object to the Organ . . . is received . . . by virtue of the Soul's presence there'; and neither they nor the brain play

[38] More, *Immortality of the Soul*, II, 8, 1–7, pp. 95–6; II, 8, 13, p. 98.

any part in memory which is categorically said to be 'in the Soul'. Forgetfulness, on the other hand, occurs when the spirits are 'not in a due temper', possibly because of 'overmuch coolness or Waterishness in the Head'.[39]

Perhaps the clearest indication of the philosophical and theological gulf between More and these other dualist thinkers is provided by their respective views on animal souls. One of the major assumptions of the mechanical philosophy was that matter could be organized in such a way that it could constitute a 'living' system. Life did not depend, therefore, upon an immaterial principle or soul. The immaterial soul was held to be unique to humankind, the distinguishing feature between men and beasts, and this meant that its realm of operation could not be extended very far. Anatomists were well aware that men and animals displayed, as Thomas Willis put it, 'altogether the same Conformation of the animal Organs'. Since there was nothing in the brain and 'nervous stock' of man that did not have its counterpart in other animal brains, it seemed obvious that vital and mental functions common to both men and animals could be attributed to the same organs or parts of organs in each. Hence only the higher mental faculties of ratiocination, abstract thought and understanding were attributed to the animal soul. To assign common animal faculties such as sensory perception, certain 'passions' or emotions, and memory to the immaterial soul would be tantamount to denying the significance of the anatomical structure of the brain and nervous system and would require that animals too have immaterial souls. This is precisely what More did.[40]

More's somewhat facile insistence that the brain was as amorphous as a 'Bowl of Curds' should be seen, therefore, not simply as stupidity or mere ignorance of contemporary medical developments in brain anatomy, but rather as a deliberate stratagem to promote his extreme soul–body dualism. By the same token, he felt it necessary to insist that animals, no less than men, have immaterial souls. Thus, when More declared that it was not spirits which kept muscles clenched but the '*Imperium* of our Souls', he immediately confronted the anticipated Cartesian criticism head-on: 'From whence it is manifest that brute Beasts must have Souls also'. Elsewhere he stood the Cartesian argument on its head as another ploy to defeat materialist monism:

[39] *Ibid.*, II, 11, 1, p. 106; II, 5, 6, p. 82, II, 11, 2, p. 106; II, 11, 4, p. 107. Thomas Willis, *Cerebri Anatome Nervorumque Descriptio et Usus* (London, 1664). Henry More, *Opera Omnia*, 2 vols. (London, 1679).

[40] Willis, *De Anima Brutorum*, p. 32. See W.F. Bynum, 'The Anatomical Method, Natural Theology and the Functions of the Brain', *Isis*, LXIV (1973), 445–68. See also Pagel, 'Medieval and Renaissance Contributions to Knowledge of the Brain and its Functions'.

For the wafting of one's hand near the Eye of a mans friend, is no sufficient proof that external Objects will necessarily and Mechanically determine the Spirits into the Muscles, no faculty of the Soul intermedling. For if one be fully assured, or rather can keep himself from the fear of any hurt, by the wafting of his friend's Hand before his Eye, he may easily abstain from winking: But if fear surprise him, the Soul is to be entitled to the action, and not the mere Mechanisme of the Body. Wherefore this is no proof that the Phaenomena of Passions, with their consequences may be salved in brute Beasts by pure Mechanicks; and therefore neither in Men.

Inversion of the Cartesian argument enabled More to use it against materialists: when animals blink it is a sign that they have (immaterial) souls; men blink, so they too must have souls.[41]

More's belief in animal souls was by no means a common theological opinion, but it was defended by his close friend, Ralph Cudworth. 'Life and mechanism', Cudworth suggested, were, until recently, 'two distinct ideas of the mind', which were not, 'confounded together': 'It being a thing that was hardly ever called into doubt or question by any before Cartesius, whether the souls of brutes had any sense, cogitation or consciousness in them or no.' In spite of this new intellectual fashion Cudworth, like More, insisted upon the incorporeal nature of the animal soul. The reason for this was not, as some historians have suggested, because they had a supposedly typical English sentimentality towards animals. It was an indispensable element in the rational theology of both men that souls must be *essentially* immaterial. For them, establishment of the immateriality of animal souls was as important as – and indeed inseparable from – the establishment of the incorporeality of human souls. As Cudworth wrote:

they who will attribute life, sense, cogitation, consciousness and self-enjoyment . . . to blood and brains, or mere organized bodies in brutes, will never be able clearly to defend the incorporeity and immortality of humane souls, as most probably they do not intend any such thing. For either all conscious and cogitative beings are incorporeal, or else nothing can be proved to be incorporeal.[42]

This stratagem was not without its drawbacks. Neither More nor Cudworth wished to assert the *immortality* of animal souls, but this

[41] More, *Immortality of the Soul*, II, 5, 6, p. 82; II, 10, 6, p. 103; cf. II, 5, 9–10, p. 83–4.

[42] Cudworth, *A Treatise Concerning Eternal and Immutable Morality*, I, pp. 73, 80. Historians have tended to regard More's and Cudworth's views as representative of English theology and as deriving simply from English sentimentality towards animals. See, for example, K. Thomas, *Man and the Natural World: Changing Attitudes in England 1500–1800* (Harmondsworth, 1983), pp. 104, and J. Rogers, 'Descartes and the English', in J.D. North and J.J. Roche (eds.), *The Light of Nature: Essays in the History and Philosophy of Science Presented to A.C. Crombie* (The Hague, 1985), 281–302, esp. 292. Thomas sees a dichotomy between Cartesianism and the views of More which are said to be 'more representative of English opinion', p. 35. In fact, most orthodox seventeenth-century theologians adopted the opinions of Pierre Gassendi, which held that animals were endowed with a lower, subtle material soul. As Thomas's sources show, More and Cudworth were in very radical and unorthodox company on this issue, pp. 138–9. See also L.D. Cohen, 'Descartes and Henry More on the Beast-Machine', *Annals of Science*, 1 (1936), 48–61.

meant that they had to sever the philosophical links between immateriality and immortality which other dualists had laboured to forge. It was axiomatic among mechanist dualists that all change was the result of matter in motion. It followed that a non-material entity could not change, and *a fortiori* could not perish. These efforts were completely undermined by More's insistence that the immaterial souls of brutes have no 'capacity of eternal life and bliss' but are only imperishable in the same sense that matter is. Similarly, while Cudworth believed that not even 'Omnipotence itself by mere will' could create a living *material* soul, there was nothing, he claimed, to prevent God from annihilating an immaterial soul 'at pleasure'. Immortality was no longer a necessary concomitant of immateriality. As one contemporary theologian, Matthew Smith, pointed out,

> incorporeity does not prove immortality, if anything incorporeal be mortal: Either, therefore, the Phenomena of Sensation do not prove Incorporeity, or incorporeity does not prove Immortality; and then there's an end of all natural arguments for Immortality or all sensible Creatures are immortal.

Smith was not the only theologian to reject the Cambridge Platonist attitude. For example, Edward Stillingfleet, the later Bishop of Worcester, also preferred the mechanistic emphasis on the natural immortality of all immaterial entities. Moreover, the fact that iatromechanism was not attacked in the bulk of the anti-atheist literature suggests that most theologians agreed with him. It can hardly be a coincidence that the only theologians to attack iatromechanism were also the only ones to insist upon the immaterial nature of the animal soul.[43]

Henry More was alone, also, in his attack on another aspect of medical theory. The widely held medical belief that life processes could be equated with some kind of fire burning in the body, was regarded by More as another materialistic threat to his concept of soul. The following account of life from Willis's *De Anima Brutorum*, for example, would have struck More as dangerously materialistic in its emphasis:

> there is not much more difference between an insensible and a sensible Body, than between a thing uninkindled, and a thing kindled; and yet we ordinarily see, this to be made from that; why therefore in a like manner, may we not judge a sensible thing, or Body to be made out of an insensible? Every matter, as it is not Burnt, so not animated

[43] More, *Immortality of the Soul*, II, 12, 1, p. 110; cf, *Appendix to the . . . Antidote against Atheism*, ch. 10, § 7, p. 171. Cudworth, *A Treatise Concerning Eternal and Immutable Morality*, I, p. 80, and III, p. 531. Matthew Smith, *A Philosophical Discourse of the Nature of Rational and Irrational Souls* (London, 1695), p. 3. Cf. Edward Stillingfleet, *Origines Sacrae*, I, p. 639, and II, pp. 385–6.

... But as soon as they have taken flame, from some incentive being put to it, by and by their Particles being rapidly moved, and as it were animated, produce a shining with Heat and Light ... In like manner, the Vital humour in an Egg, remains torpid and sluggish in the beginning, and like to uninkindled matter; but as soon as it is actuated, from the Soul being raised up, presently like an inkindled fire, it excites Life with Motion and Sense, and in the more perfect Creatures with Heat.

'Life', Willis emphasized on another occasion, 'is not so like to flame, but even a flame it self'.[44]

More dubbed this 'psychopyrism' and published an attack upon it in 1682 under the title 'A Letter to a Learned Psychopyrist'. The focus of his attack was the theory of generation-as-enkindling to which Willis and others subscribed.:

according to this, the production of a Soul, *ex Traduce*, would end in meer Materialism, and signifie only, that in Generation some matter only is newly modified, as the Tallow is or Wax [of a candle] when they pass out of their state of Wax or Tallow into that of a bright Flame. Which way of philosophizing, as it is most false, so it is most mischievous if men should be so fond as to believe it.

Significantly, More drew upon his knowledge of Cartesianism to refute this vitalistic view of fire:

it looks as if there were an *ἀυτοκινησία*, or Self-moving in Fire, it being in perpetual motion but this is a plain fallacy, for the parts of the Flame are not self-moved, but they are moved of another: As when you apply a lighted Candle to light another Candle, the parts ... are put into motion by the moved parts of the lighted Candle ...

Unfortunately for More's claims, not everyone accepted the Cartesian belief in passive matter. Willis, for example, followed Gassendi in holding matter to be active:

what is vulgarly delivered, that Matter, out of which Natural things are made, is meerly passive, and cannot be moved, unless it be moved by another thing, is not true; but rather on the contrary, Atoms, which are the matter of sublunary things, are so very active and self-moving that they never stay long, but ordinarily stray out of one subject into another[45]

Richard Baxter, the nonconformist divine, who responded directly to More's 'Letter', was utterly dismissive of More's Cartesian account

[44] Willis, *De Anima Brutorum*, pp. 33, 7. Willis cited 'Hippocrates, Plato, Pythagoras, Aristotle, Galen, Democritus, Epicurus, Laertius, Fernelius, Heurnius, Cartesius, Hogelandus, Honaratus Faber, Verulam, and George Ent' as those who believed the substance of animal souls to be 'Fire it self', p. 5, but this list could easily be extended. See Douglas McKie, 'Fire and the *Flamma Vitalis*: Boyle, Hooke and Mayow', in E.A. Underwood (ed.), *Science Medicine and History*, I, pp. 469–88; Everett Mendelsohn, *Heat and Life: The Development of the Theory of Animal Heat* (Cambridge, 1964), pp. 27–66; Audrey B. Davis, *Circulation Physiology and Medical Chemistry in England 1650–1680* (Lawrence, Kansas, 1973).

[45] More, 'Letter to a Learned Psychopyrist', in Joseph Glanvill, *Saducismus Triumphatus: Or Full and Plain Evidence concerning Witches and Apparitions*, 2nd edn (London, 1682), pp. 217, 241. This is discussed more fully in Henry, 'Medicine and Pneumatology: Henry More, Richard Baxter and Francis Glisson's *Treatise on the Energetic Nature of Substance*'. Willis, *De Anima Brutorum*, p. 33.

of fire. Far from being 'no more than motion', fire is, according to Baxter, 'the active Principle by which mental and sensitive Nature operateth on Man, and Bruits and Vegetables, and all the passive Elements, (if it be not *ipsa forma telluris*)'. More's objection to 'psychopyrism' stemmed from his own mechanistic and materialistic view of fire. Theologians who did not see fire in the same light were unlikely to share More's concern. Indeed, it would seem that analogies between fire and the soul were almost as common in theology as they were in medical writings. Jeremy Taylor, for example, drew upon an analogy with fire to explain how the soul when separated from the body would no longer need to eat to survive:

> fire can abide without matter to feed it: for itselfe is matter; it is substance. And so is the soul: and as the element of fire, and the celestial globes of fire eat nothing, but live of themselves; so can the soul when it is divested of its relative, and so would the candle's flame, if it could get to the region of fire, as the soule does to the region of spirits.

Once again we see that More's opposition to contemporary medical theories reveals more about the differences between his theology and that of his colleagues than it does about general perceptions of medicine.[46]

Similarly, More and Cudworth were the only theologians to recognize a threat to their religion in the work of Francis Glisson. The monistic vitalism of his *Treatise on the Energetic Nature of Substance* was in many respects a greater danger to their dualist endeavours than the mechanical philosophy of Cartesians and Hobbists. More and Cudworth made great play with the claim that no organization of dead matter could possibly produce life. More spoke for them both when he ridiculed the suggestion 'that *Organization* may do strange feats', as if, 'a Watch may be a living creature, though the several parts have neither Life nor Sense'. Such arguments, however, had no force against the new theories of the Regius Professor of Physick at their own University. Glisson agreed that life could not be a mere epiphenomenon of particular organizations of matter. Instead, he argued that all matter was endowed with a primitive form of life which could be seen to be manifested not only in the ability of matter to move itself but also in its apparent perceptivity and appetites.[47]

[46] Richard Baxter, *Of the Immortality of Mans Soul*, pp. 57–9. Jeremy Taylor, *The Whole Works*, I, p. lxviii.

[47] More, *Immortality of the Soul*, II, 4, 1, p. 77. On Glisson's theory of matter see: O. Temkin, 'The Classical Roots of Glisson's Doctrine of Irritation', *Bulletin of the History of Medicine*, XXXVIII (1964), 297–328; reprinted in *The Double Face of Janus*, pp. 290–316; O. Temkin, 'Francis Glisson', in C.C. Gillispie (ed.), *Dictionary of Scientific Biography*, 14 vols. (New York, 1970–80), V, pp. 425–7; Pagel, 'Harvey and Glisson'; Henry, 'Medicine and Pneumatology'.

These three powers or faculties of matter enabled Glisson to explain how parts of matter could organize and 'improve themselves' to produce sentient animals and man. For Cudworth the atheistic implications of this 'hylozoism', as he called it, were all too plain. The mechanistic proposition that matter was passive demanded some kind of external principle to account for movement and so it could be said that 'atomism . . . hath in it self a natural cognation and conjunction with *incorporealism*'. But 'hylozoism, seems to have altogether as close and intimate a correspondence with *corporealism*'. Indeed, 'there is no necessity at all left,' Cudworth went on, 'either of any incorporeal soul in men to make them rational, or of any Deity in the whole universe to solve the regularity thereof'.[48]

Originating as it did from Glisson's anatomical and physiological researches, the *Tractatus de Natura Substantiae Energetica* is a difficult work and there is little in More's and Cudworth's critiques that suggests they really understood it. Their response was one of ridicule rather than reason, frequently misrepresenting Glisson's ideas to suit their rhetoric. Here, for example, Cudworth's argument depends on the false assertion that Glisson attributed understanding to bare matter:

> If matter as such, had life, perception and understanding belonging to it, then of necessity must every atom or smallest particle therefore, be a distinct percipient by itself; from whence it will follow, that there could not possibly be, any such men and animals as now are, compounded out of them, but every man and animal, would be a heap of innumerable percipients, and have innumerable perceptions and intellections; whereas it is plain, that there is but one life and understanding, one soul or mind, one perceiver or thinker in every one.

The idea of every man being a 'commonwealth of percipients', Cudworth averred, was 'absurd and ridiculous'.[49]

The True Intellectual System of the Universe, Cudworth's massive survey of different kinds of atheism, informs us that 'hylozoick' atheism had found so few subscribers sinces its origins in Hellenistic philosophy that it had looked 'like a forlorn and deserted thing'. Since being 'urged by the writer of "The Life of Nature"', however, Cudworth believed that it 'began already to be looked upon, as the rising sun of atheism'. If Cudworth's judgement of the contemporary scene is to be relied upon, it would seem that Francis Glisson's *De Vita Naturae* had a far greater impact than historians of medicine have previously thought. Certainly, by the end of the seventeenth century

[48] Cudworth, *A Treatise Concerning Eternal and Immutable Morality*, I, p. 145; I, p. 144.
[49] *Ibid.*, III, p. 406.

and throughout the eighteenth, matter's putative ability to think was a major focus of theological polemic. Unfortunately, it is not easy to assess Glisson's role in this development. The concept of thinking matter was suggested as a possibility by John Locke in 1690, and his subsequent debate with Edward Stillingfleet made the whole issue a notorious aspect of contemporary philosophy.[50]

As a physician Locke knew Glisson's work and owned a copy of the *Tractatus*. He may even have discussed it with Lady Masham, Cudworth's learned daughter, but so far this must remain speculative. Anyway, it seems true to say that once Locke made the notion of thinking matter well known, Glisson's work, based on a recondite knowledge of anatomy and physiology, was no longer invoked. More and Cudworth, the only theologians to pay serious attention to medical theories in their anti-atheist writings, had been alone in pointing to the dangers of Glisson's 'hyloziosm'. Other theologians continued to ignore medical ideas. Hylozoism only became a focus for their attention when it had been brought out of medicine and into the realms of mainstream philosophy by John Locke.

In this chapter I have tried to offer suggestions as to why the new ideas in seventeenth-century medicine were disregarded in the otherwise comprehensive anti-atheistic literature of that time. Moreover, I have argued that the unusual interest shown in current medical theorizing by Henry More and Ralph Cudworth should alert us to the fact that their concerns were somewhat different from those of their fellow theologians. The Cambridge Platonists were concerned not only to refute atheism but also to reject the dominant Calvinistic emphasis on voluntarist theology. Repelled by the principle that 'things are good and just, because God Wills them so to be', More and Cudworth wanted to establish the claim that even God's activities are constrained by 'mutual Respects and Relations eternal and immutable, and in order of Nature antecedent to any Understanding either created or uncreated'.[51] The rationalist theology which they built up from these

[50] *Ibid.*, I, pp. 215–16; III, p. 405. John Locke, *An Essay Concerning Human Understanding* (London, 1960), book IV, ch. 3, § 6. For a full discussion of the subsequent response to Locke's suggestion from Stillingfleet and others see John W. Yolton, *Thinking Matter: Materialism in Eighteenth-Century Britain* (Oxford, 1983).

[51] [Joseph Glanvill and George Rust], *Two Choice and Useful Treatises: The One Lux Orientalis or An Enquiry into the Opinion of the Ancient Sages concerning the Praeexistence of Souls. Being a Key to unlock the Grand Mysteries of Providence, in Relation to Mans Sins and Misery. The Other a Discourse of Truth . . . With Annotations on them Both* (London, 1682), sig. N6v, and p. 45 of More's 'Annotations upon *Lux Orientalis*' (separately paginated). This edition, including More's copious annotations, was seen through the press by Henry More. The authors, by this time deceased, had both been devoted disciples of his.

beginnings depended on a strict ontological dichotomy between matter and immaterial spirit – to blur this categorical distinction was to deprive God of one of the immutable 'Respects and Relations' which was supposed to have guided His hand in creation.

One of the ways by which More and Cudworth sought to maintain this strict dichotomy was by denying any activity whatsoever to matter. For them, matter could not even fall to the ground without the intervention of the incorporeal 'Spirit of Nature'. So, although the dualism of the mechanical philosophy was perfectly acceptable to most contemporary English theologians, More and Cudworth regarded it as a threat to their rational theology – the mechanical philosophy granted too much activity to matter itself.[52] It was important for the Cambridge Platonists, therefore, to deny that any mental or vital events could be accounted for solely in material terms and it was this which led them, uniquely, to engage with medical theories. As we have seen, this, in turn, set them equally against those theologians who denied the scriptural validity of immaterial souls and the more orthodox but less severely dualistic theologians who only invoked the immaterial soul to account for the 'highest' mental faculties, such as abstract reasoning.

Although More and Cudworth were alone in taking issue with medical speculations, we should not conclude that the medical tradition was without influence. Eighteenth-century developments in theological controversy tend to suggest that the vitalistic monism which had always been present in the medical tradition came to be perceived as the major threat to the dualism of religious orthodoxy. It would seem, therefore, that More and Cudworth were right to regard medical theory as heralding 'the rising sun of atheism'.

[52] On the role of activity in mechanistic matter theory see: J. Henry, 'Occult Qualities and the Experimental Philosophy: Active Principles in pre-Newtonian Matter Theory', *History of Science*, XXIV (1986), 335–81.

VI

HENRY MORE VERSUS ROBERT BOYLE: THE SPIRIT OF NATURE AND THE NATURE OF PROVIDENCE

Pangloss enseignait la metaphysico-theologo-cosmolonigologie. Il prouvait admirablement qu'il n'y a point d'effet sans cause...

Il est démonstré, disait il, que les choses ne peuvent être autremont: car tout étant fait pour une fin, tout est nécessairement pour la meilleure fin.

Voltaire, *Candide* (1759)

I

Robert Boyle always protested his reluctance to engage in dispute about the validity of his experimental philosophy and its conclusions. Even so, of four significant attacks on his interpretation of various hydrostatical experiments he felt it necessary to respond publicly to three of them. These attacks came from the pens of Thomas Hobbes, Franciscus Linus and Henry More.[1] It is fairly easy to see why Boyle should be so disturbed by the writings of both Hobbes and Linus. Hobbes had an international reputation as a philosopher and a national reputation as an atheist. On either count Boyle could not let it be seen that Hobbes's interpretation of the physical world was more cogent than his own. Franciscus Linus, on the other hand, was a Jesuit priest and, perhaps more to the point in this context, he was a vigorous defender of scholastic Aristotelianism. Boyle and his confederates in English natural philosophy – the leading lights of the Oxford Experimental Club and later of the Royal Society – had not singled out anyone for more criticism than the contemporary Aristotelian, and it was as a representative of that breed that Linus had to be dismissed.[2]

But what about Henry More? At a superficial glance the disagreement here seems rather surprising. Boyle and More were close friends, brought together it seems by common interests in Latitudinarian Anglicanism, Cartesianism and the use of the mechanical philosophy in natural theology, and even in the empirical investigation of ghosts, witches and the like, as a means of establishing the existence of a spiritual realm.[3] So, why did Boyle feel that it was as important to refute More's interpretation of his hydrostatical experiments as it was to refute the interpretations of Hobbes and Linus? What was it about Henry More's opinions which led the fastidiously polite Boyle to say of one of More's publications, 'you had better never have printed it'?[4]

The answer to this question seems all the more important historiographically speaking since it is all too often assumed by historians of science that More and Boyle were staunch allies in the ideological struggles associated with the new philosophy. Both men were in essential agreement, it has been supposed, about the ways in which natural philosophy could be used on the one hand to combat atheism, sectarian enthusiasm, and Roman Catholicism, and on the other hand to promote Anglicanism. This assumption has given rise to a significant number of recent historical studies which have taken it for granted that the leading natural philosophers in seventeenth-century England, together with Cambridge Platonists like More and Ralph Cudworth, were committed to the belief that matter was completely inert and passive, and that all motion and activity in the world could only be brought about by the operation of an immaterial principle external to matter. According to this view, the polemic between More and Boyle was a minor disagreement about the correct interpretation of a few hydrostatic experiments.[5] The tendency to trivialize the significance of this dispute is reinforced by the extreme naivety, judged by modern standards, of Henry More's arguments. Boyle's experiments and his interpretations of them can still be seen as models of hydrostatic science, but More's claims seem almost risible. In trying to show the absurdity of Boyle's conclusions, More only succeeds in making a fool of himself. For example, More's inference, from Boyle's account of the gravitation of fluids, that the weight of water on top of a piece of wood at the bottom of a vessel would pin it down rather than make it float completely misses Boyle's point.[6]

It is the contention of this paper that the dispute between More and Boyle, in spite of the polite and careful manner in which it was conducted, represents a clash of fundamental beliefs as deep-seated as the opposing views in the more famous dispute between Samuel Clarke and Leibniz. Contrary to received historiographical wisdom, More and Boyle did not share the same theory of matter. The strict dualism of Henry More, in which matter was necessarily inert and all activity had to be attributed to a spiritual or immaterial operator, was not accepted by Robert Boyle. Conversely, Boyle's belief that God could endow matter with 'essential modifications' which included motion and 'seminal rudiments or principles' was anathema to Henry More.[7] Moreover, these differing theories of matter reflect deeper differences about the nature of God and divine Providence. More's insistence that all the activities of body, even its falling to earth when released from a height, must be attributable to an immaterial principle, the Spirit of Nature, was intimately bound up with a number of theological notions to which Boyle could not or would not subscribe. When More appropriated Boyle's hydrostatic experiments as 'proof' for the existence of the Spirit of Nature, therefore, Boyle judged it to be unsafe to turn a blind eye to More's endeavours. This was a time when religious beliefs were perceived to have marked social and political consequences, so Boyle had to register his disagreement with More and to distance himself and his new philosophy from More's unacceptable theology.[8] The aim of this paper is to indicate the theological and moral beliefs which underwrote the opposing views of these seemingly unlikely polemicists. A proper understanding of this dispute, it is suggested, reveals much not only about the nuances of seventeenth-century matter theory but also the relevance of such fine distinctions to wider issues in the intellectual history of seventeenth-century England.

II

The Spirit of Nature was the salient feature of Henry More's natural philosophy, an immaterial, universal hylarchic principle which was invoked as a moving, ordering, and animating principle in all physical phenomena – it was 'the vicarious power of God upon this great *automaton*, the world.' This concept, which seems to derive from neoplatonic notions of an *anima mundi*, appeared in More's earliest writings – his philosophical poems of the 1640s – and was successively elaborated until it achieved its mature statement in *The Immortality of the Soul* of 1659.[9] More made no significant changes to his concept of the Spirit of Nature from then until his death in 1687. However, while his ideas caused no concern to leading natural philosophers throughout the 1660s, after the reiteration of his concept in the *Divine Dialogues* of 1668 and the *Enchiridion metaphysicum* of 1671, natural philosophers suddenly felt the need to repudiate it.[10] The reveille was first sounded by the Somerset virtuoso, John Beale, in a letter to Henry Oldenburg, Secretary of the Royal Society (1671):

> Sr Did not I tell yu, wt was to be expected from Dr H.M. His confidence is as strong as Enthusiasme; & yet yu see what he does. As if he had a minde to drawe a suspicion, or (at least) to rayse ye style of Infamy agst ye RS & candid Experiment, as to be so Magicall as to call in ye ayde of Spirits & Angels. If Honble Mr Boyles health will bear it, He owes him a chastisement...[11]

Significantly, Beale links More's 'Magicall' use of 'Spirits and Angels' to an overweening confidence 'as strong as Enthusiasm'. Beale's imputation of 'enthusiasm' to the author of *Enthusiasmus triumphatus*, Henry More, may at first seem like nothing more than an attempt at facetious irony, but Beale was not the only one to see this tendency in More.[12] From the Restoration to More's death (and even beyond), a number of leading theologians objected to More's dogmatic insistence upon various more or less idiosyncratic doctrines of theology and to his enthusiastic zeal. A close reading of More's works reveals that all the doctrines opposed by these critics were present in his earliest writings but that after the Restoration More became increasingly forceful and dogmatic in his pronouncement of these doctrines.

More's dogmatism and his emphasis on the power of reason to arrive at certain truth was, in fact, a major source of disquiet for many of his contemporaries. Dogmatism was associated not only with enthusiastic sects (as Beale's statement shows) but also with Roman Catholicism and even with the atheism of rationalist materialism. The inevitable result of dogmatic rationalism, for many seventeenth-century Englishmen, was disagreement among the faithful and disruption in the Church. As Walter Charleton put it:

> the unconstant, variable and seductive imposture of Reason, hath been the only unhappy Cause, to which Religion doth owe all those wide, irreconcileable and numerous rents and schismes.[13]

Samuel Parker believed 'opinionative Zeal' to be as dangerous as 'Enthusiastick Fanaticisme' and that the 'Impudent humour of talking confidently of things uncertain' in religion was bound to lead to heresy and schism.[14]

For many of the leaders of the Anglican Church in Restoration England the safe way to salvation did not require a hard-line rationalist 'proof' of the dogmas of the true faith. On the contrary, what was called for was a broad Church based on a doctrinal minimalism to which all Protestants could freely subscribe. Anglicanism saw itself as an irenic and mediating religious movement which was the best hope for reconciling all factions to the true Christian faith. The role of reason in this was merely to show that the doctrines of revealed religion were not incompatible with 'common sense' or 'reasonableness'. Rational chains of reasoning, characteristic of Roman Catholic Scholastic 'Disputations', were not regarded as a reliable means of determining the true faith. The theology of Anglicanism was, therefore, neither dogmatic nor rationalist but was derived from the intellectual movement towards the mitigated scepticism which has been described by Richard Popkin and others.[15] If this trend is easily discerned in seventeenth-century Anglicanism, it is equally easy to see that More was swimming against the tide. For More, reason was not just a means of removing common-sense objections to the articles of faith, it was an essential means of establishing them: 'For I conceive Christian Religion rational throughout, and I think I have proved it to be so in my *Mystery of Godliness*.'[16]

The major influences on More's philosophy were, of course, Plato and Descartes, both of whom offered 'the best assistance to Religion that Reason... can afford'. And he drew from both of them to develop a philosophical theology which purported to be thoroughly rationalistic. So, according to More, the concept of the 'Spirit of Nature' was 'forced upon me by inevitable evidence of Reason... Against which I know no imaginable exception'. To doubt the existence of the Spirit of Nature, he wrote, would be 'as ridiculous, as to doubt of the truth of any one plain and easy Demonstration in the first book of Euclide.'[17]

More's invocation of Euclid was not merely a rhetorical flourish in his Preface; the argument of the whole book proceeded in a quasi-Euclidean way, based on thirty six 'self-evident' axioms. The announcement of his first ten axioms (in Chapter II) sets the tone for More's dogmatic and hectoring approach:

> And to stop all Creep-holes, and leave no place for the subterfuges and evasions of confused and cavilling spirits, I shall prefix some few Axiomes, of that plainness and evidence, that no man in his wits but will be ashamed to deny them...[18]

By adopting what he took to be the Euclidean method, then, More believed that he had demonstrated the existence of souls and the Spirit of Nature 'with evidence no less than Mathematical', and 'so evidently that nothing can be more evident in Philosophy.'[19]

More's confidence in his own powers of reasoning never shows any signs of wavering, and he defended the truth of his concept of the Spirit of Nature with the utmost tenacity throughout his career. Although in some cases his defence of the concept is best seen not so much as More putting his foot down but rather as More petulantly stamping his feet. Consider, for example, his response to one early objector:

> There will be a Spirit of Nature for all this, and I think for anything that ever will be alledg'd to the contrary or excogitated to evade the unrelish of that principle.[20]

Even after Boyle's *Hydrostatical Discourse*, More claimed that his *Enchiridion* 'proved the existence of a *Principium hylarchicum*, distinct from the matter of the Universe, against all evasions and tergiversations whatsoever.'[21] The reason for this recalcitrance, of course, was More's conviction that only his rational approach could provide 'the succours that Philosophy affords to Religion in the points of the Existence of a God and Immortality of the soul.' More's method, he alway insisted, was not only 'most True' but also 'most usefull.' It is 'of such huge importance for men to have a right notion' of spirit, he wrote, that the Spirit of Nature could not fail to be 'a very useful Dogma for assuring the souls personal subsistence after death.'[22]

Robert Boyle was not just unconvinced by all this, like Beale and others, he felt it was dangerous. Furthermore, the fact that Boyle's own experiments were being used to bolster More's dubious scheme ('delivered for a demonstration', as Boyle put it) made it imperative to give More a public telling-off.[23] There can be no doubt that Boyle knew what More was up to. News of Boyle's disapprobation of *Enchiridion metaphysicum* prompted More to write him an apologetic letter in December 1671. Speaking of Boyle in the third person, More wrote: 'the experiments of his, which I drew into my discourse, they were of that excellent importance for the design of my book, and the demonstration of the grand point in hand, that I could not by any means omit them.' The 'grand point in hand' was, of course, the proof of the existence of the Spirit of Nature, God's vicegerent in natural affairs.[24] In spite of Boyle's awareness and disapproval of More's ulterior motives, he displayed his commitment to the irenic, non-disputative Anglican methodology by confining himself, in his *Hydrostatical Discourse* of 1672 to a detailed discussion of his experiments and the epistemological burden that could be placed upon them. Implicit in Boyle's response to More was the presupposition that experimental facts were always amenable to a number of different interpretations. As Boyle made clear elsewhere,

> many Atomists and other Naturalists, presume to know the true and genuine causes of the things they attempt to explicate; yet very often the utmost they can attain to, in their explications, is, that the explicated phenomena may be produced after such a manner, as they deliver, but not that they really are so.[25]

Accordingly, experiments themselves could not, in all cases, be used to decide between rival hypotheses and Boyle's professed aim in the *Discourse* was 'not to prove that no angel or other immaterial creature could interpose in these cases', nor 'to demonstrate in general, that there can be no such thing, as the learned doctor's *principium hylarchicum*'. Instead, Boyle confined himself to showing that 'we have no need to recur to them', and that More's concept of the Spirit of Nature could be excised from consideration by wielding Ockham's razor (*entia non sunt multiplicanda absque necessitate*).[26]

The most significant aspect of Boyle's conclusions, however, was the fact that he did not confine his use of Ockham's principle of parsimony to natural philosophy. The Spirit of Nature is not only an unnecessary hypothetical construct in hydrostatics, Boyle averred, it is also superfluous in religion.

> Nor do I think it necessary [Boyle wrote] to the Doctors grand and laudable design, wherein I heartily wish him much success, of proving the existence of an Incorporeal substance. For as I think, Truth ought to be pleaded for only by Truth; so I take that, which the Doctor contends for, to be evincible in the rightest way of proceeding by a person of far less learning than He, without introducing any precarious Principle...[27]

Boyle makes it clear in this statement that he does not regard the Spirit of Nature as a 'Truth' which can be used to plead for a more general 'Truth' like the existence of immaterial substance. Furthermore he suggests that the existence of immaterial spirit or souls need not be demonstrated by rationalist arguments, since they can be evinced by men of little learning by recourse only to faith. Faith, not philosophy, is 'the rightest way of proceeding' towards religious truths.[28]

In general terms Boyle's response to More's *Enchiridion metaphysicum* can be seen as just one manifestation of Boyle's consistent scepticism and anti-dogmatism. In his *Considerations about the Reconcileableness of Reason and Religion* of 1675, Boyle made if perfectly clear that he did not believe any rationalist system of philosophy or theology could be trusted to generate new knowledge:

> there are very few conclusions, that we make, or opinions, that we espouse,

> that are so much the results of our reason, that no personal disability, prejudice, or fault, has any interest in them... the very body of mankind may be imbued with prejudices, and errors, that from their childhood, and some also even from their birth, by which means they continue undiscerned, and consequently unreformed.[29]

So much, then, for any attempt to establish the truth on rationalist grounds, be it by Plato, Descartes, Hobbes, or Henry More.

But More's dogmatic rationalism was only part of the problem. Boyle was also aware that More's dogmatism was inextricably bound up with his views on God and the nature of Providence, and that these views were totally unacceptable to Boyle himself. Boyle believed that the usefulness of his natural philosophy lay in its ability to demonstrate the existence and omnipotence of God. If we suppose, Boyle wrote in the *Hydrostatical Discourse*,

> the World to have been at first made, and to be continually preserved by God's divine power and wisdome; ...the phenomena, I strive to explicate, may be solved mechanically, that is, by the mechanicall affections of matter.[30]

Contrary to popular prejudices against the mechanical philosophy, Boyle wished to imply, the fact that the world can be explained in mechanical terms points ineluctably to the existence of an omnipotent God. Indeed, it is only our belief in an omnipotent creator that enables us to see the world as a mechanical system. Even 'Heathen Philosophers' were convinced of the existence of God, Boyle pointed out, but this did not lead them to suppose the existence of a separate Spirit of Nature. On the contrary,

> yet taking no notice of an immaterial *principium hylarchicum*, they believed things to be managed in a meer physical way according to the General Laws, settled among things Corporeal, acting upon one another.[31]

Conversely, Boyle declared that his own endeavours to account for physical phenomena in terms of 'the motion, bigness, gravity, shape and other mechanical affections of the small parts of liquors' were intended to prove that the world was 'at first made' and 'continually preserved by God's divine power and wisdome.'[32]

It is important to note that Boyle includes the occult quality of 'gravity' in his *Discourse* as a 'mechanical affection' of matter. This assumption gave him no pause for thought. While he dismisses the Spirit of Nature as 'a meer Hypothesis', the gravitation of water, occult though it may be, is 'directly proved by particular Experiments'. He freely admitted that he did not know what is 'the cause of gravity', nevertheless his experiments make it manifest that 'there is such a quality' in bodies.[33] The clear implication of this, as far as Boyle was concerned, is that God has the power to endow matter with active principles like gravity or simply motion, so that, as he said elsewhere, 'motion may be innate in the particles' of matter.[34] The omnipotence of God is even such that we must acknowledge that

> supposing the fabric of the greater bodies of the universe and the laws he had established in nature, some juicy and spirituous parts of ...living creatures must be fit to be turned into prolific seeds, whereby they may have a power, by generating their like, to propagate their species.[35]

The upshot of Boyle's insistence that God could endow matter with characteristics which were not part of its essence was to enable him to use his new natural philosophy in the service of theology. The experimental demonstration of the innate activity of matter thereby became a crucially important resource for establishing the existence and the omnipotence of God.

By contrast, as Boyle saw all too clearly, More's concept of the Spirit of Nature was used by its author as an important resource for demonstrating that *not even God* could endow matter with inherent principles of activity. This, in turn, was part of More's arsenal for establishing beyond any doubt (or so More believed) the truth of his philosophical theology, and its concomitant doctrines about moral order. More was convinced that the only meaningful system of morality was an absolute system which was capable of being discovered by all men by a process of pure reasoning. A traditional corollary of this view was the belief that God established the world in accordance with this essentially true moral order. It seemed to follow that God's omnipotence had to be constrained by the fact that there are necessary connections between elemental features of morality and of the world. So, underlying the differences between More and Boyle over the existence of the spirit of Nature and the capabilities of matter, as we shall now see, was a radical disagreement about the nature of God and his providential relationship to the world. Underlying this, furthermore, was a fundamental disagreement about the nature of the moral order.

III

More's providential theology was of the kind variously known as intellectualism, necessitarianism, theological optimism or simply rationalism.[36] What this means is that, for More, the dominant attribute of God is intellect or reason, and that God used intellectual powers, even before the Creation, to arrive at an understanding of certain essential features inherent in the very nature of things. These essentials included moral concepts like good and evil, justice and injustice, as well as natural concepts such as the categorical distinction, as More saw it, between body and soul. Having recognised such essential principles God, because of His supreme goodness, had no choice but to create the world in accordance with the moral demands placed upon Him by His own goodness and in accordance with the essential relationships inherent in the nature of things. Because of these constraints God could not even choose *not* to create the World, and the World that He *had* to create was the only one possible – namely, of course, *the best of all possible worlds.*[37]

There are hints of intellectualist theology even in More's earliest writings, and his commitment to it seems to have stemmed from his early revulsion from the Calvinist doctrine of predestination – a doctrine which More took to be inherently unjust and so incompatible with divine goodness.[38] I think it is true to say, however, that this aspect of More's theology might easily have been missed by his readers until 1668 when the *Divine Dialogues* appeared. The full title of the *Dialogues* announces that they comprise *Sundry Disquisitions & Instructions Concerning the Attributes of God and His Providence in the World*, and they contain a full account of More's intellectualist theology.[39] From God's 'infinite Goodness, Wisdome and Power', More tells us

> issue out all the Orders of the Creation in the whole Universe. So that ... his Goodness being so perfect, immutable and permanent, as never ... to be carried otherwise than to what is the best, and his wisedome never at a loss to discern, nor his power to execute it; ... For is there not all reason, that he that is so immutably Good, that it is repugnant that he should ever will any thing but what is *absolutely* for the best.[40]

'No Power, though never so omnipotent', he reiterated, 'can claim a right to ... do what is wrong ... no more then any Intellect, never so omniscient, can claim a right of authentickly thinking that true which is really false.'[41]

More returned directly to the defence of intellectualist theology in 1682 when he re-published Joseph Glanvill's *Lux orientalis* and George Rust's *Discourse of truth*, together with his own extensive annotations under the overall title, *Two Choice and Useful Treatises*. Glanvill's *Lux orientalis*, first published in 1660, was an argument for the pre-existence of souls which relied heavily upon intellectualist theological principles, while Rust's *Discourse of Truth*, first published in 1667, was a thoroughly intellectualist moral tract.[42] More's annotations, therefore, concentrate heavily on expounding and defending intellectualist principles. More argued, for example, that since goodness is one of God's essential attributes, 'he must necessarily act according to it'. In order to do that God must act in accordance with

> mutual Respects and Relations eternal and immutable, and in order of Nature antecedent to any Understanding either created or uncreated.[43]

In other words,

> though the Divine Goodness acts necessarily, yet it does not blindly but according to the Laws of Decorum and Justice.

The 'Laws of Decorum and Justice' being antecedent even to God's understanding.

One of the characteristic features of intellectualist theology, it should be noticed, is the underlying assumption that there is nothing arbitrary in the system of the world. The eternal and immutable set of 'mutual respects and relations' between things is inherent in the very nature of things themselves. Accordingly, the system of the World can be arrived at by purely rationalist means. Indeed, it was only by using His own powers of reason that God was able to work out exactly what kind of world to create to make this the best of all possible worlds. It follows, therefore, that man can, at least in principle, use his reason to 'think God's thoughts after him' and arrive at a perfect knowledge of the world system, including the physical world, the nature of man and all his institutions, and the nature of the spiritual realm.[44]

It is More's intellectualist theology, then, which underwrites the dogmatic rationalism in his works that we have already remarked upon. It accounts for the marked tendency in More to treat all his major concerns as though they are all of a piece, as though each of his major doctrines must stand or fall with the rest.[45] For example, More cannot readily treat even so recondite a notion as the preexistence of the soul as a thing indifferent to true religion, because for him it is 'The Golden Key to Providence', and this is why he reissued an annotated edition of Glanvill's *Lux orientalis*.[46] Similarly, his rationalist disjunction between passive matter and active spirit is inextricably bound up with his (intellectualist) belief that God can only create entities which have a clear and undeniable sanction from pure reasoning. Any attempt to blur the distinction between matter and spirit – which includes any suggestion that matter might be inherently active – is to undermine the validity of the rationalist approach, and therefore, to deny the fundamental assumption, the *sine qua non* of intellectualist theology. It is no accident, therefore, that More's final defence of his concept of the Spirit of Nature appears as a long digression in his annotations to George Rust's intellectualist moral tract, the *Discourse of Truth*.[47]

The point of More's insistence upon the undeniable truth of the existence of the Spirit of Nature and its corollary, that all matter is utterly dead and inert, is to add further strength to his claims about the 'eternal and immutable' truths which even God cannot change. The rationally established dichotomy between spirit and matter was intended to extend the realm of those logical 'contradictions' which all theologians accepted as being beyond God's powers to accomplish. In the same way that God's omnipotence did not enable him to create a weight so heavy that he could not lift it, so God could not endow matter with inherent principles of activity. To speak of this, according to More, would be a contradiction in terms.

More's claims about the apodictic and undeniable nature of his theory of matter and spirit must be seen, therefore, as an essential part of his wider philosophical and religious intentions. The strict dichotomy between matter and spirit served two functions in More's philosophy. Firstly, his proof of immaterial substance made it possible to go on to prove the existence of God and the immortality of the soul. Secondly, his supposed demonstration that the distinction between the two substances, material and immaterial, was both logical and natural provided clear confirmation of More's brand of providential theology in which even the Creator had to follow eternal and immutable laws.[48]

So, to reiterate this important point, More's belief in the inherent order in the nature of things which dictated to God the essential nature of the world that must be created, meant that all things were interconnected and that, therefore, moral considerations could not be entirely divorced from arguments about the dualist distinction between matter and spirit and about the role of the Spirit of Nature. His commitment to intellectualist theology, then, was a major factor in his dogmatic rationalism and explains his tenacity in defence of the Spirit of Nature. For him, the Spirit of nature had to be a *necessary* feature of the world. To give way on the existence of the Spirit of Nature was to give way on the concepts of eternal and immutable justice and goodness, and this in turn, as More fervently believed, was to lay 'the Foundations of Rantism, Debauchery and all Dissoluteness of Life', because it was only God's awareness of More's notion of justice which could guarantee that the true believer would be rewarded while the wicked were punished.[49]

IV

Now, More's providential theology was by no means unanimously accepted among theologians, much less among lay-men. Indeed, it seems true to say that the major theological alternative to intellectualism seemed to be enjoying a considerable vogue in late-seventeenth-century England. The contemporary prominence of this rival view – known as voluntarism – seems to have been an immediate legacy of Calvinism, but the tradition can be traced back to William of Ockham and beyond.[50] Voluntarist theology stressed as God's dominant attribute, the will. According to this perspective, the world was created by the power of God's arbitrary and inscrutable will. The world was a set of contingencies dependent only on God's fiat. What order there is in the world is not due to any constraints upon God's creative freedom by immutable essences inherent in things themselves, but is simply due to God's arbitrary decision to make it so. It follows from the arbitrary and contingent nature of the world system that it is futile to try to discover that system by a process of pure reasoning. In moral philosophy this meant that More's 'eternal and immutable' principles were replaced by the notion that 'things are good and just, because God wills them so to be.'[51] While, in natural philosophy, as is well-known, voluntarism underwrote empiricism.

The interconnectedness of More's views on Providence, moral laws, preexistence of souls, and other aspects of his philosophy, including his theory of matter, meant that his critics had likewise to respond to all three. So, we find a theologian like Richard Baxter, the leading Presbyterian divine and a voluntarist through and through, including a defence of the possibility of active matter in his own critique of More's notions of God and the soul. Notwithstanding More's claim to have proved apodictically that matter is necessarily passive and 'unactive of itself', Baxter pondered:

> I am not sure that if a stone in the Air fall down, it is by a Spirits motion, and that God hath not made Gravitation, and other aggravative motion of Passives to be an Essential self-moving Principle ... I dare not say, that God cannot make a Rock or Clod to fall by an intrinsick Principle of Gravitation ... [52]

The arbitrary and omnipotent will of God, according to Baxter, may even be able to grant matter the power of perception:

> I confess I am too dull to be sure that God cannot endue matter itself with the formal virtue of Perception ... that Almighty God cannot make perceptive living matter, ... I cannot prove, or I think you: Where is the contradiction that makes it impossible?[53]

More objected that Baxter's denial of the categorical disjunction between matter and active principles like gravity, perception and life, would cause people to disbelieve in God. Baxter's blurring of the logical distinction between passive matter and active principles making it impossible, so More believed, to prove the existence of spirits and therefore of God. As far as Baxter was concerned, however, the far greater threat to belief was More's denigration of God's power. Defending the theological innocuousness of perceptive matter, Baxter wrote,

> Nor do I believe that it giveth a man any more cause to doubt (as you add) of the Existence of God, or the Immortality of the Soul, than your Opinion that saith, *God cannot do this*.[54]

If Richard Baxter, the theologian, felt obliged by the density of More's philosophical theology to enter into a discussion of matter theory, we can see why Boyle felt the need to enter into theology. Boyle's own belief that matter may be endowed with inherent principles of activity such as the one which is 'the cause of gravity' went hand in hand, as we have seen, with his belief in the omnipotence of God. Involved in this same worldview was his rebuttal of More's Spirit of Nature.[55]

Boyle, as is well-known, was a devout voluntarist. Time and again throughout his works we come across statements to the effect that God

> is the creator of matter, and the sole introducer of local motion into it, so all the laws of motion were at first arbitrary to him and depended upon his free will.[56]

And that God 'has not bound up his own hands' by the laws of motion, 'but can invigorate, suspend, overrule, and reverse any of them as he thinks fit'.

Voluntarism had had a long pedigree in natural philosophy and seems to have won unanimous support from seventeenth-century English natural philosophers. For Robert Boyle, the voluntaristic assertion that the system of the world was contingent entirely upon God's arbitrary will meant that that system could only be discovered *a posteriori*. In short, voluntarism could be used to provide a mandate for the experimental method.[57]

Moreover, the belief that God had made the world by arbitrary *fiat* and could change it at will could also be used to license the kind of scepticism which was Robert Boyle's stock-in-trade. So, when Boyle counselled natural philosophers to remember that

> the same effect may be produced by divers causes different from one another; and that it will often times be very difficult, if not impossible, for our dim reasons to discern surely, which of these several ways ... Nature has really made use of to exhibit them ... [58]

he was not trying to undermine the very enterprise of natural philosophy. He was simply reminding himself and his readers of the inscrutability of God's wisdom and power, and the corruption of man's reason since the Fall.[59]

Just as More's intellectualist theology was bound up with his dogmatic rationalism, so Boyle's voluntarism stood or fell with his experimental method and his scepticism about ultimate causes. It was imperative, therefore, for Boyle's whole enterprise in natural philosophy and natural theology to reject all manifestations of theological intellectualism. Although Boyle confined himself in the *Hydrostatical Discourse* to rejecting More's claims by the presentation of further experiments, together with some passing reiterations of his voluntarist view of Providence, he was not content to leave it there. It is easy to find further attacks on intellectualist theology in Boyle's more overtly theological works.

In his *Considerations about the Reconcileableness of Reason and Religion* of 1675, for example, Boyle entered briefly into what he called 'the controversy *de jure Dei in creaturas*'. 'I think, I may safely say,' Boyle wrote,

> that God, by his right of dominion, might without any violation of the laws of justice, have destroyed and even annihilated *Adam* and *Eve*, before they had eaten of the forbidden fruit, or had been commanded to abstain from it. For man being as much and as entirely God's workmanship as any of the other creatures, unless God had obliged himself by some promise or pact, to limit the exercise of his absolute dominion over him, God was no more bound to preserve *Adam* and *Eve* long alive, than he was to preserve a lamb or a pidgeon; and therefore, as we allow, that he might justly recal the lives he had given those innocent creatures, when he pleased, (as actually he often ordered them to be killed, and burned in sacrifice to him:) so he might, for the declaration of his power to the angels, or for other reasons, have suddenly taken away the lives of *Adam* and *Eve*, though they had never offended him.[60]

Boyle went on to insist that the rules of justice and reason which *we* acknowledge must be limited within 'due bounds' and not appplied to the nature of God or to 'his absolute power and will'. He followed this up with a favourable discussion of Descartes' voluntarism and his [Descartes'] refusal to use his natural philosophy 'to limit the omnipotence of God.'[61]

But Boyle's most discursive attack on Henry More appeared in the *Free Enquiry into the Vulgarly Received Notion of Nature* of 1686. In this well-known work, Boyle fired a broadside against all those who believed there to be an architectonic or directing principle in the world, other than God, which cannot be reduced to the 'mechanical affections' of matter.[62] Boyle's critical cannons were loaded with grapeshot for this foray, so it is difficult to be specific about his targets. He himself mentions Aristotelians, Epicureans, Stoics, physicians who speak of the healing power of nature, those who speak of a God 'not really distinct from the animated and intelligent universe', (we would say 'pantheists') and last but by no means least 'the vulgar.'[63] We can be fairly sure, however, that he had Henry More in mind when he spoke of those 'Divines and other Pious Men' who believe that 'Nature' is 'an *Immaterial* Substance ... the Grand Author, of the Motion of Bodies, and that, especially in such familiar *Phaenomena*, as the Ascension of Water in Pumps, ... the running of it through Siphons, and I know not how many others.'[64]

The *Free Enquiry* is by no means unremittingly negative, however. Indeed, Boyle himself described it as his proposals 'about Providence, being written, rather to do a service to theology, than ... to justify a dissatisfaction with the received notion of nature'; the result is, as J. E. McGuire has pointed out, a protracted essay in defence of voluntarist natural theology.[65] In view of this positive aim to establish voluntarism as a 'service to theology', and in view of the fact that so notable a theologian as More was taking the opposing view, then we might expect Boyle to spend some time refuting intellectualism in theology. He does not disappoint us: among those 'Considerations ... that seem necessary to be premis'd to what I have to offer', we find Boyle denying the intellectualist view that there are *essential* forms which determine the nature of things:

> it is a mistake to imagine ... that what is called the nature of this or that body is wholly comprised in its own matter and its (I say not substantial, but) essential form; as if from that, or these only, all its operations must flow.[66]

So that, although God created the world 'according to the divine ideas which he had', he did so 'most freely.' And there are frequent repetitions that God is a free agent, unconstrained by anything in the nature of things:

> God is a free agent, and created the world, not out of necessity, but voluntarily, having framed it, as he pleased and thought fit, at the beginning of things, when there was no substance but Himself, and consequently no creature, to which he could be obliged, or by which he could be limited.[67]

Nor does Boyle miss the optimistic implications of this kind of providentialism. At one point he asks, 'Whether the World and the Creatures that compose it, are as perfect as they could be made?' or 'Whether God could make the Material World, and the Corporeal Creatures it consists of, better or more perfect than they are?'[68] Boyle leaves us in no doubt about his answer:

> I think it very unsafe to deny that God, who is almighty and omniscient, and an owner of perfections, which ... are ... more ... than we can comprehend, could not display ... them, by creating a work more excellent than this work. And his immense power and unexhausted wisdom considered, it will not follow, either, that because this world of ours is an admirable piece of workmanship, the divine architect could not have better'd it...[69]

In other words, as Boyle insisted, 'God himself *is* able to make a greater master-piece.'[70]

V

So, when Boyle decided that he had to publicly disassociate himself from and repudiate those parts of More's *Enchiridion metaphysicum* which appropriated his experiments, he was not simply concerned over the technicalities of hydrostatical principles. Moreover, as the *Free Enquiry* reveals, it does not even complete the picture to say that Boyle was concerned merely about the niceties of scientific methodology, that is to say the correct interpretation and use of experiment. In fact, Boyle wished to repudiate the very basis of More's philosophical theology. The two men did not, as some historians of science have recently assumed, share a view of matter as completely passive and inert operated on by an external immaterial principle of activity. That view was More's. Boyle's voluntarism would not allow him to insist that God *could not make* inherently active matter, and to be consistent with his self-professed experimental method he could hardly deny that matter could display active qualities like gravity, spring, magnetism or simply motion. Far from being allies with a common ideological purpose, then, the differences between Boyle and More were crucial and fundamental, extending from their epistemological and methodological views, to their beliefs about the nature of the world and of God and the relationship between them; and extending ultimately therefore to the validity, or otherwise, of what Ralph Cudworth, another intellectualist, called 'eternal and immutable morality.'[71]

A shorter version of this paper was read at the XVIIth International Congress of the History of Science, University of California, Berkeley (August, 1985). I would like to thank Drs. Michael Neve and Simon Schaffer for their help and encouragement on that occasion. Longer versions were read at the Royal Institution Centre for the History of Science and Technology, London

(April 1986) and in the History Department, University of Edinburgh (May 1986). I would like to thank the organisers of these seminars, Drs. Frank James and Nicholas Philipson respectively for inviting me to air my views, as well as those present who provided helpful comments and advice. I would also like to express my gratitude to Dr. Robert Crocker, Dr. Michael Hunter, Dr. Sarah Hutton and Mr. Malcolm Oster for their detailed and constructive criticisms of an earlier draft.

1. Thomas Hobbes, *Dialogus physicus de natura aeris, conjectura sumpta ab experimentis nuper Londini habitis in Collegio Greshamensi* (London, 1661); reprinted in Thomas Hobbes, *Opera philosophica quae Latine scripsit omnia*, ed. Sir William Molesworth, 5 vols. (London, 1839-45), 4:233-96; translated by Simon Schaffer in S. Shapin and S. Schaffer, *Leviathan and the Air Pump*, 345-91. Franciscus Linus, *Tractatus de corporum inseparabilitate; in quo experimenta de vacuo, tam Torricelliana quam Magdeburgica & Boyliana examinantur* (London, 1661). Henry More, *EM* (1671). Boyle did not bother to reply to the criticisms of Sir Matthew Hale even though Hales' strictures seem, to be as cogent as those of the other three: Sir Matthew Hale, *Essay Touching...Fluid Bodies* (1673); idem, *Difficiles nugae* (1674).
2. For an excellent account of Boyle's dispute with Hobbes see Shapin and Schaffer, *Leviathan*, which also includes briefer but very useful accounts of Boyle's disputes with Linus and More, 156-69 and 207-24. On Linus see C. Reilly, "Francis Line, Peripatetic," *Osiris*, 14 (1962): 222-53, and idem, *Francis Line S. J.: An Exiled English Scientist (1595-1675)* (Rome: Institutum Historicum, S.I., 1969).
3. Latitudinarianism is often held to have stemmed from Cambridge Platonism and More is regarded as a leading light in both movements. On the links between natural philosophy and Latitudinarianism see S[imon] P[atrick] *A Brief Account of the New Sect of Latitude-Men. Together with Some Reflections upon the New Philosophy* (London 1662). On Boyle's Latitudinarianism see for example, L. T. More, *The Life and Works of the Honourable Robert Boyle* (London: Oxford University Press, 1944), 160-87; Barbara Shapiro, "Latitudinarianism and Science in Seventeenth-Century England," *Past and Present* 40 (1968): 16-41. On More's Cartesianism, see Alan Gabbey, "Philosophia Cartesiana Triumphata." The influence of Descartes on Boyle is discussed in, for example, L. Laudan, "The Clock Metaphor and Probabilism: The Impact of Descartes on English Methodological Thought 1660-65," *AS* 22 (1966): 73-104. For More's interest in the experiential investigation of witches and ghosts see his edition of Joseph Glanvill, *Saducismus Triumphatus*, and Allison Coudert's article in this volume. Boyle's interest in the same phenomena can be surmised from his letters to Glanvill, Boyle, *Works*, 6:57-60.
4. Boyle's comment is recorded in Anne Conway to More, 4 February, 1675/6, in *Conway Letters*, p. 420. Boyle was referring to More, *Remarks upon Two Late Ingenious Discourses*.
5. This view is almost a historiographical commonplace but some influential examples are: Carolyn Merchant, *The Death of Nature*; J. R. Jacob and M. C. Jacob, "The Anglican Origins of Modern Science, the Metaphysical Foundations of the Whig Constitution," *Isis* 71 (1980): 251-67; J. R. Jacob, *Henry Stubbe, Radical Protestantism and the Early Enlightenment* (Cambridge: Cambridge University press, 1983); David Kubrin, "Newton's Inside Out: Magic, Class Struggle and the Rise of Mechanism in the West," in H. Woolf (ed.), *The Analytic Spirit: Essays in the History of Science in Honour of Henry Guerlac* (Ithaca: Cornell University Press, 1981), 96-121; and K. Hutchinson, "Supernaturalism and the Mechanical Philosophy," *History of Science* 21 (1983): 297-333.
6. More's discussion of Boyle experiments appears in chapters 12 and 13 of *EM*. Boyle replied in *An Hydrostatical Discourse* (1672), in *Works*, 3: 596-628.
7. On the Leibniz-Clarke dispute see H. G. Alexander (ed.), *The Leibniz-Clarke Correspondence* (Manchester: Manchester University Press, 1956); S. Shapin, "Of Gods and Kings: Natural

Philosophy and Politics in the Leibniz-Clark Disputes," *Isis* 72 (1981): 187-215. Shapin provides further bibilographical references, 187-8, n. On More's dualism and his concept of immaterial substance, see E. A. Burtt, *Metaphysical Foundations*, 127-44, and A. Koyré, *From the Closed World*, 125-54. For indications of Boyle's concept of active principles see below and J. Henry, "Occult Qualities and the Experimental Philosophy." The quoted phrases are from Boyle's *The Origins of Forms and Qualities* (1666) in *Works*, 3:48.

8. For another example of Boyle distancing himself from dubious theological opinions consider Boyle's reaction to Henry Stubbe's *The Miraculous conformist, or an Account of Several Marvailous Cures Performed by ...Mr. Valentine Greatarick* (Oxford, 1666). There is an excellent account of this in N. H. Steneck, "Greatrakes the Stroker," an article distinguished as one of the few studies to recognise that More and Boyle cannot be considered as part of a 'supposed unity that exists within the Latitudinarian-Anglican camp' (p. 176).

9. More, *Poems* and *IS*. The development of the concept of the Spirit of Nature can be seen in More, *CSPW*, which includes works published between 1653 and 1659.

10. More used Boyle's experiments to establish the existence of the Spirit of Nature in the 1662 edition of *AA* in *CSPW* (chap. 2, pp. 44-7). There was, however, no response in print from any natural philosopher at this time. The reasons for the very different reactions of natural philosophers before and after the appearance of *DD* will be made clear later.

11. Beale to Oldenburg, 24 June 1671, in H. Oldenburg, *Correspondence*, 8:120, 'RS' here is, of course, the Royal Society of London.

12. More, *ET*. Fear and loathing of religious 'enthusiasm' together with atheism were the main reasons which More gave for developing his philosophical theology. Nevertheless, there were marked 'enthusiastic' tendencies in More's own religious devotion. These are well brought out in Ward's *Life*, 48-50, 103. See also Sir Matthew Hale, *Observations*, 284-5; Anon [Joseph Beaumont?] *Remarks* (1690), sig. (a3)^r and (d)^v.

13. Walter Charleton, "Prolegomena to the Candid & Ingenuous Reader." in J. B. van Helmont, *A Ternary of Paradoxes...translated, illustrated and ampliated by Walter Charleton* (London, 1650), sig. F1^v.

14. Samuel Parker, *A Free and Impartial Censure*, 72, 73, 87.

15. R. H. Popkin, *The History of Scepticism from Erasmus to Spinoza* (Berkeley and Los Angeles: University of California Press, 1979); H. G. van Leeuwen, *The Problem of Certainty in English Thought, 1630-1690* (The Hague: Nijhoff, 1963); Barbara J. Shapiro, *Probability and Certainty in Seventeenth-Century England* (Princeton: Princeton University Press, 1982). Consider also, M. L. Wiley, *The Subtle Knot: Creative Scepticism in Seventeenth-Century England* (London: G. Allen and Unwin, 1972); and H. R. McAdoo, *The Spirit of Anglicanism* (London: Adam & Charles Black, 1965), 1-23. See also the discussions of 'reason' and 'reasonableness' in the studies cited in n. 28 below.

16. More, *CSPW*, "Preface General", p. iv. As far as I know there is no study of More which brings out the very clear differences between his theological method and that of his Latitudinarian contemporaries. Most studies seem to regard More as unproblematically Latitudinarian in his outlook. This assumption, erroneous in my view, probably derives from S[imon] P[atrick] *Brief Account* (1660) which equates Cambridge Platonism with Latitudinarianism. For a recent account of Latitudinarianism which avoids the usual equation with Cambridge Platonism see H. R. Trevor-Roper, "The Great Tew Circle," in his *Catholics, Anglicans and Puritans: Seventeenth-Century Essays* (London: Secker and Warburg, 1987), 166-230. I hope to return to the issue of More's relationship to Latitudinarianism in a future essay. In the meantime I hope that the present paper will provide preliminary indications by showing the nature of More's differences with the leading Latitudinarian layman, Robert Boyle. Edward Fowler, *Principles and Practices of Certain Moderate Divines of the Church of England, abusively called Latitudinarians*

(London, 1670), reports a contemporary opinion of the Latitudinarians as 'a company of men that are prepared for the embracing of any religion, and to renounce or subscribe to any doctrine...' Of course, all such opinions are polemical but it is easy to see how such charges might stick to men like Tillotson, Taylor, Stillingfleet and Wilkins. It seems unlikely to have had any force against so intransigent a dogmatist in points of faith as Henry More, however. Far from easily renouncing any doctrine, More often insisted that he had demonstrated idiosyncratic aspects of his religious system on apodictic grounds. It is perhaps significant that Ward felt it necessary to defend More against the censure that he 'was unreasonably addicted to some glorious Fancies and Opinions of his own,' *Life*, 62-3.

17. More, *CSPW*, *IS*, 13, 11, 12.

18. Ibid., 16. On reasoning *more geometrico* at this time, see H. Schüling, *Die Geschichte der axiomatischen Methode in 16 und beginnenden 17 Jahrhundert* (Hildesheim, 1969).

19. More, *CSPW*, "Preface General," pp. xv and xii, xiii.

20. More to Anne Conway, 17 March 1666, in *Conway Letters*, p. 269.

21. More, "The Digression," in *Two Choice and Useful Treatises*, 199.

22. More, "Letter to a Learned Psychopyrist," in Glanvill, *Saducismus* (1682), 239. More to Richard Baxter, 10 February, 1682, in Dr. Williams's Library, Baxter Letters vol. 3, f.284. More, "The Digression" (see n. 21 above), 186. I am grateful to the Trustees of Dr. Williams's Library for allowing me to consult and quote from the Baxter Letters.

23. Robert Boyle, *Hydrostatical Discourse*, in *Works*, 3:597.

24. More to Boyle, 4 December 1671, in Boyle, *Works*, 6:615. On the dating of this letter, see Gabbey, "Philosophia Cartesiana," 248.

25. Boyle, *The Usefulness of Natural Philosophy* (1663) in *Works*, 2:45. For a full discussion of Boyle's probabilistic experimental method see L. Laudan, "The Clock Metaphor" (see n. 3 above). For the historical context of this methodology and its wider significance in Boyle's thought see the works cited in n. 15 above and n. 29 below.

26. Boyle, *Works*, 3:627, 609.

27. Ibid., 627-8.

28. For indications of the suspicion of 'rationalist' arguments in late-seventeenth-century England, see above at notes 13 and 15, and the following. R. Hoopes, *Right Reason in the English Renaissance* (Cambridge, Mass.: Harvard University Press, 1962), 160-85; Christopher Hill, "'Reason' and 'Reasonableness'," in idem, *Change and Continuity in Seventeenth-Century England* (London: Weidenfeld and Nicholson, 1974); P. M. Rattansi, "Some Evaluations of Reason in Sixteenth- and Seventeenth-Century Natural Philosophy," in M. Teich and R. M. Young (eds.), *Changing Perspectives in the History of Science* (London: Heinemann, 1973), 148-66; L. Mulligan, "'Reason', 'Right Reason' and 'Revelation'"; and S. Schaffer, "Occultism and Reason," in A. J. Holland (ed.), *Philosophy, its History and Historiography* (Dordrecht and Lancaster: Reidel, 1985), 117-43.

29. Robert Boyle, *Some Considerations about the Reconcileableness of Reason and Religion* (London, 1675), in *Works*, 4:164. It should be noted that Boyle offers here his thoughts about how reason and religion can be reconciled. He does not argue, as we have seen More do (at n. 16 above), that religion is a rational system. On Boyle's scepticism, see L. T. More, *Life of Boyle*; van Leeuwen, *Problem of Certainty*, 91-106; J. Golinski, "Scepticism and Authority in Seventeenth-Century Chemical Discourse," in A. E. Benjamin, G. N. Cantor and J. R. R. Christie (eds.) *The Figural and the Literal: Problems of Language in the History of Science and Philosophy, 1630-1800* (Manchester: Manchester University Press, 1987), 58-82; and S. Shapin, "Robert Boyle and Mathematics: Reality, Representation and Experimental Practice" *Science in Context* 2 (1988), 23-58.

30. Boyle, *Works*, 3:608.

31. Ibid.
32. Ibid., 3:628.
33. Ibid., 3:624, 601.
34. Reported by Henry Oldenburg in a letter to Spinoza, 3 April, 1663, in Oldenburg, *Correspondence*, 2:42. Boyle's belief that 'gravity' and other 'occult qualities' are nonetheless 'mechanical' is by no means unique. Consider, for example, Robert Hooke's attack on Henry More and his rejection of the Spirit of Nature on the grounds that 'there is no need of any other Principles than the plain Mechanical principles, which supposeth every Terrestrial Body to have a Gravity in it,' in *Lampas*; reprinted in R. T. Gunther, *Early Science in Oxford*, 15 vols. (Oxford, 1923-67), 8:184. For a fuller account of these views and their use in natural theology see Henry, "Occult Qualities."
35. Boyle, *Origin of Forms and Qualities*, in *Works*, 3:48. For a fuller account of the links between Boyle's acceptance of active principles and his theological beliefs, see Henry, "Occult Qualities," 352-8.
36. I shall use the word 'intellectualist' to refer to More's providential theology as it is least likely to be confused with another philosophical position. 'Necessitarian' is in some ways preferable since it conveys the sense that God must act according to necessity. Unfortunately, however, it is all too easily confused with philosophical (as opposed to theological) determinism and equated with denials of freedom of the will. More was a fully committed believer in the freedom of man's will so he was not a philosophical necessitarian. He did believe, however, that God's supreme goodness and the 'eternal law' of justice required God to give man free will. Paradoxically, therefore, God's will was not free. Use of the word 'rationalism' is equally likely to confuse the reader. Besides, I have only seen this notion of Providence referred to as rationalist by one writer, M. B. Foster, "The Christian Doctrine of Creation and the Rise of Modern Science," *Mind* 43 (1934): 446-68.
37. The major studies of this kind of thinking are, A. O. Lovejoy, *The Great Chain of Being: A Study of the History of an Idea* (Cambridge, Mass.: Harvard University Press, 1936; reprinted, New York, 1960) and F. Oakley, *Omnipotence, Covenant and Order: An Excursion in the History of Ideas from Abelard to Leibniz* (Ithaca, NY, and London: Cornell University Press, 1984). These can now be supplemented with A. Funkenstein, *Theology and the Scientific Imagination from the Middle Ages to the Seventeenth Century* (Princeton: Princeton University Press, 1986), chap. 3, 117-201. In referring to God as 'He' I am simply following the usage of More and Boyle.
38. More's early revulsion for Calvinism on moral grounds is reported in the autobiographical part of the general preface to his *Opera*, translated in Ward, *Life*, 6-7. For early hints of intellectualist theology in More, see, for example *Psychathanasia*, bk. 3, canto 4, stanzas 21-2, in *Poems*, 178-9, *CC*, 2; and *MG*, 33.
39. Intellectualist theological views are introduced in the first dialogue, and constitute the whole of the second dialogue and much of the fourth.
40. More, *DD*, 2:24-5.
41. Ibid., 2:24.
42. The running title of Glanvill's *Lux orientalis* in More's edition is: "Praeexistence of Souls A Key for Providence," and chaps. 4, 7 and 8 are devoted to an exposition of intellectualist theology. Rust's *Discourse of Truth*, was provided with a prefatory "Letter Concerning the Subject and the Author" by Glanvill in which we read, 'the Subject is of great and weighty importance, and the Acknowledgement of the Truths here asserted and made good, will lay a Foundation for right conceptions in the Doctrines that concern the Decrees of God. For the first Errour, which is the gound of the rest, is, 'That things are good and just because God Wills them so to be; and if that be granted we are disabled from using the arguments taken from natural

Notions, and the Attributes and Perfections of the Divine Nature, against the Blackest and most Blasphemous Opinions...' sig. N6^{r-v}.

43. More, "Annotations upon *Lux orientalis*", 47, 46-7. Rust *Discourse*, 166. The pagination of the two treatises and More's two sets of annotations are all separate in this edition.

44. An important aspect of intellectualist optimism, made popularly infamous by Voltaire's *Candide* (1759), is the notion that 'everything is for the best' or, as Pope put it in his *Essay on Man* (1732), 'Whatever is, is right.' There is no shortage of similar ideas in More's writings. For instance, 'For the nature of things is such that some Particulars or Individuals must of necessity suffer for the greater good of the Whole... whatsoever designed or permitted Evil there seems in Providence, it is for a far greater good, and therefore is not properly in the summary compute of the whole affairs of the Universe to be reputed evil, the loss in particulars being so vast a gain to the Whole,' *DD*, 1:178-9. And, '...the lessening of Happiness in the one is the advancement of it in another.' Ibid., 1:309.

45. This is an example of what Lovejoy calls the 'principle of plenitude.' Pope includes the notion in his *Essay on Man*: '...all must full or not coherent be,' and 'From Nature's chain whatever link you strike,/Tenth, or ten thousandth, breaks the chain alike.' See Lovejoy, *Chain of Being*, 60 and passim.

46. More, *DD*, 1:502. Glanvill's exposition of pre-existence is also bound up with intellectualist theology (see n. 42 above). The concept of pre-existence is *indispensable* to More's theodicy. Only this concept can clear 'all the Difficulties touching the Moral evils of the world' (ibid.). The argument runs: 'Because supposing Humane Souls were created in the Morning of the World, and in such infinite Myriads, there has been time enough since that for as many and more then hitherto have peopled the Earth, to have transgressed so hainously before their entrance on this Stage, that by a just Nemesis measured and modify'd by the Divine Goodness it self they may be contrived into the worst and most horrid Circumstances, into the most squalid and disadvantageous condition and state living...' (ibid., 1:503).

47. More, "Disgression," 180-246, in *Two Choice...Treatises.*

48. It is important to note, of course, that denial of God's omnipotence would be extremely heretical except where one is speaking of logical contradictions. This, presumably, was another reason for More's repeated insistence that he had proved the major aspects of his philosophical theology, 'with evidence no less than Mathematical' (see n. 18 above). As Descartes pointed out to More, 'we do not take it as a mark of impotence when someone cannot do something we do not understand to be possible, but only when he cannot do something which we distinctly believe to be possible.' Descartes to More, 5 February 1649, in *AT (NP)*, 5:273; translation quoted from A. Kenny (ed.) *Descartes: Philosophical Letters*, 241. Needless to say, disputes raged over what we can be said to 'distinctly perceive to be possible.' On these theological debates see Oakley, *Omnipotence* (n. 37 above). More was evidently so confident of the eternal truth of his own opinions that he gave God very little leeway. Descartes, by contrast, believed that even mathematical theorems are contingent upon God's will. See Funkenstein, *Theology* (n. 37 above), 179-92.

49. Rust, *Discourse of Truth*, 181, in *Two Choice...Treatises.* More endorses Rust in his "Annotations," 45, (ibid.): 'if the measure of his [God's] Providence be his mere Power, Will or Soveraignty no man living can tell what to expect in the conclusion. All true Believers may be turned into Hell, and the wicked onely and the Blasphemer ascend into the Regions of Blis. For what can give any stop to this but God's Justice, which is a branch or mode of his Goodness?'

50. On voluntarist theology see Oakley *Omnipotence*; idem, "Christian Theology and the Newtonian Science: The Rise of the Concept of the Laws of Nature," *Church History* 30 (1961): 433-57; E. M. Klaaren, *Religious Origins of Modern Science: Belief in Creation in Seventeenth-Century Thought* (Grand Rapids: W.B. Eerdmans, 1977); R. Hooykaas, *Religion and the Rise of*

Modern Science (Edinburgh: Scottish Academic Press, and London: Chatto and Windus, 1972), 41-53; S. Shapin, "Of God's and King's" (n. 7 above); and M. J. Osler, "Descartes and Charleton on Nature and God," *JHI* 40 (1979); 445-56.

51. *Two Choice...Treatises*, sig. N6^v. More regarded this as 'a sad principle' (ibid.) and it clearly offended his moral sense. But it is perhaps worth remarking that another factor was surely involved. The notorious writings of Thomas Hobbes, in particular *Leviathan* (London, 1651), seemed to promote just the kind of arbitrary morality implicit in voluntarist theology. There is an interesting exchange between the interlocutors in *DD*, dialogue 4, which suggests a link in More's mind between voluntarism and Hobbes's ideas on sovereignty. The question is raised, 'What species of Dominion or Power it is that he [God] thus universally excercises over the Creation?' Cuphophron, who is a Cartesian, voluntarist and Latitudinarian (in so far as he holds 'indifference in the greatest points of Religion' [2:19]), suggests first of all that God rules with 'absolute and unlimited Sovereignty.' When he is accused of making God a tyrant, Cuphophron falls back on the voluntarist notion of arbitrary rule: God is 'tied to no Law' and 'acts merely according to the suggestions and sentiments of his own heart.' This notion is rejected in favour of Philotheus's suggestion that 'the Right of this absolute Sovereignty in God is founded' in 'permanent and immutable Goodness.' Ibid., 2:21-25. There can be little doubt that concepts of sovereignty provide another dimension to the political implications of More's writings and the reaction to them One might explore these implications along the lines drawn in Shapin, "Of Gods and Kings," but I do not attempt it here.

52. Richard Baxter, *Of the Immortality of Mans Soule*, 33-4. For a fuller discussion of the polemic between More and Baxter, see Henry, "A Cambridge Platonist's Materialism."

53. Baxter, *Immortality*, 28-9. See Henry, "Medicine and Pneumatology" for a fuller discussion of Baxter's argument about perceptive living matter, its source and its use in in his polemic with More.

54. Baxter, *Immortality*, 29. See also 63, 79-80, and 50-53. More repeatedly made the charge that Baxter's attitude was 'not only a Mistake but a Mischief' because it undermined the enterprise of natural theology: More, "Letter to a Learned Psychopyrist" in Glanvill, *Saducismus* (1682), 198, 217, 239. But see also Baxter, op. cit., 32.

55. See above at notes 30, 33, 34 and 35.

56. Royal Society, Boyle MSS, vol. 7, fols. 113-4, quoted from R. M. Burns, *The Great Debate on Miracles: From Joseph Glanvill to David Hume* (Lewisberg, 1981), 54. See also, for example, Boyle, *Works*, 4:161. Such examples could easily be multiplied.

57. The best account of Boyle's voluntarist theology is J. E. Maguire, "Boyle's Conception of Nature," *JHI* 33 (1972):523-42. See also Hooykaas, *Religion* (n. 50 above), 41-53; Oakley, *Omnipotence* (n. 37 above), 72-92; and Shapin, "Boyle and Mathematics" (n. 29 above). As always in matters of intellectual history there is no *necessary* connection between a voluntaristic view of Providence and empiricism. Descartes and Hobbes, the leading rationalist philosophers, also subscribed to voluntarism in theology. On Descartes, see Funkenstein, *Theology* (n. 37 above), 179-92.

58. Boyle *The Usefulness of Natural Philosophy* (1663), in *Works* 2:45.

59. Part of the difficulty with reason was held to stem from the fact that mankind was in a fallen state. Before the fall Adam might well have been able to arrive at the truth by the use of his reason. It was man's reason which made him a creature 'in God's image.' As a result of original sin, however, man's reason was never to be trusted. See Glanvill, *The Vanity of Dogmatizing: Or Confidence in Opinions Manifested in the Shortness and Uncertainty of our Knowledge* (London, 1661), 1-16; and Charleton, "Prolegomena" (n. 13 above) sig. E3^r-F1^4. Boyle even went so far as to deny the accuracy and universal applicablility of natural laws on the grounds that to affirm either might imply that God too must obey these laws. On this important aspect of Boyle's

natural philosophy see Shapin, "Boyle and Mathematics."
60. Boyle, *Works*, 4:162. See also 161.
61. Ibid., 163-4.
62. Boyle, *A Free Enquiry into the Vulgarly Received Notion of Nature* (London, 1686), in *Works*, 5:158-254. Consider, e.g., 178-9.
63. Ibid., 178, 183, 211-15, 183. Another possible target is a group of so-called 'Pagan naturalists.' See J. R. Jacob, *Robert Boyle and the English Revolution* (New York: Franklin, 1977), 161-4. 169-70, 176.
64. Ibid., 172, 191.
65. Ibid., 253. McGuire, "Boyle's Conception of Nature." McGuire provides an excellent analysis of the *Free Enquiry* and is one of the few historians to have noticed that More is one of Boyle's targets. But see also, Edwin McCann, "Was Boyle an Occasionalist?" in Holland (ed.), *Philosophy, its History and Historiography*, 229-31.
66. Boyle, *Works*, 5:179.
67. Ibid., 251.
68. Ibid., 195.
69. Ibid.
70. Ibid., 195-6.
71. Cudworth, *Treatise Concerning Eternal and Immutable Morality* (London, 1731).

Abbreviations

Abbreviations used in the notes to this paper refer to the following:

AA	Henry More, *An Antidote against Atheisme*. London, 1655.
AT	René Descartes, *Oeuvres, publiées par Charles Adam et Paul Tannery: Nouvelle presentation*, edited by B. Rochot, P. Costabel, J. Beaude. 11 vols. Paris: Vrin, 1964–1974.
CC	Henry More, *Conjectura Cabbalistica*. London, 1653.
CSPW	Henry More, *A Collection of Several Philosophical Writings*. London, 1662.
DD	Henry More, *Divine Dialogues*. London, 1671
EM	Henry More, *Enchiridion metaphysicum*. London, 1671.
ET	Henry More, *Enthusiasmus Triumphatus*. London, 1656.
IS	Henry More, *The Immortality of the Soul*. London, 1659
MG	Henry More, *An Explanation of the Grand Mystery of Godliness*. London, 1660.

Short titles in the notes to this paper refer to the following:

Anon. [Joseph Beaumont?], *Remarks upon Dr Henry More's Expositions of the Apocalypse of Daniel, and upon His Apology*... London, 1690.

Richard Baxter, *Of the Immortality of Man's Soul, and the Nature of it and Other Spirits*. London, 1682.

Robert Boyle, *The Works*, ed. T. Birch, 6 vols. London, 1772.

E.A. Burtt, *The Metaphysical Foundations of Modern Science*. London: Routledge & Kegan Paul, 1932.

Alison Coudert, 'Henry More and Witchcraft', in Sarah Hutton (ed.), *Henry More (1614–1687): Tercentenary Studies*. Dordrecht: Kluwer, 1990, pp. 115–36.

René Descartes, *Philosophical Letters*, edited and translated by A. Kenny. Oxford: Clarendon Press, 1970.

Alan Gabbey, 'Philosophia Cartesiana Triumphata: Henry More, 1646–71.' In T.M. Lennon, J.M. Nicholas, and J.W. Davis (eds), *Problems of Cartesianism*. Kingston and Montreal: McGill-Queen's University Press, 1982, pp. 171–249.

Joseph Glanvill, *Saducismus Triumphatus, or Full and plain Evidence Concerning Witches and Apparitions*. London, 1681.

Sir Matthew Hale, *Observations touching the Principles of Natural Motions*... London, 1677.

Alexandre Koyré, *From the Closed World to the Infinite Universe*. Baltimore: Johns Hopkins University Press, 1957.

A.O. Lovejoy, *The Great Chain of Being*. Cambridge, Mass.: Harvard University Press, 1936.

Carolyn Merchant, *The Death of Nature: Women, Ecology and the Scientific Revolution*. San Francisco: Harper Row, 1980.

Henry More, *Philosophical Poems*. Cambridge, 1647.

Henry More, *Remarks upon Two Late Ingenious Discourses... (by Sir Matthew Hale), so far forth as they may concern any passages in his Enchiridion Metaphysicum*. London, 1676.

Henry More, *Two Choice and Useful Treatises: The One Lux Orientalis* [by Joseph Glanvill]... *The Other, A Discourse of Truth* [by George Rust]. London, 1682.

Lotte Mulligan, '"Reason", "Right Reason", and "Revelation" in mid-seventeenth-century England'. In Brian Vickers, ed., *Occult and Scientific Mentalities in the Renaissance*. Cambridge: Cambridge University Press, 1984, pp. 375–401.

Marjorie H. Nicholson, ed., *Conway Letters: The Correspondence of Anne, Viscountess Conway, Henry More and their Friends, 1642–84*. New Haven and Oxford: Yale University Press and Oxford University Press, 1930.

Henry Oldenburg, *The Correspondence*, edited and translated by A.R. Hall and M. Boas Hall. 13 vols. Madison: University of Wisconsin Press, 1961–86.

Samuel Parker, *A Free and Impartial Censure of the Platonick Philosophie...* Oxford, 1667.

Simon] Patrick, *A Brief Account of the New Sect of Latitude-Men*. London, 1662.

George Rust, *A Discourse of the Use of Reason in Matters of Religion...* London, 1683.

Steven Shapin and Simon Schaffer, *Leviathan and the Air-Pump: Hobbes, Boyle, and the Experimental Life*. Princeton: Princeton University Press, 1985.

Nicholas Steneck, 'Greatrakes the Stroker: The Interpretations of Historians.' *Isis* 73 (1982): 161–77.

Richard Ward, *The Life of the Learned and Pious Dr Henry More*. London, 1710.

VII

Boyle and cosmical qualities

Boyle published his *Tracts about the Cosmical Qualities of Things* in 1671 (although its imprimatur is dated 3 November 1669), as a kind of sequel to the *Origin of Forms and Qualities* of 1666, but unlike the latter work these tracts have attracted very little scholarly attention. If they have been noticed at all they have been given a seemingly unproblematic mechanistic interpretation. Boyle defines 'systematical or cosmical' qualities as those qualities of a body which do not derive from the sizes, shapes and motions of its constituent particles, but

> depend upon some unheeded relations and impressions which those bodies owe to the determinate fabrick of the grand system or world they are part of.

'A system [of the world] so constituted as ours is', Boyle wrote, 'whose fabrick is such that there may be divers unheeded agents, which, by unperceived means, may have great operations upon the body we consider', can give rise to special attributes of that body.[1] It is by no means immediately clear what Boyle has in mind here, but, relying upon standard assumptions about the nature of the mechanical philosophy and about Boyle's status as a leading mechanist, it might seem reasonable to suppose that Boyle is merely extending the discussion in the *Origin of Forms and Qualities* about the power of a key to open a lock.[2]

Boyle's point in his discussion of locks and their keys is that the power of a key to open a lock becomes the 'main part of the notion and description' of that key. The result is that the ability of the key to open the lock is 'looked upon as a peculiar faculty and power in the key', even though there is no real or physical entity corresponding to this power. The key is simply a specially shaped piece of metal designed to be able to turn through the wards of a lock and draw back its bolt. The power of the key, in other words, derives from the arrangement of the system of the lock, and this power, therefore, could be said to be 'a systematic quality' of the key (though Boyle does not use this phraseology in the earlier work). A stark physical description of the shape and size of the key is, of course, perfectly true, and in a purely physical sense is all there is to be said about the qualities of the key. Nevertheless, Boyle points out, such a description misses the crucial point about it – its ability to open a particular lock. The systematic quality of the key, therefore, should be acknowledged as an important element in its make-up.[3]

According to this interpretation, then, systematic or cosmical qualities are to be seen as complex extensions of this idea, based on the lock–key analogy, which was first put forward, after all, in Boyle's preceding book. The analogy in the *Origin of Forms and Qualities* was applied to comparatively simple natural examples, the solubility of gold in *aqua regia*, the poisonous properties of ground glass and the way bodies appear to be coloured. It would seem that the aim of the tract on *Systematical or Cosmical Qualities* was merely to indicate how the mysterious qualities of some bodies may similarly be seen as epiphenomena of the way they interact with their surroundings. In spite of seemingly occult appearances and experiences, Boyle wished to explain, the real physical qualities of these bodies derive ultimately from the mechanical affections of size, shape and motion.

If Boyle's *Cosmical Qualities* has failed to attract any special attention, it is almost certainly due to the fact that he expounds his views on these qualities in such a way that there seems to be no significant advance on what was explained in the *Origin of Forms and Qualities*. The examples he gives in the course of the essay are easily explained in mechanistic terms. A wedge will not cleave unless it is hit by a hammer, a knife will not attract a needle unless excited by a magnet. A smoothly polished slab of marble can be used to pick up another piece of polished marble, even one heavier than itself, because the pieces of marble adhere 'by virtue of the fabrick of the world, which gives the ambient air fluidity and weight'. Other examples tell of the magnetisation of a piece of iron by heating it and letting it cool while the iron is orientated north–south, or of the change of a piece of paper from opacity to translucence merely by soaking it with oil or 'even a fit kind of grease'. The speed of a ship under sail can be increased by wetting the sail, we are told, and common tartar which stays dry in the air and is only dissolved in water with difficulty is changed by moderate calcining, so that it spontaneously turns into that liquour which chemists call tartar per deliquium. Glass which is heated and then cooled too quickly cracks because 'the subtle bodies that are in it' cannot find a passage of escape, but if it is cooled slowly the glass is able to settle into the appropriate texture.[4]

So, if there is nothing here that could not have been included in the earlier work on the *Origin of Forms and Qualities*, why does Boyle want to give them a special designation as cosmical qualities? A closer look makes clear that, whatever our own superficial responses to *Cosmical Qualities* might be, Boyle himself believes that there is more to be said than anything which was covered in the *Origin of Forms*:

> I have in the *Origin of Forms* touched upon this subject already, but otherwise than I am now about to do: for whereas that which I principally (and yet but transiently) take notice of is, that one body being surrounded with other bodies, is manifestly wrought on by means of those among whom it is placed; that which I chiefly in this discourse consider, is the impressions that a body may receive, or the power it may acquire from those vulgarly unknown or at least unheeded agents by which it is

> affected, not only upon the account of its own peculiar texture or disposition, but by virtue of the general fabrick of the world.[5]

This is still puzzling, however. With the exception of the example of magnetising an iron bar by cooling it in alignment with the Earth's magnetic flux, it is difficult to see why the general fabric of the world has to be invoked to explain Boyle's specific examples. Boyle himself describes the action of oil on paper in terms of its action on the pores in the paper, straightening and widening them to allow light particles through; similarly water on a sail causes the fibres of the cloth to swell and make the sail less permeable by the wind. He has already given an account of the adherence of two polished marble plates in his *Continuation of New Experiments . . . Touching the Spring and Weight of the Air* (1669) without feeling the need to explain the fluidity and weight of the air in terms of 'the fabrick of the world', and the behaviour of tartar, though inexplicable to Boyle and his contemporaries in any precise way, surely did not demand an explanation on a cosmic scale.

Certainly, all these examples cannot simply be explained in terms of what Frederick J. O'Toole has referred to as 'non-relational inherent properties', that is to say, in terms of the sizes, shapes and motions of the primary corpuscles of nature and the 'texture' or precise way in which these primary corpuscles are combined to constitute the particular bodies in question. It seems clear enough that these examples can only be explained by recourse to what O'Toole calls 'non-inherent relational properties'.[6] But these relational properties are, as O'Toole's analysis makes clear, simply those properties which Boyle did describe in the *Origin of Forms*, namely those which depend upon, or are functions of, the 'peculiar texture or disposition' of the relevant bodies, whose situations make it clear that they are 'manifestly wrought on' by one another. It seems that we have to go beyond O'Toole's account if we wish to understand what Boyle meant when he suggested that there are powers and capacities of bodies which can only be understood

> not barely upon the score of these qualities that are presumed to be evidently inherent in it, nor of the respects it has to those other particular bodies to which it seems to be manifestly related, but upon the account of a system so constituted as our world is.[7]

It is my contention that Boyle intended in *Cosmical Qualities* to provide hints towards what was effectively a new concept of qualities but failed to fully explicate what he meant by them. Furthermore, it is quite possible that Boyle's 'failure' in this regard was not accident but design. In what follows, therefore, I hope to show that there are sufficient indications in *Cosmical Qualities* that Boyle did have in mind something which is rather less compatible not only with O'Toole's analysis of the matter theory of the mechanical philosophy, but also with all standard accounts of the mechanical philosophy.[8] In addition I will try to reconstruct just what it was that Boyle only hinted at and to offer some tentative suggestions as to why he chose to obscure his own conception.

In the very opening paragraph of *Cosmical Qualities* Boyle seems to undermine any suggestion that such qualities are merely to be seen in terms of the key–lock analogy. The qualities of most bodies, he declared, 'do for the most part consist in relations, upon whose account one body is fitted to act upon others or disposed to be acted upon by them'. He then gives clear examples of the key–lock format, such as the ability of quicksilver to dissolve gold and silver. As we have already seen, however, he immediately tells us that his intention in his new book is to go beyond this to consider qualities which derive from the fact that bodies belong to 'a system so constituted as our world is'. The fabric of our world, Boyle goes on,

> is such that there may be divers unheeded agents, which, by unperceived means, may have great operations upon the body we consider, and work such changes in it, and enable it to work such changes on other bodies, as are rather to be ascribed to some unheeded agents than to those other bodies with which the body proposed is taken notice of.[9]

The real subjects of the tract, then, not brought out by the examples, are these 'unheeded agents' which work 'by unperceived means'. This is confirmed in the opening paragraph of *Cosmical Suspicions*, described as an 'Appendix' to the discussion of cosmical qualities, which suggests that 'there may be . . . peculiar sorts of corpuscles that have yet no distinct name, which may discover peculiar faculties and ways of working'.[10] It is perfectly clear from this statement that what Boyle is proposing cannot be fitted into our standard conceptions of the mechanical philosophy. The corpuscles themselves are said to have their own unique faculties and ways of working. These are obviously not inert particles which work only by impact and other contact actions to give the appearance of having powers and faculties, like the power to turn a lock, or the faculty of pulsification of the heart and arteries.

The inescapable conclusion is that these qualities are occult, and this is confirmed by Boyle's efforts to forestall criticism of his new conception:

> least you should think that under the name of Cosmicall Qualities I should introduce Chimaeras into Naturall Philosophy I must betimes advertise you, that you will meet with divers Particulars in the following Discourse, fit to show that these Qualities are not meerly fictitious qualities: but such, whose Existence I can manifest, not only by considerations not absurd, but also by real Experiments and Physical Phaenomena.[11]

The vocabulary here suggests that Boyle thought of cosmical qualities as occult qualities. In the rhetoric of the new philosophy, the occult qualities all too frequently invoked by scholastic philosophers were dismissed as 'fictitious' or as mere 'Chimaeras'. The experimental philosopher could escape these charges, however, by insisting that the qualities he discussed could be made manifest by their effects. When Leibniz described Newton's gravitational attraction as 'a chimerical thing, a scholastic occult quality', Newton rebutted the charge by insisting that gravity was a *manifest* quality whose cause only was occult.[12] It would seem from Boyle's apology that cosmical qualities were not, therefore, to be seen as mere epiphenomena of the mechanical arrangement of the world system (analogous to the power of the key to turn a lock). If they were, Boyle would scarcely have felt the need to enter into the rhetoric of occult versus manifest qualities. The fact that he did discuss cosmical qualities in these terms is a clear indication that he considered them as occult qualities whose reality could be demonstrated, experimentally, by the manifest nature of their effects.

Boyle's tract on *Cosmical Qualities* and its appendix cannot consistently be regarded as a work which adheres to strictly mechanistic precepts. Although the specific examples which Boyle gives in *Cosmical Qualities* do seem compatible with mechanistic readings, Boyle continually intimates that what he really intends to 'excite [our] curiosity and attention about', as he puts it,[13] are 'unheeded agents' which work by 'unperceived means', and which are hitherto unknown particles acting not merely by impact but by various occult means. The putative existence of such unknown particles is one of Boyle's main considerations in developing his notion of cosmical qualities. There are 'certain subtle bodies in the world', he suggests, 'that are ready to insinuate themselves into the pores of any body disposed to admit their action, or by some other way affect it', and it is the action of these subtle bodies which gives rise to the cosmical qualities of the bodies into which they have entered.[14] These agents are, Boyle proposes, 'unobserved sorts of effluvia in the air', which are as unknown now as the 'swarm of steams moving in a determinate course betwixt the north and the south' were before Gilbert demonstrated that the Earth was a giant magnet.[15] Although this last image might seem to suggest that Boyle is returning to standard mechanistic explanations in terms of streams (or steams) of effluvia, it is more likely, in view of his talk of 'unperceived means' of action and 'peculiar faculties and ways of working', that Boyle was thinking of magnetism as an exemplar of something which was not, *pace* Descartes, entirely amenable to strictly mechanistic explanations. As Boyle wrote for his *Disquisition about the Final Causes of Natural Things*,

> There are a great many things which . . . cannot with any convenience be immediately deduced from the first and simplest principles; namely, matter and motion; but must be derived from subordinate principles; such as gravity, fermentation, springiness, magnetism etc.[16]

By the same token, as he wrote in an intended 'Essay of various degrees or kinds of the knowledge of natural things':

> [it would] be backward to reject or despise all explications that are not immediately deduced from the shape, bigness and motion of atoms or other insensible particles of matter . . . [for those who] pretend to explicate every phenomenon by deducing it from the mechanical affections of atoms undertake a harder task than they imagine.[17]

Further support for this interpretation can be seen in the otherwise inexplicable fact that Boyle also speculates, in *Cosmical Suspicions*, that

> there may be a greater number even of the more general laws than have yet been distinctly enumerated, so I think that when we speak of the established laws of nature . . . they may be justly and commodiously enough distinguished; some of

> them being general rules that have a very great reach, and are of greater affinity to laws more properly so-called, and others seeming not so much to be general rules or laws, as the customs of nature in this or that peculiar part of the world; of which there may be a greater number, and those may have a greater influence on many phaenomena of nature than we are wont to imagine.

A little later Boyle reiterates this extraordinary idea, saying,

> I have some time suspected that there may be in the terrestrial globe itself, and the ambient atmosphere, divers, whether laws or customs of nature, that belong to this orb, and may be denominated from it, and seemed to have been either unknown to, or overseen by both scholastical and mathematical writers.[18]

Leaving aside the interesting proposed distinction between laws and customs of nature, it seems undeniable that Boyle is expressing dissatisfaction here with the extremely restricted, 'distinctly enumerated' laws of nature of Descartes' *Principia philosophiae*. Certainly, Descartes is the principal 'mathematical writer' that Boyle has in mind, and it seems clear that Boyle has been led by his own 'notions and observations' to doubt whether all physical phenomena can be explained by Descartes' three mechanistic laws of nature and his seven rules of impact.[19]

Whether it is true or not that Boyle's chosen examples of cosmical qualities do not adequately represent what he really had in mind, there is little we can do to explicate his meaning. What we can do, however, albeit to a limited extent, is to try to uncover what Boyle was thinking of when he wrote of those 'unheeded agents', those 'peculiar sorts of particles' with 'peculiar faculties and ways of working' which were held to produce cosmical qualities in other bodies. Boyle himself provides a clue to the nature of these agents when he tells us in chapter 2 of *Cosmical Qualities* that there are three main bodies in the fabric of the world 'whose more unobserved operations' have the most influence on cosmical qualities. These three are the subterraneal parts of the globe, the stars and aether about them and the atmosphere or air we live in.[20] He then tells us that experimental observations on each of these are to be dealt with in other tracts. A moment's inspection reveals that he does not mean the other tracts published with *Cosmical Qualities*. These deal only with the temperature of subterraneal and submarine regions, together with some hearsay reports about the bottom of the sea. If we wish to discover what these bodies are, therefore, we will have to search elsewhere in Boyle's output.

For guidance let us take those three bodies or groups of bodies in turn: the subterraneal parts of the globe, the atmosphere around us and the stars and aether. With regard to the first of these, Boyle has little to say beyond a few scattered speculations. In *Cosmical Suspicions* itself he points to the evidence for 'slow internal change' in 'the mass of the earth', such as the change of magnetic variation, and concludes that

> there may be agents that we know not of, that have a power to give the internal parts of the terrestrial globe itself a motion; of which we cannot yet certainly tell according to what laws it is regulated, or so much as whether it be constantly regulated by certain laws or no.[21]

Similar speculations are reiterated in the appendix to *Hidden Qualities of the Air*, 'Of celestial and aerial magnets':

> there may be in those vast internal parts of the earth, whose thin crust only has been here and there dug into by men, considerable masses of matter, that may have periodical revolutions, or accensions, or eustations, or fermentations, or, in short, some other notable commotions, whose effluvia and effects may have operations yet unobserved, on the atmosphere, and on some particular bodies exposed to it; though these periods may be perhaps either altogether irregular, or have some kind of regularity differing from what one would expect.[22]

Boyle also suggests that the great irregularity observed in the weight of the atmosphere must derive 'for the most part from subterraneal steams'; the very variable and inconstant phenomenon of luminosity in the sea may also derive from 'some cosmical law or custom of the terrestrial globe'.[23] 'Some subterraneal changes' are also suspected as the cause of new epidemic diseases (he mentions syphilis and rickets) which 'invade whole countries (and sometimes greater portions of the earth)'.[24] Elsewhere, however, Boyle suggests that subterranean effluvia can make a region more healthy. 'For in some places', Boyle writes,

> the air is observed to be much more healthy, than the manifest qualities of it would make one expect: and in divers of these cases I see no cause, to which such a happy constitution may more probably be ascribed, than to friendly effluvia sent up from the soil into the air.[25]

The 'subterraneal parts of the globe' only really appear in Boyle's speculations, it seems, as a source for putative effluvia in the atmosphere which are held to have otherwise inexplicable effects. This brings us, therefore, to the second of Boyle's suggested influences on cosmical qualities, the atmosphere around us. On this topic the obvious source is Boyle's *Suspicions about the Hidden Qualities of the Air*. We can tell we are on the right track from the fact that Boyle opens with another apology for discussing occult qualities. Boyle tells us that this tract was originally intended for a collection on the natural history of the air but he now published it separately, even though

> if I had more consulted my own Reputation, I should least of all have suffer'd this Title to appear, there being none of the rest, that was not less conjectural.[26]

One of the most significant of these conjectures for our purposes is the suggestion that some bodies 'may be Receptacles, if not also Attractives of the Sydereal, and other Exotic Effluviums that rove up and down in our Air'.[27] Here we have the ingredients hinted at in *Cosmical Qualities*, exotic effluvia which operate in an occult way. It would seem that Boyle wished to prove the truth of his comment in *Cosmical Suspicions* that 'it [is] not time misspent to consider whether there may not be other, and even unobserved sorts of effluvia in the air'.[28]

The first such attractive to spring to Boyle's mind is the so-called 'Philosophical Magnet' of the alchemists, which is said to attract and corporify the Universal Spirit or the Spirit of the World. Boyle dismisses these ideas as 'abstrusities', but goes on to say that we may make use of the word 'Magnet' in this context 'without avowing the receiv'd Doctrine of Attraction'.[29] What follows is a typically Boylean piece of trying to have one's cake and eat it (or denying occult qualities and recurring to them). We are first of all told that when Boyle uses the word 'Magnet' he does not mean 'a body that can properly attract our foreign Effluviums', but merely a body that is 'fitted to detain and join with them, when . . . they happen'd to accost the magnet'. But the following paragraph throws all such caution to the wind:

> And, without receding from the Corpuscularian Principles, we may allow some of the bodies we speak of, a greater resemblance to Magnets, than what I have been mentioning. For not only such a Magnet may upon the bare account of adhesion by Juxtaposition or Contact, detain the Effluviums that would glide along it, but these may be the more firmly arrested by a kind of precipitating faculty that the Magnet may have in reference to such Effluviums; which, if I had time, I could illustrate by some Instances; nay I dare not deny it to be possible, but that in some Circumstances of time or place one of our Magnets may, as it were, fetch in such steams as would indeed pass near it, but would not otherwise come to touch it. On which occasion I remember, I have in certain cases been able to make some bodies, not all of them Electrical, attract (as they speak) without being excited by rubbing, &c. far less light bodies, than the Effluviums we are speaking of.[30]

Boyle concludes that he has hereby demonstrated that without meddling in 'the mystical theories of the chymists', 'one may discourse like a Naturalist about Magnets of Celestial and other Emanations that appear not to have been consider'd, not to say, thought of, either by the Scholastic, or even the Mechanical, Philosophers'.[31]

The main theme, then, of *Hidden Qualities of the Air* is that various phenomena (usually chemical) are actually brought about by the 'operations of the Air', or rather by some previously unremarked effluvia in the air which operate not merely by force of impact, as would be required by Cartesian dogma, but by some other means. In another example, Boyle writes:

> The Difficulty we find of keeping Flame and Fire alive, though but for a little time, without Air, makes me sometimes prone to suspect, that there may be dispers'd through the rest of the Atmosphere some odd substance, either of a Solar, or Astral, or some other exotic, nature, on whose account the Air is so necessary to the subsistence of Flame; which Necessity I have found to be greater, and less dependent upon the manifest Attributes of the air, than Naturalists seem to have observed.[32]

But we do not have to leave it there. Boyle returned time and again to experiments on the nature and usefulness of the air throughout his career, and in one of his very last works, *The General History of the Air* of 1692 (finally prepared for the press by John Locke after Boyle's death), Boyle presented another speculative account of the operations of the air. Moreover, because this essay deals with 'Celestial Influences or Effluviums in the Air' it brings us to the third of Boyle's candidate bodies for producing cosmical qualities, the stars and the aether.

The essay in question, however, which is explicitly described as an 'Apology for Astrology', was not written by Boyle himself. A Latin translation of this work has been recently discovered by Antonio Clericuzio among the Hartlib Papers, and it is clear from this that the author was in fact Boyle's friend and colleague, and the man who introduced Boyle to the pleasures and usefulness of natural philosophy, Benjamin Worsley.[33] We know from a letter of John Locke's to Boyle, written in October 1691, and from Locke's 'Advertisement of the Publisher to the Reader' prefixed to the *General History of the Air*, that Boyle himself compiled the materials which were to be included in this miscellany, but we still have to be circumspect in our assessment of the significance of this particular contribution to the work.[34] We cannot assume that Boyle wholeheartedly endorsed every detail of Worsley's account, but it seems clear from the fact that Boyle chose to include it that he was, at least for the most part, in sympathy with his late friend's attempt to explain the nature of celestial influences on the Earth. Other, much shorter, extracts from other works (by Pascal and Casati, for example) suggest that Boyle's *History of the Air* is not an indiscriminate hotchpotch of all things pertaining to the air, but a presentation only of those things which Boyle took to be particularly significant in attempts to understand the nature of the air. Boyle wrote in the preface of his conviction 'that scarce any Truth, whether Historical or Doctrinal, that relates to so important a subject as the Air, is unfit to be preserv'd', so it seems reasonable to conclude that Worsley's 'Apology for Astrology' seemed to Boyle to contain at least some important truths.[35] Indeed, in the preface Boyle went so far as to say that his 'collections of particulars' were intended to afford 'some specimens of what I should have thought requisite to do upon particular subjects, if I would have ventured upon such a task, as to write a natural history of the air'.[36] Moreover, the significance of Worsley's essay for Boyle would seem to be underscored by the fact that in 1691 he remembered the relevance to a natural history of the air of a speculative essay written by his former colleague in 1657.

Although it has to be admitted that the work is very unlike anything Boyle himself wrote, it is still possible to discern resonances with some of Boyle's own more speculative pieces. In particular, I believe it is as close as we can get to gleaning what Boyle had in mind when he spoke of the role of the stars and aether about them in the operations of cosmical qualities, and what he had in mind when he wrote of 'odd substance[s], either of a Solar, or Astral, or some other exotic, nature' dispersed through the atmosphere. In *Cosmical Suspicions* Boyle wrote that we have very little knowledge of 'what communication' the Earth has 'with the other Globes we call Stars, and with the Interstellar parts of Heaven', but he expressed an intention to 'elsewhere make it probable that there may be some Commerce or other'.[37] Boyle never did return to this topic, but perhaps it is true to say that, at the end of his life, he allowed Worsley to speak for him. Accordingly, we shall now turn to a brief look at the main themes of Worsley's posthumous contribution to Boyle's *General History of the Air*.

Worsley begins by insisting that 'it may . . . still be certain, that these celestial bodies . . . may have a power to cause such and such motions, changes, and alterations [on the Earth]' in spite of the

> superstition and paganism incident to this kind of doctrine . . . the imposture, ignorance and want of learning generally observed in the persons professing this kind of knowledge, . . . the manifest mistakes and uncertainty that there is in predictions of this nature, . . . and the inexplicableness of the way or manner they come to affect one another.[38]

Predictably perhaps, Worsley's professed certainty is based upon 'undeniable experiments' and 'undoubted observations', but he clearly believes there is scope for development of these ideas, since he calls for a theory of the planets 'upon such grounds as are indubitably demonstrable', without which 'it is impossible we should assert their [the planets'] several aspects, and the mutual influences and virtues they have . . . one upon another'.[39]

The principal means by which the planets and stars operate on the Earth is light, but there are evidently many different kinds of light:

> we say, that every planet hath its own proper light: and as the light of the sun is one thing, the light of the moon another, so every planet hath its distinct light, differing from all the other. Now we must either say, that this light is a bare quality and that the utmost use and end of it is only to illuminate; or that there is no light but is accompanied further with some power, virtue or tincture, that is proper to it.; which if granted it will inform us then, that every light hath its own property, its own tincture and colour, its own specific virtue and power.

It follows, therefore, that

> those eminent stars and planets, that are in the heavens, are not to be considered by us as sluggish, inenergetical bodies, or as if they were set only to be as bare candles to us, but as bodies full of proper motion, of peculiar operation, and of life.

There is no suggestion of action at a distance in this, Worsley points out, because the virtue or power of the planet is transmitted with 'its light, and is the real property of its light'.[40]

Worsley's next move is to explain how these virtues affect the Earth. Firstly, the air is affected, being 'moved, stirred, altered, and impressed by these properties, virtues and lights, as penetrating each part of it'.[41] Secondly, various spirits which are held to be constituents of all mixed bodies, are also affected:

> as our spirits, and the spirits likewise of all mixed bodies, are really of an aerious, etherial, luminous production and composition; these spirits therefore of ours, and the spirits of all other bodies, must necessarily no less suffer an impression from the same lights, and cannot be less subject to an alteration, motion, agitation, and infection through them and by them, than the other, *viz* the air: but rather as our spirits are more near and more analogous to the nature of light than the air, so they must be more prone and easy to be impressed than it.[42]

These spirits now take on an indispensable role in the workings of the world, and are said to be 'the only principles of energy, power, force and life, in all bodies wherein they are', but these in their turn, as we have just seen, derive their properties from the truly cosmic phenomenon of light. Here then we seem to have a clear statement of what Boyle only hinted at when he wrote of 'the impressions that a body may receive, or the power it may acquire from those vulgarly unknown or at least unheeded agents by which it is thus affected . . . by virtue of the general fabrick of the world'.[43] Magnetism, for example, can be seen as a cosmic quality because

> the earth, which is not only enlightened, warmed, cherished and fructified by the power, virtue, and influence of the sun, but hath its proper magnetical planetary virtue also fermented, stirred, agitated and awakened in it; . . . and together with this magnetic planetary property of the earth, which is stirred and raised by the sun, are awakened also the seminal dispositions, odours and ferments that are lodged in and proper unto, particular regions or places.[44]

Again, this reference to properties of the Earth which are seemingly confined to particular regions or places, echoes the concern in *Cosmical Suspicions* that, as well as laws of nature, we should also consider 'the customs of nature in this or that peculiar part of the world'.[45]

It is perhaps worth considering briefly some of the likely sources for Worsley's 'Celestial Influences or Effluviums in the Air'. One possible source is Francis Bacon's speculative philosophy, described by its modern rediscoverer, Graham Rees, as Bacon's semi-Paracelsian cosmology. It is a feature of this philosophy that all mixed bodies contain a spirit of an aerious, aetherial or luminous nature, and it is these spirits which are solely responsible for the activity and operations of those bodies, their matter being totally inert. Rees himself has argued that Bacon's speculative philosophy was almost entirely overlooked by Bacon's contemporaries and immediately succeeding generations, and it is possible that the similarities between Bacon and Worsley simply derive from a common interest in, and knowledge of, Paracelsian ideas. But, equally, it may be that Rees has been too hasty in reaching his conclusion.[46] *Thema coeli*, the work in which Bacon most clearly expounded these ideas, was published by Isaac Gruter in 1653, and it seems unlikely, given their admiration for Bacon, that Worsley and others in Hartlib's circle would not have seized the opportunity to read it.

The idea that the air is penetrated throughout by these celestial lights, and that these in turn affect our spirits, is also reminiscent of ideas in the medical tradition. William Harvey, for example, wrote in his *De generatione animalium* (1651) that there was in blood a spirit which corresponded to the element of the stars. 'The blood therefore is a spirit', he went on,

> by reason of its most excellent powers and virtues. It is also celestial because in that spirit is housed a nature that is the soul, analogous to the element of the stars, and this is something bearing an analogy with the heavens as being the instrument and deputy of heaven.[47]

If there is a similarity here, there is also a clear parallel with Boyle's talk of 'sydereal', 'celestial', 'solar, or astral' effluvia in the air in his *Hidden Qualities of the Air*.

Perhaps the most obvious source of Worsley's speculations in his 'Apology for Astrology' is the Neoplatonic tradition usually referred to as 'light metaphysics'. In this tradition light was regarded as the formal and efficient cause by which God brought about the Creation. The ability of light to diffuse itself instantaneously in all directions was invoked by light metaphysicians as the fundamental formative action in the universe. As light spread itself out from the centre it took with it the unformed *prima materia* which God had created. Since the diffusion of light defined the three dimensions of space, it also gave form to the unformed matter, so producing all the individual and separate bodies of the universe. Light, then, is the first corporeal form of the universe, from which all other forms are derived, and it is the original physical cause of all development and change in the universe. It seemed to follow, for those in this tradition, that the behaviour of light was the model of all causation in the physical universe. This belief gave rise to the rich tradition of geometrical optics and to the commonplace belief in the validity of astrological influence. The universal formative agency of the autodiffusive light ensured that all things in the universe were unified with everything else. Everything in the universe contains light within it, and this internal light is ultimately the origin and principle of all motion. Robert Grosseteste, a leading writer in this tradition, even described all qualitative change in terms of light within a body being emitted and light outside a body being absorbed. Of course, this internal 'essential' light is not the same as visible light. Sensible light was regarded as only one manifestation of the emanative power of *lux*. Al-Kindi, author of *De radiis stellarum*, one of the most influential works in the tradition (also known as *De theoria artium magicarum*), believed that everything in the world produced its own rays of influence like a star. Moreover, these rays were proper to each thing and different from all others, like Worsley's lights.[48]

Benjamin Worsley's essay of 'Celestial Influences or Effluviums in the Air' fits easily into the light metaphysical tradition, and Boyle himself may well have been sympathetic to this speculative tradition. Boyle pointed out in his *Aerial Noctiluca* (an essay about a luminescent body), for example, that 'light was the first corporeal thing the great Creator of the universe was pleased to make, and . . . he was pleased to allot the whole first day to the creation of light alone, without associating with it in that honour any other corporeal thing'.[49] Moreover, Boyle certainly entertained the possibility that the sun and planets might have specific influences 'here below' which are 'distinct from their heat and light': 'On which supposition', he wrote, 'it seems not absurd to me to suspect, that the subtil, but corporeal, emanations even of these bodies may . . . reach to our air, and mingle with those of our globe in that great receptacle or rendesvous of celestial and terrestrial influences, the atmosphere.'[50] Having said this, Boyle was quick to forestall any assumptions that he was proposing a simple mechanistic account of celestial influences:

> the very small knowledge we have of the structure and constitution of globes, so many thousands or hundreds of thousands of miles remote from us, and the great ignorance we must be in of the nature of the particular bodies, that may be presumed to be in these globes (as minerals and other bodies are in the earth) which in many things appear of kin to those that we inhabit, . . . this great imperfection, I say, of our knowledge may keep it from being unreasonable to imagine, that some,

> if not many, of those bodies and their effluxions, may be of a nature quite differing from those we take notice of here about us, and consequently may operate after a very differing and peculiar manner.[51]

When Boyle wrote in *Cosmical Qualities* that there were three bodies in the world which, by unobserved operations, affect the cosmical qualities of things and so enable them to 'work . . . changes on other bodies' which they could not otherwise do, he had in mind the subterraneal parts of the globe, the atmosphere and the stars and aether around them. I have suggested that *Hidden Qualities of the Air* and Worsley's 'Celestial Influences or Effluviums in the Air' offer us our best chance of discovering Boyle's meaning. It would seem that Boyle wished to make a plea for the possibility that there were effluvia of an exotic nature, in the air, the heavens and the Earth, 'of peculiar operation', which were capable of altering Earthly bodies to give them relational properties such as gravity, magnetism, fermentation and other chemical properties. Accordingly, Boyle wished to infer, there are more laws of nature than just the three which Descartes enumerated and which (together with the derivative seven rules of impact), the influential French philosopher insisted, could explain all physical phenomena.[52] Moreover, although some of these laws may be general in their effects, there may also be local laws, or 'customs' as Boyle calls them, where phenomena which are far from universal may nonetheless be designated 'natural' and understood in terms of secondary causes operating in a regular way.

Cosmical Qualities was intended, therefore, to provide further support for Boyle's conviction that the natural world could not easily be understood in terms of the constant and undeviating operations of a few simple laws of nature. As Marie Boas Hall and Steven Shapin have both pointed out, Boyle sought to avoid conceptions of nature which were idealised and subject to mathematical abstraction.[53] The nature of the physical world was such, according to Boyle, that it could not properly be analysed in mathematical terms. The problem was not simply one of ensuring purity of samples, say, for the replication of chemical experiments, nor was it merely a recognition of the inevitability of experimenter error, or of discrepancies in instrumentation and apparatus. As Shapin has shown, Boyle believed that reality was structured in such a way that mathematical analysis was categorically inappropriate. The air, for example, was not 'a mere elementary body' but a 'great receptacle or rendesvous of celestial and terrestrial effluviums', whose precise composition varied from time to time and place to place. Similarly, Boyle believed that ores and minerals could differ in specific gravity and other qualities because 'neither nature nor art' gave 'all the productions, that bear the same name, a mathematical preciseness, either in gravity or in other qualities'.[54] Boas Hall and Shapin concentrated in their analyses on Boyle's chemical speculations, but in *Cosmical Suspicions* we see him extending his doubts about the regularity of nature to cosmology:

> when I likewise consider the fluidity of that vast interstellar part of the world wherein these globes swim, I cannot but suspect there may be less of accurateness, and of constant regularity, then we have been taught to believe, in the structure of

> the universe, and a greater obnoxiousness to deviations than the schools, who were taught by their master Aristotle to be great admirers of the imaginary perfections of the celestial bodies, have allowed their disciples to think.[55]

Boyle's attitude to the concept of natural laws derives from two major preconceptions. First, Boyle's theology, as McGuire and others have shown, was undeviatingly voluntarist.[56] For Boyle, it was God's supreme power and His arbitrary will which were paramount. The idea that the complexity of the world system could be reduced to a few simple laws of nature was associated, in Boyle's mind, with intellectualist theologies in which, as far as Boyle and all other theological voluntarists were concerned, God's power was circumscribed by the need to conform to the dictates of man-made reason and the expectations of man's ethical assumptions. The rational reconstruction of the system of nature, which Descartes no less than scholastic philosophers indulged in, seemed to imply that God could not have done things otherwise. Boyle's anti-rationalist experimental philosophy went hand in hand with his theological voluntarism. Whatever regularity there was in the system of the world was imposed by God arbitrarily by His absolute power, not as a result of God following the dictates of reason. Accordingly, we could only discover the regularities of the world by experience.

Secondly, as Steven Shapin has argued, the idealised, rationally reconstructed world picture was liable, in Boyle's eyes, to be seen as a false authority for the truth of natural philosophical claims.[57] The sceptical chemist, like many of his countrymen, was deeply distrustful of reason. The trouble with reason, as Jeremy Taylor pointed out in his *Ductor dubitantium*, was that it

> is such a boxe of quicksilver that it abides no where; it dwells in no settled mansion; it is like a doves neck, or a changeable taffata; it looks to me otherwise than it looks to you who do not stand in the same light that I doe: and if we inquire after the laws of Nature by the rules of our reason, we shall be uncertain as the discourses of the people, or the dreams of disturbed fancies.[58]

The claim to a philosophy or a theology based on 'reason' could be made by Hobbists and other atheists, by Roman Catholics, or by some highly distasteful Protestant sect. As Boyle himself wrote in his *Considerations about the Reconcileableness of Reason and Religion*,

> there are very few conclusions, that we make, or opinions that we espouse, that are so much the pure results of our reason, that no personal disability, prejudice, or fault, has any interest in them, . . . the very body of mankind may be embued with prejudices, and errors, that from their childhood, and some also even from their birth, by which means they continue undiscerned, and consequently unreformed.[59]

The way to avoid bogus claims, based on reason, was to develop an epistemology in which truth claims were sanctioned only by immediately accessible observations and experiments. This is precisely what Boyle, to say nothing of a number of his colleagues in the Royal Society, attempted to do.[60]

It should be realised, of course, that these two preoccupations of Boyle's have been expounded separately, and somewhat artificially, for the sake of elucidation. As far as Boyle was concerned, these two considerations were inextricably bound up with one another. The fact that voluntarist theology sanctioned, indeed demanded, an empiricist epistemology in which rationalist claims about the operation of a few laws of nature were illegitimate was hardly separable from the belief that rationalist arguments could be, and were, used by Roman Catholics and other groups (whether heterodox religious, or downright irreligious) merely to promote their own dubious claims. Whichever way you looked at it, as far as Boyle was concerned, rationalism was incompatible with sound religion.[61]

Whatever else may be said of Boyle's speculations in *Cosmical Qualities*, it seems clear that he was seeking to go beyond the bounds of the mechanical philosophy as it was set out by Descartes. Not only did he deny the competence of Descartes' laws in terms of the variability of things in themselves, and in terms of the variability of their interactions with other things to which they are 'manifestly related', but he also denied their ability to account for the variations brought about by the fact that all things are part of a system 'so constituted as our world is'. The world system as a whole is such, Boyle suggested, 'that there may be divers unheeded agents' operating 'by unperceived means' to directly affect bodies and to affect the way those bodies interact with other bodies.[62] The cosmical qualities themselves (whatever they might be) are to be attributed to the ordinary bodies of the world. These qualities are evidently not held to be inherent properties of the bodies in question, but relational properties, like the property of a key to open a lock, and may well be reducible to mechanistic terms.[63] Nevertheless, the 'unheeded agents' which produce the cosmical qualities in things do not seem to be so reducible. Particles of matter which operate only by impact cannot be held to affect the relational properties of other things in the way that Boyle seems to hint at. The answer, therefore, is to suppose, as Boyle does, that there must be 'peculiar sorts of corpuscles' which have 'peculiar faculties and ways of working'.[64]

One final question remains. If my reconstruction of the nature of the 'unheeded agents' capable of operating in peculiar ways to produce cosmical qualities is correct, why did Boyle not clearly say so in 1671? Why did he hold back his real thoughts on the matter? Why did he only hint obscurely at these ideas in 1671 and provide examples which, at least according to my reading, are extremely misleading because they deal with phenomena which seem to be easily explicable in terms of ordinary interactions or relationships between things, without invoking cosmic phenomena?

There are two or three closely connected reasons for Boyle's obscurity on this issue. First, Boyle did not want to be misunderstood. He did not want his readers in 1671 to think he was forsaking and rejecting the principles of the mechanical philosophy, in favour of a more occult philosophy. His point was that the mechanical philosophy was too restricted in the principles it allowed as legitimate. He did not deny the notion of natural law; he simply doubted that Descartes' three laws of nature could be used to explain all physical phenomena. By the same token, he doubted that all the complexity of interactions in the world could be explained in terms of the motions of particles of inert matter, colliding with one another and transferring their motions in accordance with Descartes' rules of impact. Nevertheless, as he wrote in *Hidden Qualities of the Air*, it should be possible to discourse like a naturalist about even such things as 'Magnets of Celestial Emanations'. Cosmical qualities should not be dismissed as chimerical, because they can be evinced by real experiments and physical phenomena.

It was important not to be misunderstood on this point, of course, because Boyle did not wish to disrupt the carefully forged links between the mechanical philosophy and Christian religion. The boast of the devout mechanical philosopher was that, by showing how all physical phenomena could be explained in terms of matter and motion, he simultaneously revealed that whatever cannot be explained in those terms cannot be a bodily or material phenomenon but must be spiritual. Any innovations in natural philosophy which threatened to blur the clear distinction between mechanical interactions and non-mechanical interactions (as the unheeded agents which were held to produce Boyle's cosmical qualities did) had to be very carefully managed. It is my belief that Boyle was not sure that his cosmical qualities, or rather the agents held to produce them, might not have been seized by irreligious thinkers to undermine the standard religious tropes of the mechanical philosophy, and he wished to be circumspect.

Boyle's concept of cosmical qualities was not merely an extension of the mechanistic ideas fully described in the *Origin of Formes and Qualities*, but was, in fact, an addendum to those ideas which Boyle regarded as essential to account for all the physical phenomena of nature. To interpret the cosmical qualities in such a way as to make them fit in with standard accounts of the mechanical philosophy is to see only part of the story. It is easy to see that the power of a key to open a lock does not have to correspond to any real or physical entity in the key, and Boyle's cosmical qualities were obviously intended to be seen in the same way. The qualities themselves, however, were produced by external agents operating in ways which were not allowed for in the standard contemporary accounts of the mechanical philosophy. It would seem that Robert Boyle might have said to Descartes or to scholastic philosophers what Hamlet said to Horatio:

> There are more things in heaven and earth . . .
> Than are dreamt of in your philosophy.[65]

Acknowledgements

I wish to thank the other contributors to the Stalbridge Boyle Symposium, all of whom made helpful comments on this paper, but I owe particular thanks to Antonio Clericuzio, Malcolm Oster and Michael Hunter for their comments upon an earlier draft.

NOTES

1 *Cosmical Qualities*, *Works*, iii, 306.
2 *Forms and Qualities*, *Works*, iii, 18.
3 *Ibid.*; Frederick J. O'Toole, 'Qualities and powers in the corpuscular philosophy of Robert Boyle', *Journal of the History of Philosophy* 12 (1974): 295–315, on pp. 308–9; Keith Hutchison, 'Dormitive virtues, scholastic qualities, and the new philosophies', *Hist. Sci.* 29 (1991): 245–78, on pp. 259–60.
4 *Cosmical Qualities*, *Works*, iii, 308, 309, 312, 313, 315.
5 *Ibid.* p. 307.
6 O'Toole, 'Qualities and powers' (n. 3), pp. 302–3.
7 *Cosmical Qualities*, *Works*, iii, 306. For a full account of what is meant here by powers and capacities, see O'Toole, 'Qualities and powers' (n. 3), pp. 310–15; and R. Harré, 'Powers', *British Journal for the Philosophy of Science* 21 (1970): 81–101.
8 O'Toole, 'Qualities and powers' (n. 3), p. 313, suggests that his analysis of Boyle's matter theory in terms of inherent non-relational and non-inherent relational properties provides us with a complete analysis of the ascriptions of powers and capacities to corporeal objects 'according to the corpuscular theory of matter'. I wish to suggest that Boyle himself would have wanted to insist that the category of non-inherent relational properties must be subdivided into those which O'Toole has discerned and others which, in Boyle's words (*Works*, iii, 306), 'are rather to be ascribed to some unheeded agents, which, by unperceived means, may have great operations'.
9 *Cosmical Qualities*, *Works*, iii, 306.
10 *Cosmical Qualities*, *Works*, iii, 316.
11 *Cosmical Qualities*, *Works*, iii, 307.
12 H. G. Alexander (ed.), *The Leibniz–Clarke Correspondence* (Manchester University Press, 1956), p. 94; Isaac Newton, *Opticks, or A Treatise of the Reflections, Refractions, Inflections and Colours of Light*, based on the 4th edn, London, 1730 (New York: Dover, 1952), p. 401. For a fuller discussion, see J. Henry, 'Occult qualities and the experimental philosophy: active principles in pre-Newtonian matter theory', *Hist. Sci.* 24 (1986): 335–81.
13 *Cosmical Suspicions*, *Works*, iii, 317–18.
14 *Cosmical Qualities*, *Works*, iii, 308.
15 *Cosmical Suspicions*, *Works*, iii, 316.
16 BP 9, fol. 40. I am very grateful to Prof. Steven Shapin for bringing this and the immediately following quotation to my attention. It may be objected to my interpretation that Boyle refers to these occult phenomena (gravity, fermentation, springiness, magnetism) as 'subordinate' and thereby implies that, ultimately, they are explicable in terms of matter in motion. I cannot answer this except by special pleading. In view of the fact that Boyle describes matter and motion as 'first and simplest principles', I believe that he might have meant to suggest that gravity, magnetism and the like were secondary and more complex explanatory principles. His use of the word subordinate was meant, therefore, to convey the notion only of an ordering of heuristic precepts without intending to imply dependence of one kind on the other. The only support I can give for this interpretation is to be found in the other quotations from Boyle's works which I discuss in this paper.
17 BP 8, fol. 166.
18 *Cosmical Suspicions*, *Works*, iii, 318.
19 *Ibid.* Descartes expounded his three laws of motion, or nature, and his seven rules of impact in his *Principia philosophiae* (Paris, 1644), Part II, sections 37–42 and 46–52. See R. Descartes, *Principles of Philosophy*, trans. V. R. Miller and R. P. Miller (Dordrecht: D. Reidel, 1983), pp. 59–62.
20 *Cosmical Qualities*, *Works*, iii, 307.
21 *Cosmical Qualities*, *Works*, iii, 319.
22 *Hidden Qualities of the Air*, *Works*, iv, 98.
23 *Cosmical Qualities*, *Works*, iii, 318, 319, 321.
24 *Cosmical Suspicions*, *Works*, iii, 321–2.
25 *Salubrity of the Air*, *Works*, v, 40.
26 *Hidden Qualities of Air*, *Works*, iv, 79.
27 *Ibid.* p. 95.
28 *Cosmical Suspicions*, *Works*, iii, 318.

29 *Hidden Qualities of Air*, *Works*, iv, 96.

30 *Ibid.*

31 *Ibid.* Note the similarity of expression at the end there with the quotation from *Cosmical Suspicious* at note 18 above. The fact that Boyle felt the same kind of defence of his position was required suggests that we are correct in linking these two discussions.

32 *Ibid.* p. 90.

33 'Of Celestial Influences or Effluviums in the Air', *Works*, v, 638–45. Worsley writes: 'You did not expect, I am sure, I should have adventured into so particular an Apology for Astrology', p. 642. The Latin translation of this essay appears under the somewhat nondescript heading, 'Exemplar literarum Benjamini Worslaei' in the Hartlib Papers in Sheffield University Library, Hartlib Papers, 42/1/18A–25B. Confirmation of Worsley's authorship is provided by a letter from Worsley to Hartlib, 20 October 1657, Hartlib Papers, 42/1/10A. I am extremely grateful to Dr Antonio Clericuzio for bringing this information to my attention. For a fuller treatment of Worsley's essays, together with related documents from the Hartlib archive, see A. Clericuzio, 'New light on Benjamin Worsley's natural philosophy', in *The Advancement of Learning in the Seventeenth Century: The World of Samuel Hartlib*, eds. M. Greengrass, M. Leslie and T. Raylor (Cambridge University Press, forthcoming). The fullest general account of the activities of Worsley and his associates in the 1640s and 1650s is to be found in Charles Webster, *The Great Instaturation: Science, Medicine and Reform 1626–60* (London: Duckworth, 1975). Eight years Boyle's senior, Worsley was evidently the initiator of the 'Invisible College', of which, as far as we know, Boyle was the only other noted member. He died in 1677.

34 Letter from Locke to Boyle, 21 October 1691, in *The Correspondence of John Locke*, ed. E. S. De Beer, 8 vols (Oxford: Clarendon Press, 1976–89), iv, 320–2, also published in *Works*, vi, 543–4. The text appearing under the heading 'Of Celestial Influences or Effluviums in the Air' is clearly presented as a letter to Samuel Hartlib, but it is unsigned. There is nothing surreptitious about this; Locke apologises in his 'Advertisement' that 'The Negligence of Transcribers has let slip the Characters of Relators, and Names and Places of Authors from whom several of the Particulars in the following Papers were taken', *Works*, v, 609. In fact, the original of this letter in the Hartlib Papers (unlike the Latin translation discussed in the previous note), is unattributed (I owe this information to Dr Stephen Pumfrey), and it seems reasonable to suppose therefore that Boyle's copy was also unattributed.

35 *General History of Air*, *Works*, v, 611. For attributed extracts see, for example, Pascal 'in his small Tract, either De la Pesanteur de l'air, or in that Del'Equilibre des liqueurs', p. 694; 'Casati, Mechanicorum, lib. VIII, c. 5, p. 792, 793, &c.' (i.e. Paulo Casati, *Mechanicorum, libri octo*, Lyons, 1684), p. 616; 'Agricola, de re metallica, lib. duodec.', p. 634; and 'Jacobi Zabarelli *de Regionibus Aeris*, c. 8' (i.e. G. Zabarella, in his *De rebus naturalibus libri XXX*, Cologne, 1590), p. 695. For unattributed contributions, see, for example, letters from Mr J.T., pp. 624, 625, 634; or from 'Fort St. George', dated 23 January 1668, pp. 645–6.

36 *General History of the Air*, *Works*, v, 612.

37 *Cosmical Suspicions*, *Works*, iii, 318.

38 *General History of Air*, *Works*, v, 638–9.

39 *Ibid.* p. 638.

40 *Ibid.* p. 640.

41 *Ibid.* p. 641.

42 *Ibid.*

43 *Cosmical Qualities*, *Works*, iii, 307.

44 *General History of Air*, *Works*, v, 641.

45 *Cosmical Suspicions*, *Works*, iii, 318.

46 Graham Rees, 'Francis Bacon's semi-Paracelsian cosmology', *Ambix* 22 (1975): 81–101; 'Francis Bacon's semi-Paracelsian cosmology and the *Great Instauration*', *Ambix* 22 (1975): 161–73; 'Matter theory: a unifying factor in Bacon's natural philosophy', *Ambix* 24 (1977): 110–25; 'The fate of Bacon's cosmology in the seventeenth century', *Ambix* 24 (1977): 27–38.

47 William Harvey, *Disputations touching the Generation of Animals*, trans. Gweneth Whitteridge (Oxford: Blackwell, 1981), p. 378, see also pp. 374–83. For a full discussion of these ideas, see D. P. Walker, 'The astral body in Renaissance medicine', *Journal of the Warburg and Courtauld Institutes* 21 (1958): 199–33, reprinted in *Music, Spirit and Language in the Renaissance*, ed. P. Gouk (London: Variorum, 1985).

48 See A. C. Crombie, *Robert Grosseteste and the Origins of Experimental Science, 1100–1700* (Oxford: Clarendon Press, 1953); James McEvoy, *The Philosophy of Robert Grosseteste* (Oxford: Clarendon Press, 1982); David C. Lindberg, *Roger Bacon's Philosophy of Nature* (Oxford: Clarendon Press, 1983); and 'The genesis of Kepler's theory of light: light metaphysics from Plotinus to Kepler', *Osiris*, second series, 2 (1986): 5–42. On Al-Kindi, see Lynn Thorndike, *A History of Magic and Experimental Science*, 8 vols (New York: Columbia University Press, 1923–58), i, 642–6. For an indication of the role of light metaphysics in the medical tradition, see Walter Pagel, *New Light on William Harvey* (Basel: Karger, 1976), pp. 161–7.

49 *Aerial Noctiluca*, *Works*, iv, 384.

50 *Hidden Qualities of the Air*, *Works*, iv, 85.

51 *Ibid.* p. 86. For another example of Boyle's consideration of the possibility of astrological influences, see also p. 98.

52 Descartes, *Principles of Philosophy* (n. 19), part IV, sect. 199, pp. 282–3.

53 Marie Boas [Hall], *Robert Boyle and Seventeenth-Century Chemistry* (Cambridge University Press, 1958), p. 216; Steven Shapin, 'Robert Boyle and mathematics: reality, representation and experimental practice', *Science in Context* 2 (1988): 23–58.

54 Shapin, 'Boyle and mathematics' (n. 53), pp. 47–50. *Hidden Qualities of Air*, *Works*, iv, 91. See also *Two Essays concerning the Unsuccessfulness of Experiments*, in *CPE*, *Works*, i, 318–54. This aspect of Boyle's work is also discussed briefly in Paul B. Wood, 'Methodology and apologetics: Thomas Sprat's *History of the Royal Society*', *BJHS* 13 (1980): 1–26, on pp. 6–7; and Michael Hunter, 'Latitudinarianism and the "ideology" of the early Royal Society: Thomas Sprat's *History of the Royal Society* (1667) reconsidered', in *Establishing the New Science: The Experience of the Early Royal Society* (Woodbridge, Suffolk: Boydell Press, 1989), pp. 45–71, esp. pp. 53, 62, 68–9, where Hunter publishes a manuscript draft by Henry Oldenburg from which Sprat evidently drew for pp. 243–5 of his *History*.

55 *Cosmical Suspicions*, *Works*, iii, 322. Boyle's use of the word 'obnoxiousness' here should be taken in the sense of the fourth meaning given for 'obnoxious' by the *Oxford English Dictionary*: 'with *to*: Exposed to the (physical) action or influence of, liable to be affected by . . . '. Hence, the planetary and stellar bodies are in a more pronounced state of being liable to be influenced by variations than the schools would have us believe. See also Hunter, 'Latitudinarianism and the Society's "ideology"' (n. 54), p. 68, where Oldenburg talks of 'Experiments [being] very obnoxious to contingencies'. It is my belief, though only speculative, that Boyle closely supervised this document of Oldenburg's. Boyle does introduce non-chemical considerations elsewhere when discussing this theme: in his *Two Essays concerning the Unsuccessfulness of Experiments* (1661), Boyle had previously mentioned the variation of the compass and some anatomical observations. But most of his examples, as he mentioned in the full title were 'chiefly Chymical'; *Works*, i, 318, 343, 345.

56 J. E. McGuire, 'Boyle's conception of nature', *JHI* 33 (1972): 523–42; Eugene M. Klaaren, *Religious Origins of Modern Science: Belief in Creation in Seventeenth-Century Thought* (Grand Rapids, Mich.: William B. Eerdmans, 1977); J. Henry, 'Henry More versus Robert Boyle: the Spirit of Nature and the nature of providence', in *Henry More (1614–87): Tercentenary Studies*, ed. Sarah Hutton (Dordrecht: Kluwer Academic, 1990), pp. 55–76.

57 Shapin, 'Boyle and mathematics' (n. 53), pp. 50–2. See also Shapin and Schaffer, *Leviathan and the Air-Pump*, for a fuller account of Boyle's concerns with the authority for natural philosophical pronouncements.

58 Jeremy Taylor, *Works*, ed. R. Heber, 15 vols (London, 1828), xi, 356.

59 *Works*, iv, 164. For fuller discussions of the suspicion of 'reason' among English thinkers at this time, see H. G. van Leeuwen, *The Problem of Certainty in English Thought, 1630–1690* (The Hague: Martinus Nijhoff, 1963); Barbara J. Shapiro, *Probability and Certainty in Seventeenth-Century England* (Princeton University Press, 1983); Christopher Hill, '"Reason" and "reasonableness"', in *Change and Continuity in Seventeenth-Century England* (London: Weidenfeld and Nicolson, 1974), pp. 103–26; Lotte Mulligan, '"Reason', "right reason", and "revelation" in mid-seventeenth-century England', in *Occult and Scientific Mentalities in the Renaissance*, ed. Brian Vickers (Cambridge University Press, 1984), pp. 375–401; J. Henry, 'The Scientific Revolution in England', in *The Scientific Revolution in National Context*, eds. R. Porter and M. Teich (Cambridge University Press, 1992), pp. 178–209.

60 See Shapin and Schaffer, *Leviathan and the Air-Pump*; Henry, 'Occult qualities and the experimental philosophy' (n. 12); *ibid.* 'Scientific revolution in England' (n. 59).

61 For another treatment of this theme from a different perspective, see Peter Dear, 'Miracles, experiments and the ordinary course of nature', *Isis* 81 (1990): 663–83.

62 *Cosmical Qualities*, *Works*, iii, 306.

63 Here again I am referring to the analysis of O'Toole, 'Qualities and powers' (n. 3), in which he distinguishes between non-relational inherent properties of bodies and non-inherent relational properties. See notes 6 and 8 above.

64 *Cosmical Suspicions*, *Works*, iii, 316. My conclusion should be taken to endorse and support, from a non-chemical perspective, Antonio Clericuzio's recent attempt to redefine Boyle's corpuscular philosophy in terms of his chemical philosophy. See A. Clericuzio, 'A redefinition of Boyle's chemistry and corpuscular philosophy', *Ann. Sci.* 47 (1990): 561–89. See also Henry, 'Occult qualities' (n. 12).

65 *Hamlet*, 1. 5. 166–7.

Additional note

Works refers to *The Works of the Honourable Robert Boyle*, ed. Thomas Birch, 2nd edn, 6 vols (London, 1772). This edition has since been superseded by *The Works of Robert Boyle*, eds Michael Hunter and Edward B. Davis, 14 vols (London: Pickering and Chatto, 1999–2000). There is a concordance in volume 14 of the new edition, which facilitates conversion of a citation in the 1772 edition, to the new edition.

VIII

ROBERT HOOKE, THE INCONGRUOUS MECHANIST

1. Incongruity

In his *De Corpore* of 1655, Thomas Hobbes dismissed all those who 'say anything may be moved or produced by itself, by species, by its own power, by substantial forms, by incorporeal substances, by instinct, by anti-peristasis, by antipathy, sympathy, occult quality, and other empty words of the schoolmen . . .'[1] Historians of science have tended to see such sentiments as entirely representative of the 'new philosophy'. The Scientific Revolution was made possible (so the story goes) because Scholastic notions of occult qualities, including antipathies and sympathies between inanimate objects, gave way to a firm conviction that everything could be explained in terms of changes brought about by contact actions between passive particles of matter moving inertially. Marie Boas Hall, for example, has said that before the establishment of the mechanical philosophy 'there were sympathies and antipathies, congruities and incongruities, attraction and hostility', but after the air-pump experiments of Boyle and Hooke, all such notions 'were banished forever from pneumatics'.[2] However, we do not have to read very far into Robert Hooke's major scientific work, the *Micrographia*, before we find him explaining certain physical phenomena in terms of what he calls the 'congruities' and 'incongruities' of inanimate matter.

At first sight these two concepts look perilously close to sympathy and antipathy, which Hooke's contemporary, Henry More, called the 'Sanctuary of Fools'.[3] Indeed, Hooke freely admitted that 'Congruity seems nothing else but a Sympathy, and Incongruity an Antipathy of bodies'. Subsequently, congruity is variously referred to as a 'property of cohaesion', 'a kind of

Acknowledgement: I should like to thank the editors of this volume and Steven Shapin and Chris Lawrence for their helpful comments on this paper.

1 Thomas Hobbes, *The Elements of Philosophy. The First Section, Concerning Body,* in *English Works*, edited by William Molesworth, 11 vols. in 12 (London, 1839 – 45), i, 531.

2 Marie Boas, 'The Establishment of the Mechanical Philosophy', *Osiris*, 10 (1952), 412 – 541, pp. 417, 422.

3 Alazonomastix Philalethes [ie. Henry More], *The Second Lash of Alazonomastix: Concerning a Solid and Serious Reply to a very uncivill Answer to Certain Observations upon Anthroposophia Theomagica, and Anima Magica Abscondita* (Cambridge, 1651) p. 165.

attraction and tenacity', 'a Tenaceous and Attractive power', and an 'attractive virtue'. His first definition of these two terms reads like this:

> By Congruity I mean a property of a fluid Body, whereby any part of it is readily united with any other part, or of any other similar fluid or solid body: And by Incongruity a property of a fluid by which it is hindered from uniting with any dissimilar fluid or solid body.[4]

Now, this cannot be dismissed as a mere aberration or a passing fancy which Hooke was unwise to allow to appear in his published work. He went on to suggest that 'this Principle' might be

> a co-efficient in the most considerable Operations of Nature . . . As in those of Heat, and Light, and consequently, of Rarefaction and Condensation, Hardness, and Fluidness, Perspicuity and Opacousness, Refractions and Colours, &c. Nay, I know not whether there may be many things done in Nature, in which this may not (be said to) have a Finger? This I have in some other passages of this Treatise further enquired into and shewn, that as well Light as Heat may be caused by corrosion, which is applicable to congruity, and consequently all the rest will be but subsequents.[5]

In view of this, it might seem that Hooke was rather an incongruous mechanist and the aim of this paper is to draw attention to, and consider the role of, congruity and incongruity and other 'active principles' in Hooke's efforts to develop a complete system of natural philosophy. It is usually accepted by the same historians of science who characterise the Scientific Revolution (at least partially) in terms of the new theory of passive matter in inertial motion, that this strict mechanistic philosophy is ultimately unworkable and a further revolution was required — the 'Newtonian Revolution', in which unexplained active principles (now dressed up in some mean and thoroughly modern mathemathetics) were reintroduced into natural philosophy — only then could modern science really take-off. It is my contention that, even before Newton, mechanical philosophers in England were trying to develop a system (or systems) of mechanical philosophy, as comprehensive as those of Descartes and Hobbes purported to be, but which frequently relied upon notions of inherent activity in matter.[6] In this perspective, therefore,

4 Hooke, *Micrographia*, pp. 16, 21, 15, 31, 32, 12.

5 Ibid., p. 28.

6 See, for example, R. S. Westfall 'Newton and Alchemy', in Brian Vickers, ed., *Occult and Scientific Mentalities in the Renaissance* (Cambridge, 1984), 315 – 35, p. 324; I. B. Cohen, 'The *Principia*, Universal Gravitation, and the "Newtonian Style", in Relation to the Newtonian Revolution in Science', in Zev Bechler, ed., *Contemporary Newtonian Research* (Dordrecht, 1982), 21 – 108; and idem, *The Newtonian Revolution, with Illustrations of the Transformation of Scientific Ideas* (Cambridge, 1980). For a further discussion of this issue see John Henry, 'Occult Qualities and the Experimental Philosophy: Active Principles in pre-Newtonian Matter Theory', *History of Science*, 24 (1986), 335 – 81, pp. 358 – 66.

there is nothing incongruous about Hooke's use of active principles, rather, we should revise our historiographical preconceptions about what constituted the mechanical philosophy in seventeenth-century England.

1.1. Hooke's Cosmology: The 'Principal of the Powers of the great System of the World'

First of all, let us run through the various major active principles which appear throughout Hooke's works. Active principles are to be understood as 'occult qualities' of bodies which were invoked by natural philosophers to account for particular kinds of observed physical activity. Congruity and incongruity were the first of such qualities to make an appearance in Hooke's natural philosophy. Indeed, Hooke's first published work, the *Attempt for the Explication of the Phenomena* of 1661, relied on these notions, and they were invoked again very early in the *Micrographia.*[7]

But, perhaps we should begin our survey at what might be regarded as the metaphysical beginning and consider Hooke's pronouncements on 'Matter and Motion', the fundamental principles of the mechanical philosophy. In his *Discourse of the Nature of Comets* in 1682 Hooke referred to body and motion as the 'two single Powers, which co-operate in effecting the most of the sensible and insensible Effects of the World'. They are, Hooke went on, 'the Male and Female of Nature'. The 'Female or Mother Principle' is Body or Matter which is 'without Life . . . a Power in itself wholly unactive, until it be, as it were, impregnated by the second Principle . . . without whose Conjunction nothing or no Alteration can be produced'. Motion itself, then, is regarded here as an active principle which must be 'impregnated' into matter before any physical phenomena can take place. Motion is not just an active principle but even a seminal principle. It should be clear, anyway, that Hooke's conception of matter being 'impregnated' with motion has a rather different connotation from the Cartesian view in which the motions of matter are continually imposed from beyond the realm of matter itself, which is defined only in terms of passive extension. According to Hooke, once God had impregnated matter with 'the second Power Motion', he made 'every Minim of it to move with infinite Varieties of Motions'. The Mater or Mother Principle 'was now no longer dead, and unactive'.[8]

7 Hooke, *An Attempt for the Explication of the Phaenomena, Observable in an Experiment Published by the Honourable Robert Boyle, Esq.; in the XXXV. Experiment of his Epistolical Discourse Touching the Aire* (London, 1661); reprinted in facsimile in Gunther, *Early Science in Oxford*, x, 1 – 50. A slightly different version of this work was incorporated into *Micrographia* pp. 11 – 32.

8 Hooke, *Posthumous Works*, pp. 172, 174. Insufficient attention has been paid to the problematic nature of Cartesian matter theory for many seventeenth-century natural philosophers, but see Alan Gabbey, 'The Mechanical Philosophy and its Problems: Mechanical Explanations, Impenetrability, and Perpetual Motion', in J. C. Pitt, ed., *Change and Progress in Modern Science* (Dordrecht, 1985), 9 – 84; and Henry, 'Occult Qualities' (n. 6) pp. 335 – 9.

Hooke even went so far as to say that his first active power might be called 'Spiritus, or hylarchic Spirit, as some call it', a remarkable comment in view of the fact that some five years before he had taken the trouble to refute Henry More's claims that a hylarchic spirit was required to explain hydrostatic phenomena.[9] Moreover, Hooke soon proceeded to discuss the regulated and ordered forms of motion in matter, the 'two great Laws of Motion', as he called them, Light and Gravity, in the same spiritualist terms. If motion is 'most properly called a Spirit, signifying its Power of moving', then

> These two Powers [Light and Gravity, Hooke wrote] seem to constitute the Souls of the greater Bodies of the World . . . and both these are to be found in every such Body in the World; but in some more, in some less; in some one is predominant, in others the other, but no one without some Degree of both: For as there is none without the Principle of Gravitation, so there is none without some degree of Light. And though some doe not show its Effects in producing Light immediately, yet I shall make it probable that it has that Motion blended with others, which hinder it from producing Effects, but yet do not wholly destroy the Principle.[10]

Once again, we can be reasonably sure that this was not just a passing rhetorical flourish. Hooke had made similar claims for light in his 1680 lectures on light:

> This being that we call light, sure if any thing may be call'd the *Anima Mundi*: Its action being so near of Kin to that of a Spirit, the whole Mass being in an instant acted by it, and made sensible as I may so speak, of

9 Hooke, *Posthumous Works*, p. 172; *Lampas: Or, Descriptions of Some Mechanical Improvements of Lamps & Waterpoises. Together with Some Other Physical and Mechanical Discoveries* (London, 1677); reprinted in Gunther, *Early Science in Oxford* viii, 153 – 208. The refutation of More's concept of hylarchic spirit appears pp. 182 – 95. More had appropriated various of Boyle's and Hooke's pneumatic and hydrostatic experiments in his *Enchiridion metaphysicum: sive, de rebus incorporeis succincta & luculenta dissertatio* (London, 1671) and compounded the felony in his *Remarks upon Two late ingenious Discourses: The One an Essay touching the Gravitation and Non-Gravitation of Fluid Bodies: The other, Observations touching the Torricellian Experiment; so far forth as They May Concern Any Passages in His Enchiridion Metaphysicum* (London, 1676). The More-Hooke polemic has not received separate attention but, for a discussion of the related polemic between More and Robert Boyle, see Steven Shapin and Simon Schaffer, *Leviathan and the Air-Pump*, pp. 207 – 24; and John Henry, 'Henry More versus Robert Boyle: The Spirit of Nature and the Nature of Providence', in Sarah Hutton, ed., *Of Mysticism and Mechanism: Tercentenary Studies of Henry More, 1614 – 1687* (Dordrecht, 1989), 55 – 76.

10 Hooke, *Posthumous Works*, pp. 174, 175.

what is done in any one Point: So that light may be said to be *tota in toto & tota in qualibet parte*, possibly with some kind of Plausibleness.[11]

Even the extraordinary suggestion in this quotation that light makes the whole mass of the universe 'sensible . . . of what is done in any one Point', is repeated elsewhere in Hooke's lectures:

> Now if we consider the Sun in the great World, we shall find it first a Being which has in itself a Power of radiating or dispersing Light into the Whole of Nature, and (consequently by this its Influence) of being as it were every where present, and of being sensible of all those other Bodies that are placed any where throughout the whole Expansum: for as it doth by its Radiation influence and affect every Point of the Universe, so must there be a kind of reflex balance upon it self from every such Point: For as any one, and every one Ray it sends forth, does meet with and affect any Body in its way, so consequently must that End of the Ray that touches the Sun, have a greater or less Resistance to be moved forwards; and consequently if there were Understanding in the Sun itself, it must be sensible that this or that Ray does somewhere in its Progress receive such or such an Impediment to its Propagation or moving forwards that way.[12]

Accordingly, Hooke concluded that the presence of light as 'one of the first and principal of the Powers of the World' unites and makes all things one, so that every part of the universe is affected by every other part. This even provided him with a justification for astrology, though it seems that he never did anything with it once he had pointed it out:

> This may inform us also, how even the very remotest star, and every one of those indefinite Number of Stars may have an influence every Moment upon this Ball of the Earth on which we tread, and every one upon every other, and all in Proportion, Measure and Harmony . . .[13]

It is important to notice, however, that Hooke's concept of Spirit, in spite of his seeming concession to the notion of 'hylarchic spirit', was not derived from the immaterialist preoccupations of Henry More and other Cambridge Platonists. No sooner had Hooke suggested that light may be *tota in toto & tota in qualibet parte* than he insisted that

11 Hooke, *Posthumous Works*, p. 79. The maxim (in Latin) that 'the whole is in the whole and the whole is (also) in any given part', is a familiar theological dictum about the human soul, see John Henry, 'A Cambridge Platonist's Materialism: Henry More and the Concept of Soul', *Journal of the Warburg and Courtauld Institutes*, 49 (1986), 172 – 95, p. 192. Elsewhere Hooke reverses the direction of this analogy by trying to explain the nature of the soul, 'the Primum movens, the self-moving Principle', by comparison with the activity of the Sun in 'the Great World' (*Posthumous Works*, p. 146). Hooke's concept of the soul has yet to be properly considered but in the meantime see B. R. Singer, 'Hooke on Memory, Association and Perception'; Oldroyd, 'Some "Philosophical Scribbles" attributed to Robert Hooke'; and Simon Schaffer, 'Godly Men and Mechanical Philosophers: Souls and Spirits in Restoration Natural Philosophy', *Science in Context*, 1 (1987), 55 – 85.

12 Hooke, *Posthumous Works*, p. 146.

13 Hooke, *Posthumous Works*, pp. 96 and 79.

> after all this we may prove it to be purely corporeal, and subjected to the same Laws that bulky, tangible and gross Bodies are subject to.[14]

Hooke's concept of spirit, therefore, was closer to that of Henry Power which, as Charles Webster has pointed out, stemmed from a different Neoplatonist tradition via the chemical philosophies of thinkers like Paracelsus and Van Helmont.[15] In this tradition there was a continuity in God's creation from the purely incorporeal to the densest matter, with concepts like light or fire and the rational soul representing the half-way mark. (This was very different from the Cambridge Platonist view, shared with the Cartesians, in which there was a strict and necessary distinction between soul or spirit on the one hand and body on the other.) For Hooke it seemed possible to arrive at an understanding of the soul by a gradual progression

> beginning from fluidity, or body without any form, . . . till we arrive at the highest form of a bruite Animals Soul, making the steps or foundations of our Enquiry, Fluidity, Orbiculation, Fixation, Angulization, or Crystallization Germination, or Ebullition, Vegetation, Plantanimation, Animation, Sensation, Imagination.

'Nor do I imagine', he added, 'that the skips from one to another will be found very great'.[16]

Accordingly, in this intellectual tradition it is perfectly possible to say, as Hooke does, that an active principle is 'most properly called a Spirit, signifying its Power of moving', while at the same time insisting that this spirit is corporeal and 'subjected to the same Laws that bulky, tangible and gross Bodies are subject to'.[17] Indeed, Hooke frequently has recourse, in his efforts to explain physical phenomena, to the notion that the finest and most subtle fluids naturally move faster than denser substances. He explains the behaviour of fire, for example, simply by asserting it to be a 'Supreme Fluid' and

14 Hooke, *Posthumous Works*, p. 79, see also pp. 115 – 6. Henry More, like Descartes, was a strict philosophical dualist, insisting upon a categorical distinction between matter (passive) and spirit (active). See Henry, 'A Cambridge Platonist's Materialism' (n. 11), and Henry, 'More versus Boyle' (n. 9).

15 Charles Webster, 'Henry Power's Experimental Philosophy', *Ambix* 14 (1967), 150 – 78; Walter Pagel, *Paracelsus: An Introduction to Philosophical Medicine in the Era of the Renaissance* (Basle, 1958); idem, *Joan Baptista van Helmont, Reformer of Science and Medicine* (Cambridge, 1982); A. G. Debus, 'Chemistry and the Quest for a Material Spirit of Life in the Seventeenth Century', in M. Fattori and E. Bianchi, eds., *Spiritus: IV° Colloquio Internazionale Roma 7 – 9 gennaio 1983* (Rome, 1984), pp. 245 – 63; D. P. Walker, 'The Astral Body in Renaissance Medicine', *Journal of the Warburg and Courtauld Institutes* 21 (1958), 119 – 33; idem, 'Medical Spirits in Philosophy and Theology from Ficino to Newton', in *Arts du spectacle et histoire des idées, Recueil offert en hommage à Jean Jacquot* (Tours, 1984), 287 – 300 (both now reprinted in D. P. Walker, *Music, Spirit and Language in the Renaissance*, London, 1985).

16 Hooke, *Micrographia*, p. 127. For an introduction to Renaissance concepts of continuity from spirit to matter see Paul Oskar Kristeller, *The Philosophy of Marsilio Ficino*, translated by V. Conant (New York, 1943); Benjamin Brickman, *An Introduction to Francesco Patrizi's 'Nova de universis philosophia'* (New York, 1941); and Walker, 'Medical Spirits' (n. 15).

17 Hooke, *Posthumous Works*, p. 174.

'Supreme Fluids always recede from the Center radiating'. Hooke took this Neoplatonic tradition to its logical conclusion when he wrote that Body and Motion 'may be found hereafter to be only differing conceptions of one and the same essence'.[18]

This brings us to another active principle in Hooke's natural philosophy, the concept of vital principles capable of producing life from putrifying matter. When discussing these matters Hooke always emphasised their speculative nature but, even so, in *Micrographia* and elsewhere he asserted that plants and even animate bodies like insects might be produced spontaneously by 'a certain instinct of nature', a 'vegetative principle', or 'seminal principle'. Admitting that he had found it impossible to be sure that the different kinds of insects which arise from 'Vinegar, Meal, musty Cakes, &c.' were not produced by unobserved parent insects, Hooke still insisted that there was nothing intrinsically difficult or absurd about the concept of spontaneous generation. He even took the trouble to dismiss one of the common objections to spontaneous generation. The fact that some of the creatures supposed to have been generated spontaneously were subsequently found to propagate themselves by more familiar means was of no consequence:

> I see not any great reason to question, but that, notwithstanding its own production was as 'twere casual, yet it may germinate and produce seed, and by it propagate its own, that is, a new Species. For we do not know, but that the Omnipotent and All-wise Creator might as directly design the structure of such a Vegetable, or such an Animal to be produc'd out of such or such a putrifaction . . . Nor need we therefore to suppose it the more imperfect in its kind . . . for he might as compleatly furnish it with all kinds of contrivances necessary for its own existence, and the propagation of its own Species . . .[19]

To sum up: at different times, in different writings, Hooke referred to 'Congruity and Incongruity' as a 'co-efficient in the most considerable Operations of Nature', with a finger in every pie; to Gravity as 'one of the most universal active principles in the world'; and to light as 'the most

18 Hooke, *Posthumous Works*, pp. 191, 115 – 6; and Gunther, *Early Science in Oxford*, viii, 339.

19 Hooke, *Micrographia*, pp. 123, 134. It should be noted that Hooke's acceptance of the possibility of spontaneous generation does not signify an abandonment of the natural philosophical enterprise. The clear implication of his pronouncements on this subject is that naturalists may yet discover the 'structure' and 'contrivances' which enable the production of living creatures from 'putrifaction'. Hooke's suggestion in *Lampas* that the existence of Henry More's hylarchic spirit may be proved by studying 'the generation, nutrition, vegetation, and propagating of Vegetables and animal substances' is meant to be taken ironically. Because such things were 'further removed beyond the reach of our senses and understandings', Hooke wrote, 'one may more boldly assert strange things of this Hylarchic Spirit without fear of controul or contradiction, and from whence possibly it may never lie within the power of Reasoning to banish him'. It seems clear, nevertheless, that Hooke held out for the hope that experimental philosophy would one day discover the principles which caused generation, vegetation, etc. and the laws which governed their operations.

operative and most considerable Ingredient of the Universe'. Motion itself was likened to a spirit and both light and gravity, being 'the two great Laws of Motion' were likened to the 'Souls of the greater Bodies of the World'. Moreover, Hooke readily accepted that there may be vital principles which supervise the genesis and production of living creatures. All discussion of these matters is couched in the language of active principles which historians of science usually associate with Isaac Newton. Perhaps the clearest example of similarity between Hooke's and Newton's speculations is afforded by comparing Hooke's congruity and incongruity with Newton's 'certain secret principle in nature by which liquors are sociable to some things, and unsociable to others'.[20] But, unlike Newton's case, Hooke's use of active principles has scarcely been noticed by historians, and before proceeding any further we should consider briefly why that might be so.

An obvious factor in the failure to take sufficient account of Hooke's speculations about occult active principles is the prevailing historiography. The cherished notion that mechanical philosophers eschewed all unexplained 'vegetative faculties', 'congruities and incongruities', and the like, and relied entirely on the heuristic possibilities of the inertial movements of passive matter, has tended to blind historians to the significance of such supposedly non-mechanistic hypothesizing. Even R. S. Westfall, one of the few scholars to recognise the importance of 'sympathies' and 'antipathies' and of occult notions of 'attraction' in Hooke's work (and, moreover, a historian well-known for emphasising the occult side of Newton's work) did not see any need to redefine our notion of the mechanical philosophy. While noting that Hooke himself declared that his hypothesis of gravitational attraction at a distance 'answered to the rules of mechanics', Westfall continued to believe that this was a path which 'mechanical philosophers tended to shun'. Once again, then, Hooke is seen as the incongruous mechanist.[21]

Less obvious but, perhaps, more cogent reasons for disregarding Hooke's discourse of active principles arise from his own equivocation and ambivalence. There are a number of places in Hooke's work where he can perhaps be seen to be striving for a strictly mechanistic account of all physical phenomena, including those which he has declared to depend upon congruity and incongruity, gravitational attraction, and so forth. It might be suggested,

20 Gunther, *Early Science in Oxford*, viii, p. 256; Hooke, *Posthumous Works*, p. 90; Letter from Isaac Newton to Robert Boyle, February 28, 1678/9, in Robert Boyle, *Works*, edited by Thomas Birch, 6 vols. (London, 1772) vi, 71. For a discussion of Newton's borrowings from Hooke see Westfall, *Force in Newton's Physics*, pp. 332, 368 – 9, 457.

21 Westfall, *Force in Newton's Physics*, p. 269, see also pp. 264, 265, 267. Westfall has produced a series of essays arguing for the alchemical antecedents of Newton's concept of force. The first of these, 'Newton and the Hermetic Tradition', in A. G. Debus, ed., *Science, Medicine and Society in the Renaissance* (New York, 1972) 183 – 93, even points out that Descartes' three forms of matter 'faithfully . . . repeated the characterisitics of the Paracelsian tria prima, and in the case of his subtle matter . . . he cast the active principles of Hermetic philosophy into mechanistic form' (p. 186). Unfortunately, Westfall chooses to see this as a sign of the inadequacy of Descartes' mechanics (p. 187); it did not occur to him that we might need to recast our historiographical conception of what counted as a mechanical explanation.

therefore, that the pronouncements which I have singled out were merely literary embellishments and flourishes, or that, in some cases, they were written absent-mindedly. In short, what we have been considering so far is only loose talk, but really Hooke saw all these things as being explicable in purely mechanical terms, that is to say, in terms of the inertial motions of categorically passive particles of matter. The *sine qua non* of all such speculations in Hooke's writings is a notion of particles of matter in incessant vibration and, before proceeding any further, we must consider these ideas in detail.

1.2. Hooke's Mechanism: Vibrations and Aethers

It is certainly true that Hooke tried to give what he regarded as a mechanistic account of the cause of congruity and incongruity. The cause of fluidity, he said, was 'a certain pulse or shake of heat' in the particles of the body so that the particles become loose from one another and fluid. Congruity and Incongruity are explained in terms of a harmony or disharmony of such vibrations. So, in a mixture of fluids or powders Hooke supposed there was a

> pulse of heat to agitate the small parcels of matter, and those that are of a like bigness, and figure, and matter will hold, or dance together, and those which are of a differing kind will be thrust or shov'd out from between them.

Congruity, then, was not, as Hooke said elsewhere, 'a first Principle, but rather of a second, third, or fourth Rank'.[22]

Similarly, the efficient cause of light, according to Hooke, was a 'Vibrative motion', which is 'exceedingly quick', and short. So, light in the medium, a ray of light, is a vibrative motion in the medium caused initially by the vibrative motions of a luminous body. A 'very short and very quick' vibrative motion was also invoked to explain gravity:

> Suppose then there is in the Ball of the Earth such a Motion, as I, for distinction sake, will call a *Globular Motion*, whereby all the parts thereof have a vibration towards and fromwards the Center, or of Expansion and Contraction; and that this vibrative Motion is very short and very quick, as it is in all very hard and very compact bodies.

Here again this vibrative motion of the Earth is passed on to the interspersed fluid of the universe which vibrates 'every way *in Orbem*, from and towards the Center, in lines radiating from the same'. Furthermore, both magnetic and electrical phenomena are explained in the same way.[23]

Now, it may be wondered how it is that vibrations in the aether, all of a very short and quick amplitude, can be said in the case of light to push things outwards, in the case of gravity to draw things inwards, and in the case of magnetism and electricity to sometimes repel and sometimes attract. No problem, Hooke simply assumed that there must be different aethers for each

22 Hooke, *Micrographia*, p. 15, Hooke, *Posthumous Works*, p. 368.
23 Hooke, *Posthumous Works*, p. 184.

of these (or, at least, different parts of the aether doing different jobs):

> every one of these distinct internal Motions of Bodies, as that of Light, and that of Sound, have distinct and differing Mediums, by which those Motions are communicated from the affecting to the affected Body. And so I conceive also that the Medium of Gravity may be distinct and differing both from that of Light, and from that of Sound . . .[24]

Hooke refers to these aethers as essentially 'stagnant', a choice of word which is presumably meant to alert the reader to the clear differences between Hooke's aether and the swirling vortexes of the Cartesian aether.[25] Strictly speaking, therefore, it is not correct to say that the aether is the cause of gravity or light or whatever. The cause is the vibrative motion of the gravitating mass, the luminous body, the magnet and so on. It is a reasonable question to ask, therefore, what is the cause of these vibrative motions in such bodies? Once again the evidence suggests that Hooke simply resorted to a concept of inherently active matter:

> Now that the parts of all bodies, though never so solid do yet vibrate, I think we need go no further for proof, then that all bodies have some degrees of heat in them, and that there has not been yet found any thing perfectly cold: Nor can I believe indeed that there is any such thing in Nature, as a body whose particles are at rest, or lazy and unactive in the great Theatre of the World, it being quite contrary to the grand Oeconomy of the Universe.

Similarly, on a loose sheet published by Waller in the *Posthumous Works*, Hooke gave a series of propositions designed to provide the philosophical underpinning for a theory 'Of Gravity'. The eleventh proposition is that 'All bodies have a tremulous Motion, as Bells, &c.' The seventeenth reads, 'The vibrating Motion of all Globular Bodies is from the Center to the Superficies, and vice versa. This shown by the Bell, Water in a Glass, &c .'[26]

So, on these accounts of light, gravity, and other active principles, in which the aether merely transmits the vibrations of a central body, we must conclude that Hooke has not succeeded in developing a strictly mechanical philosophy in the manner of Descartes or Hobbes. However, like most philosophers, Hooke was nothing if he was not inconsistent, and we had better consider, if only for completeness, the alternative speculations which he occasionally developed.

There is, for example, an early version of an aether theory in which the aether itself is the direct cause of gravitational 'attractions' between bodies. In the *Micrographia* Hooke suggested that gravity might be explained by assuming (or proving) that there is a universal aether, pervading all bodies and all

24 Hooke, *Posthumous Works*, p. 184.

25 Boas Hall, 'Establishment of the Mechanical Philosophy' (n. 2) regards Hooke's aether as essentially Cartesian, p. 454.

26 Hooke, *Micrographia*, p. 16, Hooke, *Posthumous Works*, p. 191.

interstices in the world, which is 'heterogeneous' or incongruous to all the bodies of the world. The incongruity of this aether to all other substances would, we are to assume, set up density gradients in the aether; it being denser in empty space than in a body to which it is incongruous. Accordingly, the aether would 'endeavour to detrude all earthly bodies as far from it as it can'.[27] Judging from the frequency with which Hooke recurred to his different hypotheses, he seems to have dropped this idea in favour of the merely transmissive aethers already described. It is worth pointing out, however, that this theory was taken up by Isaac Newton who made a note of it when he read the *Micrographia* sometime in the late 1660s, and offered it as an explanation of gravity in Query 21 of the 3rd edition of the *Opticks* (1718). Clearly, this theory cannot be used to suggest that Hooke held all matter to be passive and free from unexplained occult qualities.[28]

Before leaving the subject of congruity and incongruity, it is perhaps worth pointing to another occult aspect of Hooke's account of this 'co-efficient in the most considerable Operations of Nature'. His earliest attempt to explain the separation of immiscible fluids in terms of the differing vibrations of their constituent particles drew upon an ingenious analogy with vibrating musical strings:

> particles that are all similar, will, like so many equal musical strings equally stretcht, vibrate together in a kind of Harmony or unison; whereas others that are dissimilar, upon what account soever, unless the disproportion be otherwise counter-ballanc'd, will, like so many strings out of tune to those unisons, though they have the same agitating pulse, yet make quite differing kinds of vibrations and repercussions . . .[29]

Subsequently, in the *Discourse of Earthquakes*, Hooke proffered a hypothesis of magnetism which depended on the assumption that 'all magnetical Bodies have the constituent Parts of them of equal Magnitude and *equal Tone*'.[30] Ingenious as this might be, it surely would not have convinced all of Hooke's contemporaries as an account entirely free of any occult principles. Henry More, for example, had pointed out in 1659 that sympathetic resonance could not be explained 'by the mere Vibration of the Aire . . . for if it could, these Vibrations would not faile to move other Bodies more movable by farre then the string itself that is thus moved'.[31] for most of Hooke's contemporaries sympathetic resonance remained a magical phenomenon. In fact, as we shall see below (in Section 1.3.), Hooke claimed to have a perfect justification for the use of such occult phenomena in his natural philosophy, but the point to

27 Hooke, *Micrographia*, p. 22.

28 Newton's notes 'Out of Mr Hooks Micrographia' have been published in A. R. Hall and M. Boas Hall, eds., *Unpublished Scientific Papers of Isaac Newton* (Cambridge, 1962) pp. 400 – 13.

29 Hooke, *Micrographia*, p. 15.

30 Hooke, *Posthumous Works*, p. 364.

31 Henry More, *The Immortality of the Soul* (London, 1659) quoted from 2nd edition, in *A Collection of Several Philosophical Writings* (London, 1662) p. 193.

notice here is that sympathetic resonance was still regarded as an unexplained or occult phenomenon.

There is yet another aether hypothesis, presented in *De potentia restitutiva* (1678), which we must finally consider. Unlike the passive transmissive aethers, this seems to have some vibrative power of its own. This short but remarkable work attempts to explain the cause of another notoriously problematic active principle operating in nature — the springiness of bodies. Once again Hooke's account relies upon the ingenious use of presupposed vibrations in all particles of matter. Perhaps this hypothesis can be seen as a purely kinetic version of what has come to be known as Newton's 'nut-shell' theory of matter. Just as Newton, in the 'Queries' to the *Opticks*, suggested that bodies need not be conceived as entirely solid but as arrays of minute particles invested with repulsive forces which prevented penetration, so Hooke spoke of a body as 'a determinate Space . . . defended from being penetrated . . . by a power from within'. He illustrated this by imagining a thin plate of iron a foot square moving rapidly to and fro 'the length of a foot'. Providing the motion was fast enough this would compose 'such an essence as I call in my sense a Cubick foot of sensible Body'.[32]

Here, though, we get an unequivocal, absolutely explicit claim that this power of vibration is not an active principle but merely the result of impacts from other bodies:

> This Vibrative motion I do not suppose inherent or inseperable from the Particles of body, but communicated by Impulses given from other bodies in the Universe.

This is entirely in keeping with the kind of strict mechanist philosophy devised by Descartes, and familiar to historians of science. The communication of impulses, which Hooke postulated, once again required an aether:

> I do further suppose, A subtil matter that incompasseth and pervades all other bodies, which is the Menstruum in which they swim which maintains and continues all such bodies in their motion, and which is the medium that conveys all Homogenious or Harmonical motions from body to body.[33]

Leaving aside the fact that the operation of this aether seems to depend in part upon the occult notion of sympathetic resonance, we still need to know what Hooke means by saying that this aether 'maintains and continues all bodies in their motion'. Either it has its own inherent vibrative motion, in the nature of a subtle corporeal spirit (in which case Hooke can no longer be seen as a mechanist in the Cartesian mode), or we have to assume that the aether is

32 Gunther, *Early Science in Oxford*, viii, 340. See Arnold Thackray, ' "Matter in a Nut-shell": Newton's *Opticks* and Eighteenth-Century Chemistry', *Ambix* 15 (1968), 29 – 53.
33 Gunther, *Early Science in Oxford*, viii, 340, 341.

once again, as in other versions of Hooke's philosophy, a passive entity and this whole account is meant to be seen as a completely inertial system, like that of Descartes and Hobbes. Well, this certainly may be the case. It is perfectly possible that Hooke hoped to develop a system to rival that of Descartes, in which nothing was left to unexplained principles of activity. But the fact remains that throughout his career, as we have seen, Hooke wrote in a way that suggested he was willing to entertain notions of matter endowed with occult principles of activity. *De potentia restitutiva* is, at best, a mere hint towards a completely inertial system of mechanical philosophy.[34]

Even so, it might still be argued that I am simply placing too much emphasis on aspects of Hooke's philosophy which were merely tentative, speculative, and deliberately presented only as queries. And that, in fact, this only provides evidence that Hooke was not a very cogent or effective natural philosopher, but rather one who consistently failed to develop a plausible system of mechanical philosophy. This would be a serious underestimation of Hooke's powers as a natural philosopher. For, we can see from Hooke's concern with scientific methodology that he was always concerned about the validity of his hypotheses and the epistemological status of his hypothetical constructs.

A number of historians have already paid due tribute to Hooke's efforts to develop a sophisticated methodology of science but less emphasis has been placed on the reasons why Hooke felt such considerations to be so important to the presentation of his case. As recent work has indicated, we should not take it for granted that a 'good scientist' will automatically review his method as he proceeds. Methodological pronouncements within scientific texts are there to serve a purpose. That purpose, of course, is to convince the reader that the scientific theories proposed must be granted a high degree of credibility and plausibility.[35] In Hooke's case, as I now want to show, his method is carefully constructed to justify the use of unexplained principles of activity in nature.

34 For earlier speculations in which the aether itself seems to maintain 'all bodies in their motion' see *Micrographia*, pp. 13 – 4. Here Hooke seeks to explain the ability of air to undergo extreme rarefaction and condensation by analogy with a 'tincture'; the air is a 'solution of terrestrial and aqueous particles dissolv'd into' the aether. The 'agil Aether', therefore, keeps the air 'dissolv'd and agitated' but it is not clear what makes the aether agile. In his 'Discourse of the Nature of Comets', for example, he distinguishes between an 'almost fluid aether' and a 'quite fluid aether'. The 'agil aether' of the *Micrographia* must have an 'almost fluid' nature, being composed of little particles. The 'quite fluid aether' by contrast, 'hath no one Part that may be called a Solid, but its parts are infinitely or indefinitely fluid' (*Posthumous Works*, p. 165) The meaning of this is obscure but Hooke seems to be implying that the 'quite fluid aether' is a continuum which is, by its very nature, in a constant state of agitation or (which seems to mean the same thing) fluidity.

35 The importance of Hooke's methodological pronouncements has been discussed in Hesse, 'Hooke's Development of Bacon's Method'; idem, 'Hooke's Philosophical Algebra'; Oldroyd, 'Hooke's Methodology'; idem, 'Some Writings of Robert Hooke on Procedures for the Prosecution of Scientific Enquiry'; Hunter and Wood, 'Towards Solomon's House'; and Hunter, 'Science, Technology and Patronage', *Establishing the New Science*. On the rhetorical import of methodological claims see John Schuster, 'Methodologies as Mythic Structures: A Preface to the Future Historiography of Method', *Metascience* 1/2 (1984) 15 – 36; and idem and R. R. Yeo, 'Introduction', in Schuster and Yeo, eds., *The Politics and Rhetoric of Scientific Method: Historical Studies* (Dordrecht, 1986) pp. ix – xxxvii.

1.3. Hooke's Method: Experimentalism and Active Principles

One of the first things Hooke does after introducing his concept of Congruity and Incongruity is to insist that his ideas are all firmly grounded on experimental evidence. Not just on a few, narrow and obscure, experiments but on

> divers kinds of Experiment, and all and every one of those Trials, I make the Standards or Touchstones, by which I try all my former Notions, whether they hold out in weight, and measure, and touch, &c.[36]

Surprising though it may seem, Hooke even manages to contrive an experiment to show the separation of sand and fine gravel solely by the agency of vibrating. He simply places a mixture of the two on a 'nimbly beaten' drum-head and 'hey presto' they separate out. Later, when he attributes light to the insensible vibrations of the luminous body, he outlines the experimental evidence for this, and concludes that, thereby, these insensible motions are made 'very manifest'. Similarly, he supports his speculations about the internal sub-microscopic structure of various bodies by claiming to have demonstrated them *ad oculem* with a company of bullets:

> so that there was not any regular Figure . . . that I could not with the composition of bullets or globules, . . . imitate, even almost by shaking them together.[37]

Even his brief speculations about the existence of a universally incongruous aether (to explain gravity) are accompanied with the claim: 'that there is some such fluid, I could produce many Experiments and Reasons, that do seem to prove it'. The same could be said, evidently, for all the different kinds of aether which Hooke invoked. In *Remarks about Comets* (1678), for example, Hooke partially developed a rather more complex notion of aether but he defended it in precisely the same terms:

> though the supposition even of the Aether, may seem to be a Chimera and groundless; yet had I now time, I could by many very sensible and undeniable experiments, prove the existence and reality thereof.[38]

The important thing to note here is that Hooke claims to be able to use the experimental method, if not always to make insensible phenomena manifest, at least to make them plausible, and perhaps even probable. On every

36 Hooke, *Micrographia*, p. 28.
37 Hooke, *Micrographia*, pp. 12, 15, 54, 85.
38 Hooke, *Micrographia*, p. 22; Gunther, *Early Science in Oxford*, viii, 269; Hooke, *Posthumous Works*, p. 165.

occasion in his work when he has recourse to a hypothetical concept beyond the reach of the senses, Hooke interjects a claim that he can make the existence of this concept probable by analogy with experimentally amenable phenomena. And, the reverse side of this intellectual coin, whenever he discusses the correct method of science, he always emphasises the usefulness of experimentation for establishing the behaviour of 'secret', 'subtle', 'abstruse', and 'hidden' principles of nature. Consider, for example, Hooke's analogy between the universal active principle, light, and other putative cosmic principles, which he develops in the *Lectures on Light* (c. 1680):

> possibly there may be many other Motions and Operations of Bodies at a distance, and several other ways by which the Bodies of the World may influence one another, though it has pleased God not to give us Organs or Senses to discover them, and thereby many things that are accounted Sympathetick or Magical may be done by Natural Causes and Powers of which we have no Organs to make us sensible. Now from divers late Discoveries about the influencing Powers of Gravity and Magnetism, . . . of which we have no Sense that does immediately inform us; but we become knowing and assur'd of them by other means . . .[39]

The trick, of course, is to study the effects of these hidden powers and to draw an analogy between them and sensible powers. In order to understand 'Causes, Principles, and Operations . . . far removed from the reach of our Senses', therefore,

> the best and utmost we can do towards the discovery of them, is only accurately to observe and examine all those Effects produced by them, which fall within the Power of our Senses, and comparing them with like Effects produced by Causes that fall within the reach of our Senses . . . and so from Sensibles to argue the Similitude of the nature of Causes that are wholly insensible.[40]

In Hooke's *General Scheme, or Idea of the Present State of Natural Philosophy*, he suggested that one of the defects of all previous natural philosophy was that its practitioners did not try to discover the true nature of bodies, to whit: 'what the inward texture and Constitution of them is, and what the internal Motions, Powers and Energies are'. The trouble is, Hooke went on, that the favoured method, relying solely on the use of logic or reason is 'wholly deficient' for discovering the kinds of secret and subtile Actors, and what the abstruse and hidden Instruments and Engines there made use of, may be'. The remedy is to develop what Hooke calls a Sixth Sense which will enable us to discover all those

39 Hooke, *Posthumous Works*, p. 79.
40 Hooke, *Posthumous Works*, p. 165.

> Effects of Nature as seem to be more secret and reserv'd working on Bodies remov'd at some distance, such strange Effects as our Senses are wholly unable to show us any probable cause thereof.

The way to develop this sixth sense, of course, is to follow Hooke's experimental method.[41]

What follows from Hooke's belief 'in the Similitude of the nature of Cause' — a notion which seems indistinguishable from Newton's concept, similarly deployed, of the 'Analogy of Nature'— is that acceptable hypotheses about occult or insensible powers must all be

> uniform and similar to Powers, Operations, Effects or Motions, which are within our reach and command, which we daily try, see, and find the regular working of.

Even a superficial glance at Hooke's work reveals that he practised what he preached. All his experimental investigations of vibrations and oscillations in springs and pendulums can be seen as attempts to refine his system of incessantly vibrating matter, and so enabling him to say, as he did in his *Discourse of the Nature of Comets*, 'what I here suppose to be in all Bodies, I can prove to be in some, and so suppose nothing absurd or impossible'.[42]

I said above that Hooke did not need to be concerned about objections to mechanistic accounts of sympathetic resonance. We can now see why. Even if Hooke were to accept the impossibility of conceiving how the motions of a mechanical aether could transfer the vibrations of one musical string to a distant heavy string while not causing the least tremor in a dangling piece of the lightest cotton or a near-by feather, it would not affect his case. The important thing for Hooke was that sympathetic resonance could consistently be shown to occur by clear and simple demonstrations. In supposing that all bodies had a particular tone, therefore, which readily resonated with bodies of the same or harmonic tones he was proposing nothing impossible, even if it was inexplicable. Sympathetic resonance, no matter how occult it might be, still came into the category of 'powers . . . which we daily try, see, and find the regular working of'. The same could be said, of course, of the other occult qualities which Hooke considered in his work: gravity, magnetism, spring, spontaneous generation, and so on.

41 Hooke, *Posthumous Works*, pp. 3, 46, see also p. 61.

42 Hooke, *Posthumous Works*, p. 179. On Newton's methodological stratagem see J. E. McGuire, 'Atoms and the "Analogy of Nature": Newton's Rules of Philosophizing', *Studies in the History and Philosophy of Science* 1 (1970) 3 – 58. For further examples of Hooke's efforts to establish the existence and attributes of active principles by experimental means see Schaffer, 'Godly Men and Mechanical Philosophers' (n. 11). For more general surveys of the links between the perceived need to account for and deal with occult qualities and the experimental method see Henry, 'Occult Qualities and the Experimental Philosophy' (n. 6) and Ron Millen, 'The Manifestation of Occult Qualities in the Scientific Revolution', in M. J. Osler and P. L. Farber, eds., *Religion Science, and Worldview: Essays in Honor of R. S. Westfall* (Cambridge, 1985), 185 – 216, especially pp. 201 – 16.

Of course, not all Hooke's hypotheses provided such extensive fields for experimental investigation. His proposal that gravitational attraction might be caused by a radiating shock wave sent out from a pulsating body was not easy to establish. One of the few ways in which he could make this 'intelligible' was 'an Observation very commonly known amongst Tradesmen', in which a loose hammer or axe head can be driven more firmly on to the helve by striking the opposite end of the helve from the head. The outward pulse along the handle does not make the axe-head fly off, but brings it closer to the initial point of impact, just as in Hooke's model of gravity.[43] Hooke's inability to come up with anything else that showed the same effect led him to use that old rhetorical trick, making a virtue of a necessity:

> some one plain but pertinent Experiment [he wrote], apply'd with judgment, may be more significant than thousands of such as are pompous, amusing, and excite Admiration. And I am satisfied that more Discoveries in Nature may be made by the most plain, obvious and trivial Experiments to be everywhere met with, than by the far-fetcht and dear-bought Experiments which some seek after.[44]

Talk is cheap, but I suppose if any one earned the right to say this, it was Hooke.

We can see now, therefore, why for Hooke the real natural philosophy, as he made clear in the Preface to *Micrographia*, was not simply 'mechanical' but also 'experimental'. A mechanical philosophy which tried to proceed independently of experimental investigation could say nothing about the hidden and 'secret workings of Nature'. The 'inlargement of the dominion of the Senses' which the *Micrographia* did so much to make possible was not confined to magnification. It also pointed the way towards developing the 'Sixth Sense' which Hooke believed was necessary to reveal the causes of 'those effects of Bodies which have been commonly attributed to Qualities . . . confess'd to be occult'.[45] It can hardly be maintained that Hooke's references to active principles in nature are merely loose talk in view of the fact that his sophisticated methodology, his 'philosophick Algebra' as he called it, was largely aimed at defending the use of such active principles.

One of the major claims of the mechanical philosophy was that it could give intelligible accounts even of those phenomena which Scholastics simply referred to as 'occult'. Descartes and Hobbes believed they could do this simply by extending their metaphor of continuous impacts between particles of matter in inertial motions. For Hooke, and other English philosophers, this could not work. There were too many instances where the phenomena eluded such crude attempts to explain them. The answer was to allow the existence of principles of activity. But these principles differed from the old scholastic

43 Hooke, *Posthumous Works*, p. 185.
44 Hooke, *Posthumous Works*, p. 184.
45 Hooke, *Micrographia* sig. g1.

notions, like that of the dormitive power of opium, in so far as their behaviour, their precise way of working, could be investigated and established by experimental procedures; the reality of them could be made to hold out, as Hooke put it, 'in weight, and measure, and touch'. Admittedly, the cause of these active principles might have to remain inscrutable. Man's senses and reason are limited and ultimately he may simply have to admit, as Hooke himself did,

> That they are what they are; Powers created by the Omnipotent to be what they are, and to operate as they do.[46]

2. *Congruity*

Robert Hooke must surely rank high on every historian's list of the major mechanical philosophers of the seventeenth century.[47] The point of the foregoing is not to suggest that Hooke must no longer be seen as a mechanical philosopher. Rather, it is intended to show that, if we wish to understand what Hooke meant when he wrote of 'the *real*, the *mechanical*, the *experimental* Philosophy', we must refine our conception of what constituted the mechanical philosophy. Sucked in by the vortex of Cartesian studies, historians of science have tended to see the mechanical philosophy in terms of Descartes' own claims about his new philosophical system.[48] According to the standard characterization of the mechanical philosophy, matter was always held to be passive, and moved only as a result of impact with other inertially moving particles of matter, and all unexplained or occult properties of matter were banished from discourse. The conclusion that Hooke's matter theory cannot properly be comprehended in such simplistic terms should not lead us to suppose that he was not a mechanical philosopher — for if he was not, who was? — but should alert us to the possibility that there are more things in the mechanical philosophy than are dreamt of in previous historiographies.[49]

46 Hooke, *Posthumous Works*, p. 173, see also p. 165.

47 R. S. Westfall has said that he would not wish to include Hooke in the very first rank alongside Kepler, Galileo and Newton, but is willing to concede that he is a figure of 'major importance' in the history of science. See R. S. Westfall, 'Robert Hooke, Mechanical Technology and Scientific Investigation', p. 89.

48 Descartes' own account of his system may well sound more strictly mechanical than it was in practice. See note 21 above.

49 Hooke, *Micrographia* sig. a2r.

2.1. Mechanism and Magic: Occult Qualities and the Natural Magic Tradition

The significance of occult qualities in the development of early modern science is currently being reassessed. In a pioneering article Keith Hutchison has shown that, contrary to what has formerly been supposed, the mechanical philosophers did not eschew occult qualities from their worldview but deliberately embraced them as one means of showing the superior heuristic possibilities of the new philosophy. Ron Millen has endorsed Hutchison's thesis while arguing that the beginnings of the reassessment of occult qualities can be traced back to a number of Renaissance Aristotelian or Galenic thinkers.[50] The story that is sketched-out is one in which natural philosophers became increasingly dissatisfied with Aristotelian notions of what constituted *scientia* — true scientific knowledge. *Scientia* required a knowledge of causes, but Aristotelian epistemology was sensationalist. The only causes which could be claimed to be certainly true, therefore, were those that derived from the sensible qualities of heat, cold, wetness and dryness. The inevitable result, particularly in the medical tradition but increasingly in natural philosophy, was a proliferation of so-called occult qualities, invoked to 'explain' the behaviour of natural objects which could not be attributed to the four sensible qualities. This set in train the shift away from the emphasis on sensibilist epistemology which is famously illustrated in Descartes' opening words to *Le Monde*:

> . . . the first point I want to draw to your attention is that there may be a difference between the sensation we have of light (ie. the idea of light which is formed in our imagination by the medium of our eyes) and what it is in the objects that produces this sensation within us (i.e. what it is in a flame or the sun that we call by the name 'light'). For although everyone is commonly convinced that the ideas we have in our mind are wholly similar to the objects from which they proceed, nevertheless I cannot see any reason which assures us that this is so. On the contrary, I note many observations which should make us doubt it.[51]

The fact that the corpuscular philosophy aims to explain all physical phenomena in terms of the interactions of invisible and insensible particles of matter suggests, as Hutchison has pointed out, that the mechanical philosophy

50 Keith Hutchison, 'What Happened to Occult Qualities in the Scientific Revolution?', *Isis*, 73 (1982), 233 – 53; Millen, 'The Manifestation of Occult Qualities' (n. 42). Both authors claim that the use of occult powers in scientific explanation is now taken for granted (Hutchison, p. 235; Millen, p. 216). For an excellent discussion of this see Rom Harré, 'Powers', *British Journal for the Philosophy of Science*, 21 (1970) 81 – 101.

51 C. Adam and P. Tannery, eds., *Oeuvres de Descartes*, 12 vols. (Paris, 1897 – 1913; reprinted, Paris, 1957 – 8), xi, 3. Quoted from J. Cottingham, R. Stoothoff, D. Murdoch, eds., *The Philosophical Writings of Descartes*, 2 vols. (Cambridge, 1985), i, 81.

rejected the manifest qualities of Aristotelianism (as mere illusions) while accepting the occult qualities as a valid part of *scientia*.[52]

Both Hutchison and Millen are quick to point out that this new historiographical perspective has important implications for our assessment of the role of magical traditions in the formation of early modern science. Hitherto even the most forceful scholarly accounts of the continued presence of magical assumptions, attitudes and tropes in the works of seventeenth-century natural philosophers, have merely led to a consensus that the magical and the mechanical worldviews or 'mentalities' persisted side by side for a time before the magical tradition finally gave place to modern (mechanistic) science. It should now be recognised, however, that the new philosophers' seemingly paradoxical adherence to earlier occult modes of thought stemmed from the fact that their 'new' philosophy grew out of the natural magic tradition.[53]

Many of the characterising features of the new philosophy can now be seen to reflect the outlook of the Renaissance magus. In particular, the empiricist methodology and the utilitarian impulse of the new science have been convincingly shown, by a number of historians, to derive from attitudes which were commonplace among magicians. What had always distinguished magic from the academic subject of natural philosophy was its preoccupation with the practical benefits to be gained from an understanding of the powers and forces of nature. The magician was not concerned, as was the Aristotelian philosopher, with *scientia* or knowledge of causes for its own sake. Being satisfied with a pragmatic knowledge of how one object interacted with another, the magician was not overly concerned with the precise causal chain, or mechanism, by which objects interacted with one another. The important point, for the practical magician, was the ability to reproduce the same effects. The upshot was a far greater emphasis on empiricism in the magical tradition than would have been appropriate for academic Aristotelian philosophy. Furthermore, this empiricism, unlike Scholasticism, easily accommodated the notion of occult powers and qualities. The magician was

52 Hutchison, 'What Happened to Occult Qualities?' (n. 50) pp. 242 – 3; Millen, 'The Manifestation of Occult Qualities' (n. 42) pp. 198, 202, 205. It is perhaps worth remarking that the distinction between our sensations and the way things really are, as exemplified in the quotation from Descartes, was always a fundamental assumption of the magical tradition. See Georg Luck, *Arcana Mundi Magic and the Occult in the Greek and Roman Worlds* (Baltimore, 1985) pp. 8 – 9; D. P.Walker, *Spiritual and Demonic Magic from Ficino to Campanella* (London, 1958) pp. 76 – 7; and Bert Hansen, 'The Complementarity of Science and Magic before the Scientific Revolution', *American Scientist*, 74 (1986) 128 – 36, pp. 130 – 1. For a very fine literary illustration of this aspect of magic consider 'The Franklin's Tale' in Chaucer's *Canterbury Tales*.

53 Hutchison, 'What Happened to Occult Qualities? (n. 50) pp. 249-50; Millen, 'The Manifestation of Occult Qualities' (n. 42) p. 185.

able to defend his claims about the existence of occult powers at work in the universe simply by pointing to his experiential results.[54]

For the most part the new philosophy was formulated and propagated by thinkers who not only worked outside the confines of the university system but also took pains to eschew the fundamentals of Aristotelian natural philosophy. A characteristic boast of the early mechanical philosophers was that their philosophy was more easily intelligible than the Aristotelian philosophy, and yet the new philosophy aimed to replace a sensibilist account of natural phenomena with one which was based on the occult behaviour of invisible and insensible corpuscles. The only way this rhetorical strategy could succeed was to concentrate attention on the occult qualities and insist that their philosophy made better sense of these than Aristotelianism. Increasingly, therefore, the mechanical philosophers, particularly in England, adopted the same empiricist methodolgy as the natural magicians.[55]

The utilitarian impulse of the new philosophy derived most directly, perhaps, from the forceful advocacy of Francis Bacon, but Paolo Rossi has shown the influence of the natural magic tradition on this aspect of Bacon's thought.[56] The premiss of this utilitarianism was that artificial devices and machines could be made to perform seemingly extraordinary feats merely by exploiting the laws of nature in a special way. A corollary of this was that study of the operations of such machines could lead to an improved understanding of those same natural laws. Here again we are in the realms of natural magic. The traditional natural philosophy of the universities concerned itself only with the so-called natural motions: the rectilinear motion of the earthly elements and the circular motions of the heavenly spheres. If the 'unnatural' motions of mechanical devices were beyond the purview of Scholastic philosophy, they were the stock-in-trade of the natural magician. Strange as it may seem to our technological age, throughout the Middle Ages and the Renaissance the construction of 'miracle-working' machinery was the province of the magician. Moreover, it is important to note that such machinery was not simply deemed magical by the ignorant, because they were unaware of the internal mechanical workings of such machines. On the contrary, devices whose mechanisms were described and even illustrated in written sources were still designated as magical. The crucial point for these earlier thinkers was that any artificial means of harnessing natural laws was a magical, not a natural philosophical, enterprise.[57]

54 See Lynn Thorndike, *A History of Magic and Experimental Science*, 8 vols. (New York, 1923 – 58); Bert Hansen, 'Science and Magic', in David C. Lindberg, ed., *Science in the Middle Ages* (Chicago and London, 1978), 483 – 506; Paolo Rossi, *Francis Bacon: from Magic to Science*, translated by S. Rabinovitch (London, 1968).

55 On the rhetoric of 'intelligibility' see Hutchison, 'What Happened to Occult Qualities' (n. 50).

56 Rossi, *Francis Bacon* (n. 54).

57 A. G. Molland, 'Roger Bacon as a Magician', *Traditio*, 30 (1974) 445 – 60; William Eamon, 'Technology as Magic in the Late Middle Ages and the Renaissance', *Janus*, 70 (1983) 171 – 212; Bert Hansen, 'The Complementarity of Science and Magic' (n. 52); Peter French, *John Dee: The World of an Elizabethan Magus* (London, 1972), pp. 160 – 87.

It seems impossible to deny, therefore, that a fundamental initial premiss of the mechanical philosophy — that natural processes can be understood by analogy with the artificial operations of mechanical devices — derived from the natural magic tradition. The expansion of the scope of natural philosophy to include, for the first time, explanations of the workings of machinery testifies, surely, to an increased recognition of the validity and usefulness of the natural magicians' approach to the natural world and its laws. It should not be regarded, whiggishly, as the inexorable triumph of a 'scientific mentality'.[58]

As D. P. Walker has pointed out, the age-old tradition of natural magic effectively began to dissappear during the Renaissance, at precisely the moment when, thanks to the activites of a number of leading Renaissance 'Magi', it was achieving its acme. There is nothing mysterious about this disappearance. Major aspects of natural magic became absorbed into what had previously been academic natural philosophy, to form the hybrid which was the new philosophy. What made the new philosophy so different from previous natural philosophy, so much more 'modern' to our eyes, was the fact that it had appropriated many of the methods, aims, and assumptions of natural magic. Those other aspects of magic which were not appropriated in this way, spiritual and demonic magic, divination, the more religious side of alchemy, numerology and many other beliefs, remained as so many straw men to bear the brunt of anti-magical polemic.[59]

Magic had always had its detractors among the clergy and the academics, to say nothing of the superstitious masses, and what it gained from patronage at the various European courts it could often lose when political circumstances seemed to require it.[60] Our image of magic today is conditioned largely by attitudes to what was left after the most powerful part of the tradition was

58 One recent scholar has tried to see the development of the Scientific Revolution in terms of the (inevitable) triumph of the 'scientific' over the 'occult' mentality. See Brian Vickers, 'Introduction', in Vickers, ed., *Occult and Scientific Mentalities* (n. 6) pp. 1 – 55. For a corrective see Charles Webster, *From Paracelsus to Newton* (Cambridge 1982).

59 Walker, *Spiritual and Demonic Magic* (n. 52) pp. 75 – 84.

60 Such opposition led some noted magicians to deny that they were magicians. See, for example, Roger Bacon, 'Letter to William of Paris on the Secret Works of Art and Nature and on the Nullity of Magic', in *Opera quaedam hactenus inedita*, ed. by J. S. Brewer (London, 1859) and John Dee, *The Mathematicall Praeface to the Elements of Geometrie of Euclid Megara* (1570), edited by Allen G. Debus (New York, 1975). On patronage of magic and magicians by Royal Courts see, for examples, French, *John Dee* (n. 57); R. J. W. Evans, *Rudolf II and His World: A Study in Intellectual History 1576 – 1612* (Oxford, 1973); idem, *The Making of the Habsburg Monarchy 1550 – 1700* (Oxford, 1979); and David Goodman, *Power and Penury: Government, Technology and Science in Philip II's Spain* (Cambridge, 1988). The best example of the use of magical accusations for socio-political purposes is in Peter Partner, *The Murdered Magicians: The Templars and Their Myth* (Oxford, 1981).

absorbed into early modern science.[61] Among today's well-educated Westerners magic is nothing more than a body of entertaining or fraudulent tricks, or of ignorant absurdities associated with a fantastic and untenable worldview. But it was not always so. Through the Renaissance and well into the seventeenth-century, a natural magician like Agrippa, Paracelsus, Della Porta, or John Dee could be regarded as a supreme master of natural law and lore. Moreover, if modes of thought and concepts which we now consider to be 'magical' can be discerned in otherwise 'modern' mechanical philosophers, it should not be taken as a sign of momentary weakness and inability to shake off out-moded ways of thought. It should be regarded as all part of the process of taking what was useful from natural magic — the counterpart of the well-known rejection of supposedly moribund aspects of traditional natural philosophy — to form the 'new philosophy'.

2.2. Hooke and Magic: The Natural Philosopher's Image

Hooke's use of magical modes of thought have been pointed out before. In particular, attention has been focussed on Hooke's easy acceptance, long before the appearance of Newton's *Principia mathematica*, of attractive forces capable of acting between bodies at a distance from one another. J. A. Bennett has argued that the provenance of this idea is to be found in what he has called the 'magnetical philosophy'. Although this magnetical philosophy was taken up by Kepler, Bennett sees it as an essentially English tradition deriving from the animate and magical philosophy of William Gilbert, author of *De magnete*.[62] The 'animate', as opposed to 'mechanical', nature of heavenly bodies in this tradition gave rise to notions of 'spheres of activity' surrounding bodies and within which their influence, 'magnetical' or gravitational, could be felt. Significantly, Gilbert himself realised that the only way to achieve an understanding of the occult power of magnetism and the animate behaviour of magnets was through careful experimental study. Furthermore, the continued

61 The reader may think that, in calling natural magic 'the most powerful part of the tradition' of magic, I am guilty of whiggism: that I simply define anything which was absorbed into early modern science, ipso-facto as 'powerful'. This is not so. Natural magic was always the dominant form of magic from the Middle Ages through the Renaissance and into the early modern period. Evidence for this is afforded, for example, by Thorndike, *History of Magic* (n. 54), and J. Tester, *A History of Western Astrology* (Woodbridge, Suffolk, 1987). Furthermore, natural magic was held to be more powerful than demonic magic because demons had no supernatural power, merely a superior knowledge of natural laws. 'They [demons] have no advantage over us, but operate only by applying active things to passive, like as men do', wrote John Webster in *The Displaying of Supposed Witchcraft* (London, 1677) p. 224. It follows that the supreme magical art was not necromancy but natural magic. On this see Stuart Clark, 'The Scientific Status of Demonology', in Vickers, ed., *Occult and Scientific Mentalities* (n. 6), 351 – 74.

62 J. A. Bennett, 'Cosmology and the Magnetical Philosophy'; William Gilbert, *De magnete, magnetisque corporibus et de magno magnete tellure; Physiologia nova . . .* (London, 1600) translated by P. Fleury Mottelay (New York, 1958). Not only did Hooke freely discuss attraction at a distance before Newton, but the historical consensus is that Hooke actually suggested the idea to Newton. See Bennett p. 176; R. S. Westfall, *Force in Newton's Physics*, pp. 210, 272, 426 – 7.

influence of the magnetical philosophy was made possible, Bennett suggests, by the collaborative experimentalism of groups like that centred on Gresham College. Experiments could demonstrate beyond any doubt that there is an attractive force between iron and the loadstone, even if they could say nothing about the causal nature of that force. The conclusions of such investigations could then be extended, by analogy, to an analysis of the behaviour of heavenly bodies.[63]

Hooke's appeals to magnetism as an active principle, and his experimentalist defence of the use of such an occult quality in his natural philosophy, is undeniable but, as we have seen above, it is far from being the only occult active principle in his system. Besides, Gilbert's initial work on magnetism can itself be seen as a special case of the Neoplatonist worldview of the Renaissance magus. It seems necessary, therefore, to extend Bennett's claims about the tradition of magnetical philosophy. The 'alternative' tradition to mechanicism, which Hooke drew upon and developed in his natural philosophy, was not founded *ex nihilo* by Gilbert and confined to speculations about magnetism and gravity. It was not, in short, a magnetical tradition but a magical one.[64]

R. S. Westfall's studies of the antecedents of Newton's theory of gravitational attraction have led him too to see Hooke as a thinker in the magical or 'Hermetic' tradition.[65] Hooke played a crucial role in the development of modern dynamics by switching the problem of the orbital motions of planets. Descartes presented the problem as one of balancing a centripetal and a centrifugal force and it is evident that Newton accepted this formulation of the problem until Hooke gave him an alternative view in December 1679. As is well-known, Hooke suggested that the elliptical orbit of the planets would result when a tangentially, and inertially, moving planet was deflected by an appropriate attractive force acting towards the centre. The appropriate force, furthermore, would obey the inverse-square law. For all this, Hooke can still only be said to have pointed the way towards a satisfactory dynamics of circular motion. Moreover, he cannot be said to have anticipated Newton's theory of universal gravitation. Hooke seems to

63 Bennett, 'Cosmology and the Magnetical Philosophy', pp. 165 – 8 (on Gilbert) pp. 167 – 8 (on empirical continuation of Gilbertian tradition). On the analogy between magnetism and gravity see pp. 172 – 3 and 174.

64 Bennett writes of an 'alternative' or 'different' tradition, Bennett, 'Cosmology and the Magnetical Philosophy', pp. 167, 172, 175. I have tried to deal with the kind of issues picked out by Bennett as characteristic of 'magnetical philosophy' and demonstrate their broader significance in seventeenth-century English natural philosophy in Henry, 'Occult Qualities and the Experimental Philosophy' (n. 6).

65 Westfall, *Force in Newton's Physics*, pp. 210, 269 – 70, 425 – 7. The designation 'Hermetic', where 'magical' would perhaps be better, is a legacy of the remarkably influential work of Dame Frances Yates; especially her *Giordano Bruno and the Hermetic Tradition* (London, 1964). The tendency has been to see Hermeticism as the broad category into which magic is subsumed instead of the other way round. For correctives see Charles B. Schmitt, 'Reappraisals in Renaissance Science', *History of Science* 16 (1978), 200 – 214; and Nicholas H. Clulee, 'At the Crossroads of Magic and Science: John Dee's Archemastrie', in Vickers, ed., *Occult and Scientific Mentalities* (n. 6), pp. 57 – 71.

have considered gravity to be an attraction of similars: a notion that depends heavily upon his belief in the cosmic significance of congruities and incongruities. According to Westfall, therefore, Hooke's theory of gravitation was not universal but specific to different kinds of matter and depended on 'the conception, Hermetic but not uniquely Hermetic, of the attraction of like for like'.[66]

If Hooke's acceptance of attraction at a distance and his repeated use of occult active principles like gravity, magnetism, light and incessant vibrations are the most prominent examples of his use of magical traditions of thought, there are a number of lesser but still significant indications of the same. For example, it is worth comparing the opening words of Hooke's 'General Scheme, or Idea of the Present State of Natural Philosophy' with some well-known pronouncements of representative Renaissance magi. This is what Hooke writes:

> The Business of Philosophy is to find out a perfect Knowledge of the Nature and Proprieties of Bodies, and of the Causes of Natural Productions, and this Knowledge is not barely acquir'd for it self, but in order to the inabling a Man to understand how by the joyning of fit Agents to Patients according to the Orders, Laws, Times and Methods of Nature, he may be able to produce and bring to pass such Effects, as may very much conduce to his well being in this World, both for *satisfying his Desires*, and the relieving of his *Necessities* . . .[67]

By now we hardly need to remark upon the utilitarian ethos expressed in this passage, but what might not stand out is the precise method by which the promised perfect knowledge and practical improvements are to be brought about. In fact, everything is to be achieved by the 'joyning of fit Agents to Patients'. Cornelius Agrippa's description of natural magic in *De incertitudine et vanitate scientiarum* makes the art dependent on 'uniting actives to passives'; magic can only succeed when 'inferior and superior things are joined by an interchanging application of each to each'. Giambattista della Porta, author of the compendious *Magia naturalis*, sums up the work of the magician as 'the mutual and fit applying of one thing to another': 'strange works' are achieved 'by the agreement and disagreement of things, either so to sunder them, or else to lay them so together'. Similarly, for Pietro Pomponazzi the magician achieves his effects by 'applying the active to the passive'. Finally, when Francis Bacon insisted that magic should be 'restored to its ancient and honourable meaning', namely 'sublime wisdom', he described it in the same terms:

66 Westfall, *Force in Newton's Physics*, pp. 269 – 72, 457. The quotation is from p. 267 and is said in connection with Roberval's views but Hooke's opinion is said to be 'essentially similar' on p. 268. See also Westfall, 'Newton and the Hermetic Tradition' (n. 21), p. 194.
67 Hooke, *Posthumous Works*, p. 3, Hooke's emphasis.

> I . . . understand it as the science which applies the knowledge of hidden forms to the productions of wonderful operations; and by uniting (as they say) actives with passives, displays the wonderful works of nature.[68]

In the folklore of magic the power of flight always ranked high as a mark of the true adept. Daedalus was regarded as one of the greatest of ancient magicians because he was able to construct an artificial means of flying. Roger Bacon, the most renowned English 'magician' wrote of the possibility of constructing 'instruments by which to fly, so that one sitting in the instrument, and turning about [a device] by which the wings being artificially produced may beat the air after the manner of a flying bird'.[69] Bearing this in mind, we might ask ourselves what kind of public image Hooke conjured up when he claimed to have devised thirty different ways of flying while he was at Westminster School. Such boasts were not confined to his childhood. He made a number of claims about his ability to fly to various colleagues during the 1670s. On October 8, 1674, for example, Hooke recorded in his Diary: 'At Councell at Arundell told Sir Robert Southwell that I could fly, not how'. Unfortunately we do not know how Southwell responded but when Hooke 'told Wild and Aubery of flying' (December 18, 1675), Wild was 'cold'. Christopher Wren, by contrast, 'liked it' when he learned of Hooke's schemes for a 'flying chariott by horses'. It is important to note that Hooke's means of flying were always, as far as we can tell, dependent on his technological wizardry, his mastery of mechanics. The ability to fly was not gained by smearing oneself with the fat of a 'new murder'd babe', but by harnessing the power of horses, 'by vanes', or by 'bending Springs by gunpowder for flying'. Even so, as we saw earlier, such machinery had hitherto been regarded as part of the magician's art and was only slowly beginning to be seen as a legitimate part of natural philosophy.[70]

Hooke's interest in the psychological and physiological effects of music has recently been shown to be an important element in his natural philosophy. Providing an obvious source for the experimental investigation of sympathy

68 Cornelius Agrippa of Nettesheim, *De incertitudine et vanitate omnium scientiarum et artium* (n.p., 1531), ch. 42. John Baptista Porta, *Natural Magick* (London, 1658), pp. 1 – 2. Pietro Pomponazzi, *Les causes des merveilles de la nature ou les enchantements* (Basle, 1556), reprinted, ed. by Henri Busson (Paris, 1930), p. 124; see Millen, 'The Manifestation of Occult Qualities' (n. 42), p. 193. Francis Bacon, *Of the Advancement of Learning* in J. Spedding, R. L. Ellis, and D. D. Heath, eds., *The Works of Francis Bacon*, 9 vols. (London, 1857 – 1862) iv, pp. 366 – 7 (Bk III, ch. 5). See also the quotation from John Webster in note 61 above.

69 Roger Bacon, 'Letter to William of Paris' (n. 60), quoted from Hansen, 'Science and Magic' (n. 54), p. 486. See also Molland, 'Roger Bacon as Magician' (n. 57).

70 'Espinasse, *Robert Hooke*, p. 3 (see also pp. 106 – 9, 116 – 7); Hooke, *Diary. 1672 – 80*, pp. 125, 202, 233, 146, 124. It should be perfectly clear from the above quotation from Bacon (see previous note) and the well-known story of Daedalus that the magical power to fly was known to depend on mechanical devices. See also Eamon, 'Technology as Magic' (n. 57). While such notions became absorbed into the mechanical philosophy, the rather different tradition of witches flying by means of some totally occult power were rapidly being consigned to the realm of folk- tale and fairy story. See Barbara J. Shapiro, *Probability and Certainty in Seventeenth-Century England: A Study of the Relationships between Natural Science, Religion, History, Law, and Literature* (Princeton, 1983), pp. 194 – 226.

and antipathy (or congruity and incongruity) and action at a distance, music had always been redolent of magic. As Vitruvius had written, the musician must be a magician, for the sympathies and antipathies of sounds had profound thaumaturgical effects.[71] Even something as seemingly idiosyncratic as Hooke's secretiveness may reflect an approach inspired, at least partially, by the traditions of magic. The belief that if something is worth knowing it is also worth keeping it secret was virtually axiomatic among magicians. Hooke himself can be seen advocating secrecy about new discoveries in his proposals for reform of the Royal Society written in the 1670s, but he also assumed that earlier writers must have consigned profound truths to cryptic writings. For example, according to Hooke, the ancients,

> that they might the better conceale their Knowledge, and keep it to themselves, and abscond it from the Vulgar, and such as were not initiated and admitted into their Fraternities, they had contrived and digested [their wisdom] into fabulous Stories, which, as they might serve to amuse and awe the Vulgar by the Daemonology they had thereupon superstructed, so they might serve to instruct and inform the Adepti, or such as were admitted to the true interpretation and understanding of what they knew, of the real History that was concealed thereby, as also of their Philosophical or Physical Hypothesis for the Explication and Solution thereof.[72]

It is perhaps significant, also, that the subject matter of Hooke's *Cutlerian Lectures* was not the history of 'Trades', as Sir John Cutler had wished, but the 'History of Nature and Art'. Hooke's defence of this approach to 'usefull Philosophy' took for granted the magical assumption that artificial processes obey and reveal natural laws. Michael Hunter's analysis of the background to the Cutlerian Lectures has revealed Hooke's reluctance to concentrate on (perhaps even his distaste for) the 'operative part' of 'mechanicks'. Even if we make allowance for the rhetorical aspects of Hooke's pronouncements in this context, we can still see his concern for the 'perpetuall & universal knowledge' which is supposed to underwrite mechanical operations. The balance between the natural and artificial which Hunter sees Hooke trying to strike (for the benefit of both kinds of knowledge) is entirely typical of the natural magic tradition. Hooke perhaps recognised that not much had really changed since

71 Vitruvius, *De architectura*, Bk. 1, Sect. 3, quoted from Gouk, 'The Role of Acoustics and Music Theory', p. 594; see also pp. 587, 591. For a further consideration of the role of music in Hooke's cosmology see Kassler and Oldroyd, 'Robert Hooke's Trinity College "Music Scripts" '. For general accounts of the links between music and magic see Walker, *Spiritual and Demonic Magic* (n. 52), pp. 3 – 29.

72 Hooke, *Posthumous Works*, p. 394; the same general approach is heavily used pp. 371 – 416. On Hooke's secretiveness see Hunter and Wood, 'Towards Solomon's House', pp. 73 – 4, and Steven Shapin's article in this volume. On Hooke's belief in cryptic writings of the ancients see Gouk, 'The Role of Acoustics and Music Theory', p. 595; Rossi, *Dark Abyss of Time*, pp. 16 – 17. For the links between such beliefs and magical traditions see J. E. McGuire and P. M. Rattansi 'Newton and the "Pipes of Pan" ', *NRRS*, 21 (1966) 108 – 43; and D. P. Walker, *The Ancient Theology: Studies in Christian Platonism from the Fifteenth to the Eighteenth Century* (London, 1972).

Francis Bacon had written that 'it is esteemed a kind of dishonour upon learning for learned men to descend to inquiry or meditation upon matters mechanical; except they be such as may be thought secrets of art, or rarities and special subtleties'.[73]

It need not be supposed that Hooke was self-conscious about the similarities between his own assumptions and approaches and those of magicians. The historical process of absorbing or converting natural magic into natural philosophy can be traced back to the Renaissance (and in the case of certain individuals, like Roger Bacon, back to the Middle Ages). By Robert Hooke's day, therefore, certain aspects of the magical tradition and of what had formerly been the magician's 'image' were already beeing seen as part of the role of 'natural philosopher'. Steven Shapin has superbly shown just how difficult it was for Hooke's contemporaries to comprehend what kind of man the new philosopher should be. If, as Shapin rightly says, Hooke could not play the part of the Christian Virtuoso, he could perhaps appear in the guise which had formerly belonged to the natural magician. For one thing, part of the image of the magician was that he was a retained man. Court astrologers, alchemists and other kinds of magus were familiar figures in people's conception of the highest levels of their society. Moreover, Hooke's contemporaries in the Royal Society would have the more specific exemplars of the ninth Earl of Northumberland's 'three magi'. Of these three magi, Thomas Harriot had a particularly high reputation in the Restoration as a learned mathematician, alchemist and natural philosopher but he was well-known to have been Sir Walter Ralegh's and subsequently Henry Percy's pensioner.[74]

Hooke's positive attitude towards the traditions of natural magic is clearly brought out in his Cutlerian Lecture 'Of Dr. Dee's Book of Spirits'. In this lecture Hooke argued that John Dee's 'Treatise about Supernatural Effects or Productions', his *True and Faithful Relation of What Passed for Many Yeers Between Dr. John Dee and Some Spirits*, 'may be as properly referred to the Improvement of Natural Knowledge to understand it, as of any other Book that has plainly and expressly treated of the History of Nature and Art'. Although this notorious book ostensibly consisted of conversations between the Elizabethan magus and various angels or demons, and seemed to be 'a Rhapsody of

73 Hooke quoted from Michael Hunter's article 'Science, Technology and Patronage' (n. 35). Francis Bacon, *Of the Advancement of Learning* Bk. II ch. 2, in Spedding, Ellis, Heath, eds., *Works* (n. 68) iv, p. 296. For suggestive indications of the similarities between Hooke's Cutlerian Lectures and the magical tradition compare the contents of *Lampas* (n. 9) with Eamon's account of Arab and Medieval Western hydraulic devices inspired by the treatises of Philo and Hero, Eamon, 'Technology as Magic' (n. 57), pp. 175 – 6. Note also that when Roger Bacon wished to deny that he did anything by the aid of demons or supernatural means, he wrote of the 'Secret Works of Art and Nature', Bacon, 'Letter to William of Paris' (n. 60).

74 See Steven Shapin's article in this volume. On Court magicians see the works by Evans and Goodman cited in n. 60. On the 'Wizard' Earl of Nurthumberland's 'three magi', and Thomas Harriot, see John W. Shirley, *Thomas Harriot: A Biography* (Oxford, 1983) and G. R. Batho, *Thomas Harriot and the Northumberland Household,* Durham Thomas Harriot Seminar Occasional Paper No. 1 (London, 1983).

incoherent and unintelligible Whimsies of Prayers and Praises, Invocations and Apparitions of Spirits, strange Characters, uncouth and unintelligible Names, Words and Sentences, and Relations of incredible Occurrences', Hooke deemed it worthy of discussion before the Royal Society.[75] Meric Casaubon, the editor of this demonological work, believed Dee 'to be a Conjurer or Witch and to have dealt with the Devil all along', but Hooke, impressed by the scope of Dee's writings, saw him as 'an extraordinary Man, both for learning, Ingenuity and Industry'. Quoting from Elias Ashmole's *Theatrum Chemicum Britannicum* Hooke tried to convince his audience of Dee's genius:

> he chiefly bent his Studies to the Mathematicks, in all parts of which he was an absolute and perfect Master; witness his Mathematical Preface to *Euclid's Elements* wherein are enumerated many Arts of him wholly invented (by Name, Definition, Property and Use) more than either the Grecian or Roman Mathematicians have left to our Knowledge . . . His great Ability in Astrology, and the more secret Parts of Learning (to which he had a strong Propensity and unwearied Phansy) drew from the Envious and Vulgar, many rash, hard, and lying Scandals upon his most honest and justifiable Philosophical Studies . . .[76]

Hooke's admiration for Dee's work and his contempt for the misconceived reactions of the vulgar led him to reassess Dee's demonolgy and to find in it a 'concealed History of Nature and Art'. 'I do conceive', Hooke wrote, 'that the greatest part of the said Book, especially all that which relates to the Spirits and Apparitions, together with their Names, Speeches, Shews, Noises, Clothing, Actions, and the Prayers and Doxologies, &c. are all *Cryptography*'.[77]

If Dee was a magician, therefore, he was always a natural magician, one who achieved his effects by artificial means. Dee's claims to conjure up visions of spirits in his crystal ball were simply further attempts to use his knowledge to 'make a Fortune under the Emperor'. By his own account, Hooke said, Dee tried to make his fortune

75 'Of Dr. Dee's Book of Spirits', in Hooke, *Posthumous Works*, 203 – 10, p. 205. John Dee, *A True and Faithful Relation of What Passed for Many Yeers Between Dr. John Dee and Some Spirits* (London, 1659). On Dee see Peter French, *John Dee* (n. 57); Wayne Shumaker, *Renaissance Curiosa* (Binghamton, New York, 1982) pp. 15 – 52.

76 Hooke, *Posthumous Works*, pp. 203, 208. Elias Ashmole, *Theatrum Chemicum Britannicum* (London, 1652), Hooke gives the page reference as p. 480. Dee's 'Mathematicall Preface' was published in Sir Henry Billingsley's edition of *The Elements of Geometry of the most auncient Philosopher Euclide of Megara* (London, 1570).

77 Hooke, *Posthumous Works*, pp. 205, 206. Hooke draws an analogy (p. 207) between Dee's book and Johannes Trithemius, *Steganographia: Hoc est: Ars per occultam scripturam animi sui voluntatem absentibus aperiendi certa* (Frankfurt, 1606). This work was regarded by many as another notorious piece of demonology but Hooke was evidently one of those who recognised it as a work on cryptography. For a full exposition of Trithemius' cryptography see Shumaker, *Renaissance Curiosa* (n. 75) pp. 91 – 131. Dee's cryptography and Hooke's assessment of it are discussed also in Richard Deacon, *John Dee: Scientist, Geographer, Astrologer and Secret Agent to Elizabeth I* (London, 1968).

> by means of Chemistry, or Mathematicks, or Astrology, or Mechanicks, all which I find by his Writings he was well versed in, and especially in the Business of Opticks, and Perspective and Mechanick Contrivances; an effect of which I conceive his Chrystal or Angelical Stone, or Chrystallum Sacratum, as he terms it, to have been, for that it was of a considerable bigness and was placed upon a Pedestal or Table, which he calls a Holy Table, which might contain the Apparatus to make Apparitions, when he had a mind to be seen in it, as likewise to produce Noises and Voices, if there were occasion. All which might be done by Art, as has been shown, both formerly by Roger Bacon, and of late by the Echoing Head.[78]

Hooke's defence of Dee is a remarkable document; written in 1690 towards the end of Hooke's own career, it shows a real concern for the importance of natural magic in 'the Improving [of] Natural Knowledge'. It may be taken to be a clear and unequivocal indication, therefore, of Hooke's own belief, even in the 1690s, in the validity and usefulness of the ways of the natural magician. Hooke was well placed to recognise the similarities between his own work and that of a 'magus' like Dee, and it is hardly surprising that he should try to interpret all the work of so distinguished a predecessor, even his demonology, in terms which he found acceptable to himself. The continuity between the traditions of natural magic and the new philosophy of Robert Hooke seems hard to deny.[79]

Conclusion

If, as Lord Keynes said, Newton was 'the last of the magicians', are we to conclude that Hooke was the penultimate magician? The answer must surely be no. Historical labels cannot be dispensed with because they often serve a useful purpose. Problems arise, however, when the label obscures or distorts our understanding of the past. By calling Newton the last of the magicians, Keynes paradoxically succeeded in perpetuating the assumption that there were two distinct traditions in Western thought: the magical and the scientific. Even for Keynes himself, Newton had 'one foot in the Middle Ages and one foot treading a path for modern science'. This seems to imply that Newton was in danger of tearing himself apart. Similarly, Hugh Kearney, a scholar

78 Hooke, *Posthumous Works*, p. 206. The Emperor in question was Rudolf II, the Habsburg Emperor.

79 The date of this lecture is noted in Royal Society, Copy Journal Book, vol. 7, pp. 289 – 90. I am grateful to Michael Hunter for this reference. A number of the papers in Hooke's *Philosophical Experiments and Observations* also show marked similarities with earlier natural magic works, such as della Porta's *Natural Magick* (n. 68). Consider for example, the 'Discourse . . . shewing a Way how to communicate one's Mind at great Distances' (pp. 142-50) 'Divers curious Recepts' (pp. 190 – 5); and his due acknowledgement of Roger Bacon and della Porta as inventors of the telescope (pp. 257 – 8).

sympathetic to the 'magical tradition' in the history of science, saw Newton as a monstrous figure, a 'Great Amphibium' who managed to span two worlds even though they were 'clearly incompatible'.[80]

It would do equal damage to a proper historical understanding if we were to deny that there was any significant difference between magic and science. There are differences between these two world views today, and there were in the past. The point of the present essay has been to try to indicate that these two approaches to an understanding of the world lie along a continuous spectrum and that dynamic interchange between the two has always been possible. Instead of seeing Newton as the last of the magicians, with its suggestion of a dramatic switch from magic to science, we should try to see him, as I have tried to present Hooke, seeking to develop a new philosophy out of earlier systems of natural knowledge. The undue emphasis on Newton's use of active principles and other occult qualities in his work has tended to obscure the use of similar ideas by his older colleagues in English mechanical philosophy. In fact, the essential elements of Newton's matter theory, his use of unexplained active principles and the development of an experimental methodology in their defence, can all be found not just in Hooke but in virtually every major contributor to seventeenth-century English mechanical philosophy.[81]

I have not attempted here a detailed comparison of the shifting speculations of Hooke and Newton on active principles but even a superficial glance at Newton's 'Hypothesis of Light' of 1675 reveals striking similarities to Hooke's work. Newton suggested that light was 'something or other capable of exciting vibrations in the aether', before going on to suppose, like Hooke, that there may be more than one aether because 'the electric and magnetic effluvia, and gravitating principle, seem to argue such variety'. Furthermore, he hypothesised that these aethers may be universally 'unsociable' and this could account for gravity and capillary action in small glass-pipes. Nor did Newton subsequently transcend such speculations by arriving at what might be unanimously agreed to be the 'correct answer' to such problems. For all Newton's success in other areas of natural philosophy, it is at least arguable that the 'Queries' at the end of the *Opticks* still so reminiscent of Hooke's own queries, show no significant advance beyond the older man's work. If Newton can be said to have drawn on magical traditions, one of his sources for those traditions was surely Robert Hooke.[82]

None of this should be taken to mean that Hooke was a magician and not a

80 John Maynard, Lord Keynes, 'Newton, the Man', The Royal Society, *Newton Tercentenary Celebrations* (Cambridge, 1947), 27 – 34; pp. 27, 32. Hugh Kearney, *Science and Change 1500 – 1700* (London, 1971), p. 196.

81 See Henry, 'Occult Qualities and the Experimental Philosophy' (n. 6).

82 Isaac Newton, 'An Hypothesis explaining the Properties of Light, discoursed of in my several Papers', in I. Bernard Cohen, *Isaac Newton's Papers & Letters on Natural Philosophy* (Cambridge, Mass., 1978) pp. 179, 180, 183 – 4. Isaac Newton, *Opticks: Or A Treatise of the Reflections, Refractions, Inflections & Colours of Light*, based on the fourth edition of 1730 (New York, 1952) pp. 339 – 406.

mechanical philosopher. The categories of 'mechanical philosopher', and 'experimental philosopher' were being newly defined throughout the whole period of Hooke's career, and were forcing a redefiniton of the old category of 'natural philosopher'. In this defining process many of the fundamental ingredients of the tradition of natural magic, as this case-study of Hooke has indicated, played a crucial role. It should now be clear, therefore, that the title of this essay is meant ironically. Few seventeenth-century thinkers were better qualified than Hooke to define 'the real, the mechanical, the experimental Philosophy'. It is absurd for us to define the mechanical philosophy in such a way that Hooke seems not to fit into it. Once we are able to acknowledge that various aspects of natural magic were recognised as valid and routinely appropriated by the new philosophers, it should become clear to us that, for all his use of occult active principles and other notions from the magical tradition, there was nothing incongruous about the mechanical philosophy of Robert Hooke.

References

Short titles in the notes to this paper refer to the following:

J.A. Bennett, 'Cosmology and the Magnetic Philosophy, 1640–1680', *Journal for the History of Astronomy*, 12 (1981), 165–77.

P. Gouk, 'The Role of Acoustics and Music Theory in the Scientific Work of Robert Hooke', *Annals of Science*, 37 (1980), 573–605.

R.T. Gunther, *Early Science in Oxford*, 14 vols (Oxford, 1930).

M.B. Hesse, 'Hooke's Development of Bacon's Method', in *Proceedings of the Tenth International Congress of the History of Science* (Ithaca, 1962), 2 vols; vol. 1 (Paris, 1964), pp. 265–68.

M.B. Hesse, 'Hooke's Philosophical Algebra', *Isis*, 57 (1966), 67–83.

Robert Hooke, *Philosophical Experiments and Observations of the Late Eminent Dr Robert Hooke*, edited by W. Derham (London, 1726).

Robert Hooke, *The Posthumous Works*, edited by R. Waller (London, 1705).

Robert Hooke, *Micrographia: or Some Physiological Descriptions of Minute Bodies Made by Magnifying Glasses* (London, 1665)

M. Hunter and P.B. Wood, 'Towards Solomon's House: Rival Strategies for Reforming the Early Royal Society', *History of Science*, 24 (1986), 49–108.

J.C. Kassler and D.R. Oldroyd, 'Robert Hooke's Trinity College "Music Scripts": His Music Theory and the Role of Music in His Cosmology', *Annals of Science*, 40 (1983), 559–95.

D.R. Oldroyd, 'Some "Philosophical Scribbles" Attributed to Robert Hooke', *Notes and Records of the Royal Society*, 35 (1980), 17–32.

D.R. Oldroyd, 'Some Writings of Robert Hooke on Procedures for the Prosecution of Scientific Inquiry, Including His "Lectures of Things Requisite to a Natural History"', *Notes and Records of the Royal Society*, 41 (1987), 145–67.

P. Rossi, *The Dark Abyss of Time: The History of the Earth and the History of Nations from Hooke to Vico* (Chicago, 1984).

S. Shapin, 'Who was Robert Hooke?', in M. Hunter and S. Schaffer, eds, *Robert Hooke: New Studies* (Woodbridge, Suffolk, 1989), pp. 253–85.

B.R. Singer, 'Robert Hooke on Memory, Association and Perception', *Notes and Records of the Royal Society*, 31 (1976), 115–131.

R.S. Westfall, *Force in Newton's Physics* (London, 1971).

R.S. Westfall, 'Robert Hooke, Mechanical Technology and Scientific Investigation', in J.G, Burke, ed., *The Uses of Science in the Age of Newton* (Berkeley, 1983), pp. 85–110.

IX

"PRAY DO NOT ASCRIBE THAT NOTION TO ME": GOD AND NEWTON'S GRAVITY*

The precise nature of the force of gravitational attraction was always problematic for Isaac Newton. As is well known he was forced by the criticism of Leibniz to acknowledge in the General Scholium to the *Principia mathematica* that he did not know the *cause* of gravity. Making a virtue out of a necessity, he insisted that he was not interested in feigning explanatory hypotheses, being perfectly content to show "that gravity does really exist and act according to the laws which we have explained." Elsewhere, however, as is also well known, Newton did try out various explanatory hypotheses. Essentially, Newton's speculations derived from four earlier traditions with which he was familiar. Drawing upon the Neoplatonic tradition of light metaphysics, he suggested that light might combine with matter to give it various active powers; the alchemical tradition linked ideas of light with ideas of an active spirit, present in all things, which again might be said to give rise to various unceasing activities of matter. This active spirit, in its turn, could be linked to more recent ideas, developed in the new mechanical philosophy, in which an all pervasive aether was used as a medium of transmitting impulse from one part of the universe to another. Newton's own aether speculations were by no means purely mechanistic, since his aether consisted of particles held apart from one another, and from particles of other matter, by repulsive forces operating between them, but they clearly owed something to the mechanical as well as the alchemical traditions. The fourth tradition was Christian theology: gravitational attraction being held to be brought about by God.[1]

What I want to do in this paper is look at the fourth of these explanatory speculations. The general consensus among historians of Newton's thought is that none of the first three proved satisfactory to Newton for long. The conclusion, therefore, is that God was Newton's most favoured explanation for Gravity. However, it seems to me that the precise way in which Newton was said to have invoked God to explain gravity is based on a misreading and a misunderstanding of one or two crucially important pronouncements of Newton's. Let's begin, therefore, by looking at what some of the leading Newtonian scholars have said about God and Newton's gravity.

A. Rupert Hall in his *The Scientific Revolution* (1954) noted that Newton, on a superficial view of his writings, seems to believe in gravitational attraction as an inherent property of matter which was capable of acting at a distance, but Hall simply denies that Newton could ever have thought this way:

> That Newton seemed, by the theory of universal gravitation, to contravene the principles of mechanism, was due to misapprehension. Though certain phrases in the *Principia* might seem to indicate the contrary he did not believe that gravity was an innate property of matter, nor that two masses could attract each other at a distance without some relationship.

Thirty years later, when Hall rewrote this book under the title *The Revolution in Science*, he found himself able to say that Newton's comments upon gravity took him to an "impasse":

> On the one hand Newton says that no aetherial means, or presumably any kind of material mechanism occupying the celestial spaces, can exist; on the other he says that the existence of gravitational force within a void without "mediation" between the masses is absurd.

The way out of this impasse, Hall declares, referring for support to Newton's friends Fatio de Duillier and David Gregory, was to invoke God as the cause of gravitational attraction. More recently, Hall has insisted that "Newton was... certain... that though these powers, virtues or forces [including gravity] are invariably associated with matter, they are not produced by matter." Newton's subsequent attempts to explain the principle of gravity by alternative natural means are described by Hall as confusing. He then goes on to discuss the role of God in Newton's gravitational theory, suggesting at one point that God is the agent "continually at work" to produce gravitational effects, and at another that the "certain very subtle spirit penetrating dense bodies," which Newton refers to in the General Scholium appended to the second edition of the *Principia*, it to be identified with Newton's "idea of ubiquitous, active God."[2]

Alexandre Koyré, in his influential *Newtonian Studies* of 1965, claimed that,

> It is well known that Newton did not believe gravity to be an "innate, essential and inherent property of matter"... attraction as action at a distance through vacuum without mediation... was an utter absurdity that nobody could believed in...

Like Hall, Koyré also talks of a misapprehension by Newton's contemporaries of what he meant. Readers of the first edition of the *Principia* "could hardly fail to misunderstand Newton's position and to ascribe to him just those opinions which he so vehemently rejected... ", he wrote. And a bit later he said, "It may seem rather surprising, in spite of these very decided and definite declarations by

Newton, ... that his teaching could be, and was, interpreted as positing action at a distance by an attractive force residing in bodies." Koyré, however, insists that Newton had in fact stated quite clearly that gravity had to be "performed by something which is not material, that is, by God,"[3]

Hall's and Koyré's belief that Newton relied more or less directly upon God to explain the force of gravity is bound up with their belief that he did not believe action at a distance to be possible and that gravitational attraction could not, therefore, be a property of matter. This same view of Newton's thinking has been most forcefully stated by I. Bernard Cohen, in a recent article on 'Newton's Third Law and Universal Gravity'. This is what Cohen writes:

> From the time of his *Principia* until his death Newton was deeply troubled by the concept he had introduced: a universal gravity. He had been brought up in the "received" philosophy, sometimes known as the "mechanical philosophy," centering around the ideas of Descartes, which held that all explanations in natural philosophy must be couched in terms of what Robert Boyle called "those two grand and most catholick principles of bodies, matter and motion." At first sight it would seem that to these Newton had added a third principle, force. Not only was this a departure from accepted norms, it also introduced a kind of force that was astonishing in its primary characteristics or qualities. For this force had to have the power of extending itself over many hundreds of millions of miles as a kind of grasping entity which could affect huge bodies. For instance, the gravitational force could extend far beyond the reaches of the solar system ... to turn a comet around and cause it to return to the visible regions of the neighbourhood of the sun. Newton again and again sought for some explanation of how universal gravity might act. That is, he attempted to reduce universal gravity to the action of something else, a shower of aether particles, electrical effluvia, variations in an all-pervading aether. All of these attempted "explanations" or reductions of universal gravity to some accepted kind of mechanism failed – because none could fulfill two major requirements: that the resultant force vary inversely as the square of the distance and that this force act mutually on every pair of bodies so as to attempt to bring them together.

Now, one of the crucial factors in this account of an embarrassed Newton, is that it is beyond doubt because it is fully chronicled. As Cohen goes on to say, "It is well known that by the time of the second edition of the *Principia*, in 1713, Newton publicly acknowledged the difficulty ... in accepting the action-at-a-distance of universal gravity ... " And a little later: "Newton's aversion to accepting a force that acts at a distance without any secondary explanatory mechanism is well documented ... ." This being so, Cohen goes on to ask, "How, then, could Newton have initially been able to consider the properties and actions of a force of "attraction," of a type in which he could not fully believe?"[4] We need not pursue the details of Cohen's attempt to answer this question because it is my contention that it is based on a misunderstanding of Newton's views on the nature of gravitational attraction.

So, according to Hall, Koyré and Cohen, Isaac Newton did not believe that gravity could be an inherent property of matter, and he did not believe that action at a distance was possible "without mediation." However, in spite of his own very clear statements to this effect, Newton was completely misunderstood by most of his contemporaries. Furthermore, the fullest modern attempt to understand Newton's meaning on these matters, that of Ernan McMullin in his book *Newton on Matter and Activity* (1978), also found it extremely difficult to understand Newton. McMullin confessed that he had "failed to identify any single principle, consistently pursued, as the driving force behind Newton's search for an explanatory account of gravitational motion." McMullin's failure can be attributed to the fact that he could not reconcile Newton's clear and unabashed pronouncements on the reality of actions at a distance in a number of places in his writings, with a couple of famous, and very insistent, statements by Newton that actions at a distance are impossible.[5]

All of these writers have been very influential and their reading of Newton on this important issue has virtually become canonical. So much so, that nobody now makes any serious attempt to resolve these difficulties. It is simply assumed that, as McMullin implies, Newton's problems were intractable and that he never did reach a consistent position; or that, as Hall says, Newton merely speculated "confusingly."[6] It is my contention, however, that this standard account is almost entirely based on a misreading of two crucial statements where Newton is supposed to have denied that gravity is inherent in matter, and that action at a distance is impossible.

These crucial statements appear in two passages in Newton's letters to Richard Bentley, written between December 1692 and February 1693 when Bentley was preparing his Boyle Lectures for the press. In the second of his letters to Bentley, Newton wrote:

> You sometimes speak of Gravity as essential and inherent to Matter. Pray do not ascribe that Notion to me; for the Cause of Gravity is What I do not pretend to know, and therefore would take more Time to consider it.[7]

This is quoted in every recension of the canonical account. With good reason since it seems to me to be literally a *sine qua non*. In the following letter Newton partially explains his meaning:

> The last Clause of the second Position I like very well. It is inconceivable, that inanimate brute Matter should, without the Mediation of something else, which is not material, operate upon, and effect other Matter without mutual Contact, as it must be, if Gravitation in the Sense of Epicurus, be essential and inherent in it. And this is one Reason why I desired you would not ascribe innate Gravity to

> me. That Gravity should be innate, inherent and essential to Matter, so that one Body may act upon another at a distance thro' a Vacuum, without the Mediation of any thing else, by and through which their Action and Force may be conveyed from one to another, is to me so great an Absurdity, that I believe no Man who has in philosophical Matters a competent Faculty of thinking can ever fall into it. Gravity must be caused by an Agent acting constantly according to certain Laws; but whether this Agent be material or immaterial, I have left to the Consideration of my Readers.[8]

There can be no doubt, then, about the vigour of Newton's denials. Small wonder that Koyré, Hall, McMullin, Cohen and all those who have followed them found it hard to reconcile the suggestion that no man with a competent faculty of thinking could ever believe in action at a distance with all those statements in the Queries appended to the *Opticks* where Newton talks glibly of actions at a distance. Consider for example, Query 29:

> Pellucid Substances act upon the Rays of Light at a distance in refracting, reflecting, and inflecting them, and the Rays mutually agitate the Parts of those Substances at a distance for heating them; and this Action and Re-action at a distance very much resembles an attractive Force between Bodies.

And, of course, Query 31:

> Have not the small Particles of Bodies certain Powers, Virtues, or Forces, by which they act at a distance, not only upon the Rays of Light for reflecting, refracting, and inflecting them, but also upon one another for producing a great Part of the Phenomena of Nature? For it's well known, that Bodies act one upon another by the Attractions of Gravity, Magnetism and Electricity; and these Instances shew the Tenor and Course of Nature, and make it not improbable but that there may be more attractive Powers than these.[9]

But there is an easy way out of this dilemma. What we have to ask ourselves is, what exactly is Newton denying so vigorously in his letters to Bentley? Is Newton saying, for example, that gravity is not a property of matter at all, since matter is inert and passive? In which case McMullin is correct in supposing that Newton had set himself "the thankless task of finding a home for forces somewhere else than in matter".[10] Is Newton implying that, in fact, gravitational attraction should be taken to be a Spirit – spirits after all are active principles – which is all-pervasive but independent of matter? This is the line taken by R. S. Westfall and Betty Jo Dobbs.[11] Or, is he perhaps saying that gravitational attraction must be directly performed by God – "an Agent acting constantly"?

In fact, we do not have to accept any of these interpretations. There is another reading in which these pronouncements in the letters to Bentley can be seen to be perfectly compatible with all his other statements or speculations on the nature of gravity and other forces, including those optical Queries where he calmly proposes the operation of actions at a distance.

An obvious place to seek help in interpreting Newton's meaning is to look to see what Richard Bentley, the recipient of the letters, made of these particular passages. At the relevant point in his Seventh Boyle Lecture ('A Confutation of Atheism from the Origin and Frame of the World'), Bentley rejected the notion that Gravity was inherent in or essential to matter, "so that several Particles placed in a Void space at any distance whatsoever would without any external impulse spontaneously convene and unite together," Still following Newton's lead, Bentley identifies this rejected view as the "Epicurean Theory of Atoms descending down an infinite space by an inherent principle of Gravitation."[12] Now this is important because none of the commentators in the canonical tradition have ever pointed to the puzzling nature of Newton's reference to Epicurus in his third letter to Bentley. "It is inconceivable," Newton said, "that... Matter should... operate upon and affect other Matter without mutual Contact, as it must be, if Gravitation in the sense of Epicurus be essential and inherent in it."[13]

It seems clear that the concern of both Newton and Bentley is to dissociate Newton's concept of gravity from the ancient system of Epicurus which was generally regarded as atheistic. Neither man wanted to provide hostages to the ranks of atheists which they perceived to be an ever-present threat to their society and their religion. If gravity and other principles of activity could be said to be essential and inherent in matter, then the universe might seem capable of running itself, without the benefit of divine guidance. So, when Newton said "Pray do not ascribe that notion to me," the notion he was objecting to was not that gravitational attraction might be a property of matter, but that gravitational attraction might be held to be an *essential* property of matter, in the way that extension was held to be. Extension was generally agreed to be an essential attribute of matter. Matter could not be conceived of without extension but it could easily be conceived of without gravitational attraction.

But what about the other notion that Newton is supposed to have rejected in his letter to Bentley: the notion of action at a distance? How did Bentley treat this in his Boyle Lecture? Surely he must have rejected it out of all countenance? Evidently not. In fact, this is what he wrote:

> Now, mutual gravitation or attraction, in our present acceptation of the words is the same thing with this, 'tis an operation, or virtue, or influence of distant bodies upon each other through an empty interval, without any *effluvia* or exhalations, or other corporeal medium to convey and transmit it.[14]

Are we to assume that Bentley simply disagreed with Newton's objections to action at a distance? Was Bentley sufficiently confident in his own powers "in philosophical Matters" that he chose to disregard Newton's strictures against this great absurdity of action at a distance? Certainly this is the line that Alexandre Koyré and, more recently, Perry Miller have taken.[15] As we shall see later, both of these commentators have blamed Bentley for the fact that most of Newton's contemporaries thought, erroneously according to Koyré and Miller, that he did believe in action at a distance.

If we read on, however, it becomes possible to see that Bentley was not ignoring Newton's advice but was putting it to good use:

> This Power therefore cannot be innate and essential to Matter. And if it be not essential; it is consequently most manifest (seeing it doth not depend upon Motion or Rest or Figure or Position of Parts, which are all the ways that Matter can diversify itself) that it could never supervene to it [matter], unless impress'd and infused into it by an immaterial and divine Power.[16]

Bentley can be seen to be providing here a very clear statement of exactly what Newton had in mind about the nature of gravity. Gravity is a power, or virtue which has been superadded to matter by God. Certainly there is nothing in Newton's letters to Bentley which directly contradicts this description, nor even so much as a hint that he dissents from it. We have no reason to suppose (although, as we shall see later, a number of scholars *have* supposed) that Bentley misunderstood Newton and misrepresented his ideas.

It is important to be clear about what Bentley is saying. The fact that he refers to mutual gravitation implies straight away that he takes gravity to be acting between bodies, not for example between a body on the one hand and the centre of rotation of a vortex, as in the Cartesian system, on the other hand. So, gravity is a virtue, or influence of one body on another at a distance ("distant bodies"), through absolutely empty space – not even a tenuous effluvium is required to transmit it. He then goes on to draw implications from this. These implication are fully in keeping with Newton's denials about gravity in his letters to Bentley. This mutual attraction cannot be essential to matter, like extension. Furthermore the gravitational attractive power of any given body cannot be shown to be an epiphenomenon of the arrangement or motions of the particles which constitute that body. The only alternative, therefore, is to suppose that this non-essential power is *superadded* to bodies by (who else but) "an immaterial and divine power."

So, Newton could be said to fully accept what Bentley says here without having a belief in *innate* gravity ascribed to him. Moreover, it is now possible to understand the otherwise deeply puzzling claim of Newton in his letter to Bentley that he has "left to the consideration of [his] Readers," whether the Agent responsible for gravity is material or immaterial This is deeply puzzling on the standard reading because Newton says just a few lines above this: "It is inconceivable, that inanimate brute Matter should, without the Mediation of something else, *which is not material*, operate upon, and effect other Matter without mutual Contact, . . . " In what sense does he leave his readers to decide whether the agent responsible for gravity is material or immaterial, when he has already told his readers that action at a distance is inconceivable without the mediation of something which is not material?

To resolve this difficulty we simply have to realise that the "Agent acting constantly according to certain laws," which is the immediate cause of gravity and which could be either material or immaterial, is different from the *immaterial* Mediator which endows matter with the non-essential power of attracting other matter. It is inconceivable, Newton is saying, that matter should have a power to attract other matter at a distance unless such power is mediated or brought about by God. Certainly, Bentley himself interpreted Newton's letter this way, as can be seen by a close paraphrase in the seventh lecture:

> 'Tis utterly unconceivable, that inanimate brute Matter (without the mediation of some Immaterial Being) should operate upon and affect other Matter without mutual Contact . . .

Or, as Bentley wrote in one of his letters to Newton:

> 'tis unconceivable, yt inanimate brute matter should (without a divine impression) operate upon & affect other matter without mutual contact: as it must, if gravitation be essential and inherent in it.[17]

Note that Bentley seems to be admitting at this point in his letter (and Newton never took him to task for this) that if gravitation is essential to matter then brute matter *must* act at a distance. What he objects to, and Newton by all accounts must be taken to agree with him, is not that matter can act at a distance, but that it can be held to do so without this ability having been conferred upon it by a divine mediator. The objection is to the notion of gravity as essential to matter, not that matter can be held to operate at a distance.

For both Newton and Bentley, God was the immaterial mediator whose omnipotence enabled him to impose upon matter an agent of gravitational attraction which acts constantly according to certain laws. Bentley, in his Lecture, seems to imply that the agent must be immaterial ("without any Effluvia or Exhalations or other corporeal Medium to convey and transmit it"). If this was Bentley's opinion, Newton evidently disagreed with him. Being a fully committed voluntarist in his theology, Newton would not presume to say whether the agent which God chose to use was immaterial or material.[18]

The clear implication of what Newton says in his correspondence with Bentley is that gravity is perfectly acceptable as long as we recognise that it must have been endowed upon matter at the creation, since it is not logically entailed by the nature of matter, or to put it another way, it isn't essential to matter. This is perfectly compatible, for example, with his statement at the end of the *Opticks*:

> All these thing being consider'd it seems probable to me, that God in the Beginning form'd Matter in solid, massy, hard, impenetrable, moveable Particles, of such Sizes and Figures, and with such other Properties and in such Proportions to Space, as most conduced to the End for which He form'd them.... It seems to me farther, that these Particles have not only a *Vis inertiae*, accompanied with such passive Laws of Motion as naturally result from that Force, but also that they are moved by certain active Principles, such as is that of Gravity, and that which causes Fermentation...[19]

Furthermore, in case any body should still be in doubt about God's ability to impose gravity upon matter, he later adds:

> it may be also allow'd that God is able to create Particles of Matter of several Sizes and Figures, and in several Proportions to Space, and perhaps of Different Densities and Forces, and thereby to vary the Laws of Nature, and make Worlds of several sorts in several Parts of the Universe. At least, I see nothing of contradiction in all this.[20]

To recapitulate, it would seem that Newton believed that matter, which is essentially passive, was endowed with various active principles by God. One of these active principles was, or gave rise to, gravitational attraction: "we must... universally allow that all bodies whatsoever are endowed with a principle of mutual gravitation."[21] Gravity could be said, therefore, to be a property inherent in matter providing it was realised that it was a superadded property. Certainly, gravity was not to be seen as a property which was logically entailed by the nature of matter itself, in the way that extension was. Newton's natural philosophy, based on induction and mathematical analysis, was able to explain in precise detail the behaviour of bodies endowed with gravitational attraction, and so reveal the regular laws which governed gravity. What Newton freely admitted he could not do was define exactly what was the secondary cause of gravitational attraction (the primary cause was, of course, God). He could not even say whether it was material or immaterial. Part of the difficulty here was surely semantic: if matter had a property of attracting other matter at a distance without any material linkage between them, might not this property still be designated "material" by virtue of the fact that it was an epiphenomenon of matter?[22] Be that as it may, Newton certainly never insisted that gravitational attraction must be brought about by some kind of *immaterial* spirit, whether divine or otherwise.

I take issue, on this point, with those commentators who have taken Newton's pronouncements in the letters to Bentley, together with some of his statements elsewhere, to imply that gravitational attraction must be brought about by the direct and constant intervention of God himself. R. S. Westfall, for example, has said that the third letter to Bentley makes it clear that Newton believed gravity was caused by an immaterial aether. He then goes on: "What could an immaterial aether be? To Newton, it was the infinite omnipotent God who by his infinity constitutes absolute space and by his omnipotence is actively present throughout it." God, Westfall reiterates elsewhere, was the "agent acting constantly according to certain laws" that "makes bodies move as though they attract each other."[23]

Betty Jo Dobbs, in a refinement of these arguments, has recently argued that Newton's gravity derived from Neostoic ideas in which the Stoic *pneuma* becomes Platonized into an immaterial principle, which Newton, in turn, equated with God:

> a powerful ever-living Agent, who being in all Places, is more able by his Will to move the bodies within his boundless uniform Sensorium, and thereby to form and reform the Parts of the Universe, than we are by our Will to move the Parts of our own Bodies.

"What we call gravitation," Dobbs writes, is nothing more than the Divine Mind forming and reforming the Parts of the Universe.[24]

There are a number of difficulties with Westfall's and Dobbs's views. In the first place their theories imply that Newton did not believe there was a secondary cause of gravity. If this is so, why did he spend time trying to find a satisfactory aetherist account? And why did he include "the cause of gravity" among the Active Principles of body (along with "the cause of Fermentation," for example)? Secondly, their immanentist account of God's activity in the world is incompatible with the transcendental theology which is more common in Newton's writings about God. Newton's talk of space as the sensorium of God in which we live and move and have our being,[25] which is emphasized in these interpretations, was always presented as an analogy and Westfall and Dobbs are taking it too literally. Newton explicitly denied that God was gravity:

> In him are all things contained and moved; yet neither affects the other: God suffers nothing from the motion of bodies; bodies find no resistance from the omnipresence of God.[26]

Furthermore, Newton's God was a god of dominion: "Lord God, Pantokrator." "Deity," Newton insists, "is the dominion of God not over his own body, as those imagine who fancy God to be the soul of the world, but over servants." We reverence and adore God, therefore,

> on account of his dominion: for we adore him as his servants; and a God without dominion, providence, and final causes, is nothing else but Fate and Nature.[27]

Westfall's and Dobbs's God of Gravity would be, in Newton's terms, merely fate and Nature, "blind metaphysical necessity." Westfall is wrong, I believe, to interpret Newton's talk of God's dominion as though it can mean God's direct involvement: "Newton's god was the Pantocrator... who holds dominion over all that is,

...not by his watchful providence but by a similar dominion over the course of events." Newton's God works indirectly, exerting dominion to establish his providence, which is brought about by his servants, the secondary causes.[28]

My disagreement with Westfall and Dobbs can be expressed theologically. Westfall and Dobbs both seem to suggest that Gravity is brought about by God's *potentia absoluta*, his absolute power. On my reading, gravity is taken care of by God's *potentia ordinata*, his ordinary power. Having superadded the power of gravitational attraction to bodies, and established the laws of interaction, God need only use his ordinary power to maintain all things in existence. This is perfectly compatible with Newton's attempts to understand the cause of gravity in terms of the alternative secondary causes of superadded active principle or an aether.[29]

So, "the Agent acting constantly according to certain Laws", which Newton refers to in his third letter to Bentley, is not God but the active principle which God has added to the matter (either to all matter, or merely to the matter of the aether, depending upon which speculation Newton favoured at the time). The "Mediation" of "something else which is not material" which then enables matter to "operate upon, and affect other Matter without mutual Contact" need only take place once, presumably at the Creation.[30] God's continuous operation is no more required to account for gravitational attraction than it is to account for the impenetrability of matter, the immortality of the soul, or the continual truthfulness of the logical law of excluded middle. Which is to say that gravity, along with everything else in the world, has to be continually maintained by God's arbitrary will. It is not a phenomenon of such a unique kind that it requires a special effort on God's part to ensure from moment to moment that it works according to plan. If Newton *had* believed that gravity was such a special part of God's creation that it required his particular attention, then Leibniz's suggestion that Newton's gravity was "the effect of a miracle" would have been correct. But, of course, Newton always vigorously denied this.[31]

The interpretation of the two crucial statements in Newton's third and fourth letters to Bentley which is presented here does little more than note a difference of emphasis. What Newton was objecting to was not that gravity should be seen to be a property of matter but that it should be seen to be essentially so. In case this seems like too nice a distinction, it is worth pointing out the implications, and advantages, of this reading for Newtonian exegesis.

Firstly, I think it is easier to see how important the Newtonian concept of gravity is for Bentley's purpose in the Boyle Lectures and, for that matter, for Newton himself who said in his very first letter to Bentley that "When I wrote my Treatise about our System, I had an Eye upon such Principles as might work with considering Men, for the Belief of a Deity, and nothing can rejoice me more than to find it useful for that Purpose."[32] As Bentley pointed out "if it be made appear, that there is really such a power of Gravity perpetually acting in the constitution of the present System." then,

> This would be a new and invincible Argument for the Being of God: being a direct and positive proof, that an immaterial living Mind doth inform and actuate the dead Matter, and support the Frame of the World.[33]

In other words, by demonstrating the existence of active principles in matter, which although universal and inseparable from matter were not essential to it, Newton's natural philosophy closed up the loophole which atheists had supposedly discovered in the Cartesian system. The Cartesian atheist only had to assume that the world was eternal to be able to conclude that God was never required to set the whole system a-rolling. This argument could work because the Cartesian system relied only upon what were taken to be essential properties of matter (like extension). The crucial role of gravity in the Newtonian system put a stop to such atheist claims. Since, as Bentley said, "We have proved, that a Power of mutual Gravitation, without contact or impulse, can in no-wise be attributed to mere Matter" as an essential property (logically entailed, as it were, in the very nature of matter), then it must have been "impress'd and infused into it by an immaterial and divine Power," even if the world has endured eternally. "As for Gravitation", Bentley wrote, "tis impossible yt That should either be coaeternal & essential to Matter, or ever acquired by it" unless it be endowed upon matter by divine intervention.[34] It could not be further from the truth, therefore, to suggest that Newton wished to dissociate himself from the view that gravity was an inherent property of matter. On the contrary, it was important for him to insist that it was because the otherwise inexplicable presence of this property in matter (since it was not a natural concomitant of matter, nor could it have been acquired by matter in some natural way) provided incontrovertible evidence for the existence of an omnipotent God.

An obvious advantage of the reading presented here, which has already been pointed out, is that it does away with all the puzzlement in the standard account caused by the fact that, in spite of Newton's "Pray, do not ascribe that notion to me" in his letter to Bentley, elsewhere he makes no bones whatsoever about talking of gravity as a property of matter and about casually discussing the evident reality of action at a distance. Small wonder that Koyré should believe that contemporaries "could hardly fail to misunderstand Newton's position and to ascribe to him just those opinions which he so vehemently rejected...," or that "in spite of these very decided and definite declarations by Newton, ...his teaching could be, and was, interpreted as positing action at a distance by an attractive force residing in the bodies." Similarly, as Rupert Hall wrote in 1954, Newton "was widely misunderstood."[35] In fact, Newton was not misunderstood; in ascribing to him belief in action at a distance by an attractive force residing in bodies his contemporaries, like Bentley, were reading him correctly. His English contemporaries in particular were used to the voluntaristic theological claim that God could, by his arbitrary will, endow matter with various qualities and powers. So, when Newton, in the final book of the *Opticks*, asks: "Do not Bodies act upon Light at a distance, and by their action bend its rays...?" he did not need to explicitly invoke God's omnipotent will at that point. Nevertheless, most English theologians, or other intellectuals raised in the traditions of the English church, would have taken Newton's voluntarism for granted. It is surely significant that the major critic of such ideas, Leibniz, was a committed intellecualist in his theology, and therefore opposed to Newton's voluntaristic approach.[36]

But the advantage of this reading is not merely confined to a better understanding of Newton's letters to Bentley. It has implications for our understanding of all of Newton's statements about gravity and action at a distance. We have already seen how there is no contradiction between his strictures to Bentley and his glib talk of actions at a distance in the Queries to the *Opticks*. We can make a precisely similar point about another problem for Newtonian exegetes, namely: What advantage over his theory of gravitational attraction at a distance did Newton see in his various aether theories, given that they were not mechanical aethers but depended on repulsive forces, operating at a distance between the particles of the aether? This issue has been raised particularly by Ernan McMullin in his *Newton on Matter and Activity*.[37]

The point here is that the aether theories were not intended to be a way of avoiding actions at a distance, which – contrary to the standard account – Newton was not averse to, but merely a way of reducing the number of occult qualities in the universe. Newton was swayed during the periods of his aether speculations by the reductionist belief that it is better to have one active matter, like the aether, responsible for all physical activity and change than to have each and every parcel of matter endowed with its own activity. The aether speculations do not, therefore, represent attempts by Newton to be more "mechanical" in his thinking than when he assumed the existence of "active principles" in matter. And they are neither more nor less in keeping with his pronouncements in the letters to Bentley. Active principles and aethers both entailed a belief in matter having virtues or powers which could not be reduced to the motions, positions, or arrangements of the particles of matter, and which could operate at a distance. If there was an aether, the mutual repelling forces between its particles were active principles which, like gravity, or the cause of fermentation and cohesion in his non-aetherist speculations, could not be essential to the aether particles.

It might be objected that there were a number of other places, besides the letters to Bentley, where Newton denied the concept of attractive force residing in bodies and the possibility of action at a distance. But it seems to me that any pronouncements of Newton's which are read this way have simply been (mis)interpreted in the light of the standard misreadings of the letters to Bentley. Indeed, Koyré's essay, 'Gravity an

Essential Property of Matter?', shows very clearly how Newton's most significant pronouncements on gravitational attraction were all interpreted by contemporaries in terms of a property of matter capable of acting at a distance. Koyré himself explains this, for him, lamentable misunderstanding by virtue of the fact that Newton's contemporaries did not have the benefit of his clear denials of this interpretation in the letters to Bentley, which were not published until 1756.[38]

I.B. Cohen has pointed to a number of places in the *Principia* where Newton seems to apologize for his use of the word "attraction," or for his inability to give a mechanistic explanation of gravity. In the Introduction to Section XI, of Book I of the *Principia*, for example, Newton writes:

> I shall therefore at present go on to treat of the motion of bodies attracting each other; considering the centripetal forces as attractions; though perhaps in a physical strictness they may more truly be called impulses. But these Propositions are to be considered as purely mathematical; and therefore, laying aside all physical considerations, I make use of a familiar way of speaking, to make myself more easily understood by the mathematical reader.[39]

Cohen sees this profession of mathematical intrumentalism, "laying aside all physical considerations," as Newton's way of freeing himself from constraints that mainstream natural philosophy would have laid upon him – such as avoiding talk of actions at a distance. This, and the famous disclaimer at the end of the General Scholium that he "feigns no hypotheses" concerning the cause of gravity, are taken by Cohen to represent a sophisticated methodological approach which enabled Newton to go further than any of his contemporaries. Cohen is surely right about this. But we do not need to suppose that Newton developed this means of presenting his ideas as a result of his own "repugnance" to action at a distance, as Cohen, citing the third letter to Bentley, seems to think.[40] We have every reason to assume that Newton developed his sceptical, intrumentalist and empiricist approach partly in accordance with the powerful contemporary tradition of empiricism associated with the Royal Society of London, and with a full awareness of the dominance of Cartesian ideas on the Continent with their vigorous strictures against actions at a distance.[41] But another important element in Newton's approach was his theological voluntarism, which required him to accept that God could, if he so willed it, create bodies capable of acting upon one another across distances. To deny this was to deny God's omnipotence. Cohen presents Newton's sophisticated methodology, which is so important an element in what he calls the "Newtonian style," in a rather negative, apologetic light. For all its intellectual power and influence, it remains for Cohen a clever escape: with one bound Newton was free of the constraints of contemporary natural philosophy. This is only part of the story. Newton's method was also a positive defence of the concept of action at a distance, and therefore of the omnipotence of God.[42]

To conclude, it is worth emphasizing that Newton does not, either in these passages in the *Principia* or anywhere else, deny the reality of actions at a distance. Not even, as we have seen, in the letters to Bentley. Similarly, his frequently cited comment in the "Advertisement" to the second English edition of the *Opticks* (1717) says nothing against actions at a distance. Once again, he simply says, "I do not take Gravity for an essential Property of Bodies." The aetherist hypothesis to which he is drawing the reader's attention nevertheless relies upon actions at a distance.[43] Moreover, the hypothetical nature of the Optical Queries encourages him to dismiss all methodological caution and explicitly discuss actions at a distance, whether between all particles or simply between the particles of a putative aether.

It is surely significant, also, that although Samuel Clarke seems to deny action at a distance during his controversy with Leibniz, in the subsequent defence of this particular statement it becomes clear, as Leibniz had seen straight away, that he was not denying it at all. In his Fourth Reply Clarke wrote,

> That one body should attract another without any intermediate means, is indeed not a miracle but a contradiction: for 'tis supposing something to act where it is not. But the means by which two bodies attract each other, may be invisible and intangible, and of a different nature from mechanism...

"What does he mean," Leibniz asked, "when he will have the sun to attract the globe of the earth through an empty space?" The German philosopher made a few suggestions, including God, immaterial substances, spiritual rays, an accident without a substance, some kind of *species intentionalis*, or a *je ne sais quoi*. Clarke opted for the latter:

> That this phenomenon is not produced *sans moyen*, that is without some cause capable of producing such effect; is undoubtedly true. Philosphers therefore may search after and discover that cause, if they can; be it mechanical or not mechanical. But if they cannot discover the cause; is therefore the effect itself, the phenomenon, or the matter of fact discovered by experience, ... ever the less true?

There are two things to notice about Clarke's final reply. Firstly he subtly shifts his ground from denying gravitational attraction can take place without an "intermediate means" to saying it cannot take place without a cause. Secondly, he does not opt for God as the direct cause, nor for any kind of immaterial entity.[44]

We have seen above how a close reading of Bentley's seventh and eight Boyle Lectures helps us to understand what Newton meant in the crucial passages of his letters to Bentley It has been more usual, however, for historians to suppose that Bentley is not just a dubious guide to Newton's meaning, but was downright wrong in his interpretation. It is important for the case presented here, therefore, to try to dispell this view of Bentley.

In his *Newtonian Studies* Koyré placed much of the blame for the subsequent misunderstanding of Newton on Bentley who, we are told,

> in spite of Newton's admonition, said in his *Confutation of Atheism* that "a constant Energy [is] infused into Matter by the Author of all things" and that "Gravity may be essential to Matter".[45]

Similarly, Perry Miller, writing in 1978, illustrated his doubts that Bentley had the mental wherewithal to "fully get the point" in his discussions with Newton by saying:

> despite his effort to make clear his agreement with Newton, Bentley still calls it [gravity] "a constant energy infused into matter by the Author of all things."[46]

The first thing to say here is that Bentley's designation of gravity as "a constant Energy infused into Matter," which appears in his eighth and final Boyle Lecture, is perfectly in accordance with Newton's own view, expressed in the letter to Bentley, that gravitational attraction at a distance is only impossible, "without the Mediation of something else, which is not material." It seems safe to assume, therefore, that Newton, pace Koyré and Miller, would have seen nothing of contradiction in the suggestion that gravitation is "a constant Energy infused into Matter by the Author of all things" On the contrary, it entirely suited his voluntarist theology.

But what of Koyré's report that Bentley also said that "Gravity may be essential to matter"? Regrettably, this is simply a very bad case of misreading by Koyré. Having said that "the concentric Revolutions of the Planets about the Sun" can be explained in terms of a gravitational attraction to the sun and a perpetual motion, tangential to the planet's particular orbit, both of which (the power of attraction and the transverse motion) being "impress'd at first by the Divine Arm," Bentley goes on to insist that even if an atheist were to claim that these two phenomena were not of divine origin, the system of the world as we know it could not be shown to arise spontaneously from this force of attraction and this tangential motion. This is what he actually wrote:

> But now admitting that Gravity may be essential to Matter; and that a transverse Impulse might be acquired too by Natural Causes, yet to make all the Planets move about the Sun in circular Orbs; there must be given to each a determinate Impulse... For had the Velocities of the several Planets been greater or less than they are now, at the same distance from the sun; or had their Distance from the Sun, or the quantity of the Sun's Matter and consequently his Attractive Power been greater or less than they are now, with the same Velocities: they would not have revolved in concentric Circles as they do, but have moved in Hyperbola's or Parabola's or in Ellipses very Eccentric.[47]

Bentley's argument against atheists is in two parts and he introduces the second part by means of a counterfactual. The first part of the argument states that gravitational attraction cannot have become a property of matter unless it had been put there by God. The second part of the argument, which relies upon the supposedly clear signs of intelligent design in the universe, suggests that even if gravity *was* part of the essence of matter the atheist can go no further without invoking a cosmic intelligence.

A. R. Hall has also taken Bentley to task for going beyond what Newton intended. Having accepted Newton's insistence that gravity is not essential and inherent to matter, Hall wrote, Bentley unwarrantably jumped to the conclusion that gravity must, therefore, have been impressed and infused into it by God. According to Hall, "This blurs Newton's clear contention" that God, acting continually, is directly responsible for gravitational effects.[48] I have already argued above that this is based on a misunderstanding of Newton's theology. Once again, I suggest, Bentley's reading of Newton can be taken to approach very close to what Newton himself intended. It is perhaps worth mentioning also, that there is a total lack of any circumstantial evidence that Newton was in any way upset by Bentley's use of his ideas. As is well known, Newton allowed Bentley to supervise the preparation of the second edition of the *Principia* for the press, an unlikely position of trust for one who had supposedly so badly misunderstood the philosophical and theologial import of Newton's work.

Beside, there is some evidence to suggest that the huge importance bestowed by historians upon the crucial statement of Newton's in his correspondence with Bentley does not reflect the attitude of the correspondents themselves. The original mistake which Newton picked upon: "You sometimes speak of Gravity as essential and inherent to Matter," was, according to Bentley, only loose talk: "If I used yt word, it was only for Brevity's sake." Perry Miller dismisses this excuse, but we have no real reason for supposing it wasn't true.[49] Bentley did not, after all, let slip in one of his published lectures that "Gravity may be essential to Matter" – except (as we have seen) when he was putting the atheists' case for them.

If Bentley was a bit slapdash in his choice of word, he was in good company. Roger Cotes, who edited the second edition of the *Principia* and who furnished it with an editorial Preface, also slipped into referring to gravity as essential to matter. He dropped it at the advice of Samuel Clarke but if we go by the evidence of Cotes's letter, which is all we can go by (since Clarke's advice seems to have been passed on orally by a Dr. Cannon), it wasn't seen as a horrendous *faux pas*:

> I return my thanks for Your corrections of the Preface, & particularly for Your advice in relation to that place where I seem'd to assert Gravity to be Essential to Bodies. I am fully of Your mind that it would have furnish'd matter for Cavilling and therefore I struck it out immediately upon Dr Cannon mentioning Your Objection to me . . .[50]

Cotes's slip becomes for Koyré "an obvious error for which he was taken to task by Clarke." It is perhaps symptomatic of Koyré's overemphasis on the importance of this issue that he failed to notice that this criticism was passed on orally by an intermediary. Koyré tells us that, "Clarke's letter has not been preserved and one can only infer its contents from Cotes's answer." It would seem that in Koyré's eyes the error that Cotes made was so bad that it was inconceivable that it could have been dealt with so casually.[51]

It is difficult to know what kind of further "Cavilling" Cotes and Clarke thought the word "essential" would have inspired, given that what they did allow to stand in the new "Preface" made a bold plea for asserting gravity to be the metaphysical equivalent of extension, mobility and impenetrability, the three mainstays of the mechanical philosophy. Taking an empiricist line, Cotes insists that the extension, mobility and impenetrability of bodies are known only by observations and experiments. But the same method teaches us that all bodies are heavy. It will not do, Cotes says, to argue that we have not yet observed the gravity of the fixed stars because we have not yet observed their extension, mobility or impenetrability. "In short," he concludes,

> either gravity must have a place among the primary qualities of all bodies, or extension, mobility, and impenetrability must not. And if the nature of things is not rightly explained by the gravity of bodies, it will not be rightly explained by their extension, mobility, and impenetrability.[52]

One thing is certain, Cotes was not riding a hobby-horse of his own making here. He was drawing directly upon his understanding of Newton's opinion, and in particular upon what he took to be the intention of Newton's Third Rule of Reasoning. In his own explanation of the significance of this rule Newton wrote:

> if it universally appears, by experiments and astronomical observations, that all bodies about the earth gravitate towards the earth, and that in proportion to the quantity of matter which they severally contain; ... and all the planets [gravitate] one towards another; and the comets in like manner towards the sun; we must, in consequence of this rule, universally allow that all bodies whatsoever are endowed with a principle of mutual gravitation. For the argument from the appearance concludes with more force for the universal gravitation of all bodies than for their impenetrability; of which, among those in the celestial regions, we have no experiments, nor any manner of observation.[53]

Newton clearly wishes to suggest that gravity has as much claim to being considered a primary, or universal, property of body as those primary qualities which are emphasized in the mechanical philosophy. This is not to say, however, as Newton immediately adds, that gravity is an essential property of body, any more than impenetrability or motion is.[54] It seems clear, therefore, that Newton and Cotes, were clearly concerned to argue that gravity was a property of bodies on a par with extension, mobility, and impenetrability. At the same time, however, it was important to maintain the theological significance of this by insisting that gravity was not an essential, logical concomitant of the very nature of body, but a property superadded by God.

I believe it is safe to say that all of Newton's pronouncements upon gravitational attraction are consistent with the view that he believed gravity to be a superadded inherent property of body which was capable of acting at a distance. His cautious, sometimes tentative, ways of expressing this view were only partly defensive. Certainly he wished, even more than Cotes, to avoid "cavilling" from Cartesian mechanical philosophers, or from the "mathematical readers" he spoke to in the introduction to Section XI of Book I of the *Principia.* But he also wished to present his philosophy in accordance with the sceptical and empiricist principles laid down by the Royal Society, which he, like the leading Fellows of the Society who struggled to establish it, believed to be the surest means of establishing the true philosophy.[55] Moreover, this same way of presenting gravitational attraction, as a causally, inexplicable matter of fact, made evident by phenomena, dovetailed perfectly with his voluntarist theology. Mysterious though it is, God could make matter act at a distance – to deny this was to deny God's omnipotence.

It may be countered to the reading of Newton offered here that it cannot be correct if it disagrees so widely with the interpretation of the foremost Newtonian scholars. This may well be true and, ultimately, I defer to the judgement of others. But, in the meantime, it may be relevant, and I hope not too impertinent, to offer some reflections on why these scholars have, according to my account, consistently misread Newton's views. I believe that the misreadings I have pointed to stem, initially, from the preconceptions of the scholars involved. I think it is surely significant, for example, that the leading proponents of the standard account, Koyré, Cohen and Hall, are the most vigorous in rejecting any suggestions that Newton was influenced by alchemy or other occult traditions. A. Rupert Hall, for example, has recently expressed his fear that scholarly interest in Newton's alchemy will "cloud the clarity of reason and intellectual integrity" in the great man's life's work. "I would have regarded Newton as a founder of reason," he wrote, "so, I think, he wished to be regarded... not as flotsam on the weltering sea of the human unconscious."[56] Given this attitude, I think it is unlikely that Hall and his like-minded colleagues could accept the passage from Bentley's lecture, which I quoted earlier, as a valid illustration of what Newton had in mind about the nature of Gravity:

> Now mutual Gravitation or Attraction (in our present acceptation of the Words) is the same thing with This; 'tis an operation or vertue or influence of distant Bodies upon each other through an empty Interval, without any Effluvia or Exhalations or other corporeal Medium to convey and transmit it.[57]

This is much too obviously an occult view of gravity. If Newton is to be preserved from the taint of occultism, then this cannot be his view but must be the mistaken view of Bentley. Fortunately, this is easy to establish, since we have clear and unequivocal evidence in the letters to Bentley that Newton rejected these very ideas – or so the old familiar story goes. But Newton's own writings do not allow much freedom of interpretation. Accordingly, these scholars have opted for what seemed to them the only alternative, as Cotes foretold in his Preface to the 1713 edition of the *Principia*: "either they will deny gravity to be in bodies, which cannot be said, or else, they will therefore call it preternatural," which is to say, directly caused by God.[58]

The mechanical philosophy is still regarded fondly by thinkers like Koyré, Hall, Cohen and many others, as the foundation upon which all subsequent scientific thinking was erected. We got where we are today, thanks to the rejection of occult ways of thinking and the development of the supremely rational ways of thinking of mechanical philosophers like Galileo, Descartes, and above all Isaac Newton. So, even though Koyré, Hall and Cohen accept that Newton did indeed introduce attractive forces seemingly capable of acting over distances of empty space, they insist on giving a non-occultist account of how he developed these notions. All three of them, therefore, emphasise the importance of Newton's methodological pronouncements in his three rules of philosophizing, in his insistence that it is enough that he has shown that gravity does really exist and that it acts according to the laws he has discovered, and in his development of an elaborate and unique "Newtonian style." All this is excellent and important work but I think there can be no doubt that it is a one-sided view of Newton, failing to see that the first modern scientist was also the last of the magi.[59]

This view is also bound up with the still dominant, late-nineteeth-century trend in physics to denounce and reject all suggestions of action at a distance. From Faraday, Thomson, and Maxwell onwards, actions at a distance have been excluded from legitimate scientific theorizing. There was no such all-pervasive prejudice in the seventeenth and eighteenth centuries as a number of historial studies have clearly shown.[60] Indeed, F. H. van Lunteren has recently pointed to the use of Newton's letters to Bentley by the nineteenth-century opponents of distance-forces:

> the phrase in the third letter to Bentley expressing the absurdity of bodies acting upon another at a distance without the mediation of anything else must have been Newton's most frequently quoted statement in the second half of the nineteenth century.[61]

Newton's words to Bentley came to be seen, therefore, as authoritative pronouncements on the impossibility of actions at a distance, and, of course, clear evidence that the great Newton never believed in them. Having been invested with this kind of authority it is hardly surprising that the original meaning of his words became hard to see. It is the authoritativeness of this reading, perhaps, which prevented R. S. Westfall and Betty Jo Dobbs, two scholars who have never ignored the occult side of Newton, from seeing Newton's comments to Bentley as a tacit acceptance of the possibility of action at a distance.

The history we write, because it depends upon our perceptions as much as the historical evidence, is bound to tell us something not only about our past but also about ourselves. It is this last aspect of history-writing which enables subsequent generations of historians (with a different set of preoccupations) to recognise what reflects the past and what reflects the recently present in the histories of the previous generation. Each generation, therefore, in spite of the handicap of its own preconceptions, can build upon the previous work to refine its reading of the past, sifting out the more "presentist" elements and approaching ever closer to the way things were. "So we beat on, boats against the current, borne back ceaselesly into the past."[62]

Notes

* Earlier versions of this paper were read to the Open University History of Philosophy Research Group Seminar, at Birkbeck College, London (May, 1990); and at a conference on "Ciència i Religió: Perspectives Històriques i Filosòfiques', held in the Department of Philosophy of the Universitat Autonoma de Barcelona, Catalonia (March 1992). I am grateful to Professor Tom Sorrell of the Open University, and Professor Antoni Malet of the Universitat Autònoma de Barcelona for inviting me to air my views, and for their generous hospitality during my visits. I owe many thanks also to the audiences on both occasions for their helpful and encouraging comments.

1. The role of light metaphysics in Newton's thinking has not received separate treatment but see Ernan McMullin, *Newton on Matter and Activity* (Notre Dame and London, 1978), pp. 84–96. The role of light is also alluded to in some of the discussions of Newton's alchemy, particularly when discussing his alchemical treatise, 'On the Vegetation of Metals', and his 'Hypothesis explaining the Properties of Light' of 1675. See, for examples, P. M. Rattansi, 'Newton's Alchemical Studies', and R. S. Westfall, 'Newton and the Hermetic Tradition', both in A. G. Debus (ed.), *Science, Medicine and Society in the Renaissance* (New York, 1972), pp. 167–82 and 183–98; and J. E. McGuire, 'Transmutation and Immutability: Newton's Doctrine of Physical Qualities', *Ambix* 14 (1967): 69–95. On Newton's aether speculations see Henry E. Guerlac, 'Newton's Optical Aether', *Notes and Records of the Royal Society of London* 22 (1967): 45–57; P. M. Harman, 'Ether and Imponderables', in G. N. Cantor and M. J. S. Hodge (eds), *Conceptions of Ether* (Cambridge, 1981), pp. 61–83; and R. W. Home, 'Newton on Electricity and the Aether', in Zev Bechler (ed.), *Contemporary Newtonian Research* (Dordrecht, 1982), pp. 191–213. Newton's pronouncements on gravity as an effect brought about by God are discussed in R. S. Westfall, *Never at Rest: A Biography of Isaac Newton* (Cambridge, 1980), pp. 505–11. Westfall's biography should be consulted for excellent discussions of all the foregoing issues as well (the index is excellent). Isaac Newton, *Mathematical Principles of Natural Philosophy*, translated by Andrew Motte, revised by Florian Cajori (Berkeley, 1946), p. 547.
2. A. Rupert Hall, *The Scientific Revolution* (London, 1962), p. 273; idem, *The Revolution in Science, 1500–1750* (London, 1983), p. 323; idem, *Henry More: Magic, Religion and Experiment* (Oxford, 1990), p. 231, 232, 238, 240. Although the first two of these books were intended for undergraduate

audiences, I have included them in my survey because that will have gained them a certain amount of influence in the field.

3. Alexandre Koyré, 'Gravity an Essential Property of Matter?', in idem, *Newtonian Studies* (London, 1965), pp. 149–63, see p. 149, 152, 149.
4. I. B. Cohen, 'Newton's Third Law and Universal Gravitation', *Journal of the History of Ideas*, 48 (1987): 571–93, pp. 587–88, 588–89. This article also appears in P. B. Scheurer and G. Debrock (eds), *Newton's Scientific and Philosophical Legacy* (Dordrecht, 1988), pp. 25–53.
5. Ernan McMullin, *Newton on Matter and Activity* (see note 1), p. 104.
6. McMullin, *Newton on Matter and Activity* (see note 1), pp. 105–6; Hall, *Henry More* (see note 2), p. 232.
7. Isaac Newton, *Four Letters from Sir Isaac Newton to Doctor Bentley containing some Arguments in Proof of a Deity* (London, 1756), Letter II, p. 20. This letter is dated January 17 1692/3. The letters are conveniently reprinted in I. B. Cohen (ed.), *Isaac Newton's Papers & Letters on Natural Philosophy*, 2nd edition (Cambridge, Mass., and London, 1978), pp. 279–312. See, p. 298. Henceforward I will refer to this edition, as *Papers & Letters*, and give page numbers of the original edition in brackets.
8. *Papers & Letters*, pp. 302–3 (25–6). This letter is dated February 25, 1692/3.
9. Isaac Newton, *Opticks, or A Treatise of the Reflections, Refractions, Inflections & Colours of Light*, based on the 4th edition London, 1730 (New York, 1979), pp. 370–1, 375–6.
10. McMullin, *Newton on Matter and Activity* (see note 1), p. 53.
11. R. S. Westfall, 'The Rise of Science and the Decline of Orthodox Christianity: A Study of Kepler, Descartes, and Newton', in D. C. Lindberg and R. L. Numbers (eds), *God and Nature: Historical Essays on the Encounter between Christianity and Science* (Berkeley, 1986), pp. 218–37, especially p. 233; B. J. T. Dobbs, 'Newton's Alchemy and his "Active Principle" of Gravitation', in P.B. Scheurer and G. Debrock (eds), *Newton's Scientific and Philosophical Legacy* (Dordrecht, 1988), pp. 55–80, p. 74. These are discussed more fully below.
12. Richard Bentley, *A Confutation of Atheism from the Origin and Frame of the World*, Part II (London, 1693), pp. 331–2 (19–20). This is conveniently reprinted in *Papers & Letters* and I will refer to that edition henceforth, giving page numbers of the original edition in brackets.
13. *Papers & Letters*, p. 302 (25).
14. *Papers & Letters*, p. 339 (27).
15. Koyré, 'Gravity an Essential Property of Matter?' (see note 3), p. 149. Perry Miller, 'Bentley and Newton', in *Papers & Letters*, pp. 271–78.
16. *Papers & Letters*, p. 341 (29). See also, pp. 332–3 (20–1), and p. 363 (p. 11 in Part III, the Eighth Lecture).
17. *Papers & Letters*, p. 340 (28). Bentley, Letter to Newton, 18 February, 1692/3, see H.W. Turnbull et al. (eds), *The Correspondence of Isaac Newton*, 7 vols. (Cambridge, 1959–71), Vol. 3, pp. 246–253, p. 249.
18. For discussions of Newton's voluntarism see Francis Oakley 'Christian Theology and the Newtonian Science: The Rise of the Concept of Laws of Nature', *Church History* 30 (1961): 433–57; and J. E. McGuire, 'Force, Active Principles and Newton's Invisible Realm', *Ambix* 15 (1968): 154–208.
19. Newton, *Opticks* (see note 9), pp. 400–01.
20. Newton, *Opticks* (see note 9), pp. 404–5.
21. Newton, *Principia* (see note 1), p. 399.
22. This is a difficult point to get across. Consider the phenomena of shadows. Would it not seem misleading to describe shadows as immaterial entities? The unenlightened might jump to the conclusion that shadows must, therefore, be spiritual beings. To avoid this, we might prefer to consider them to be material phenomena, even though they are not composed of matter, because they are produced by the behaviour of matter and cannot exist without material objects to bring them into existence. Perhaps there should be a third category for referring to such epiphenomena of matter, but in the meantime, whether shadows are material or immaterial must be left to the consideration of individuals.

Another point to bear in mind when dealing with historical concepts of matter and spirit is that the distinction between material and immaterial in seventeenth-century thought was by no means always so clear cut as it was in fully committed philosophical dualists, such as Descartes or Henry More. The status of light, for example, on a spectrum from material to immaterial never achieved a consensus, and J. E. McGuire in his detailed studies of Newton's concept of force certainly concluded that Newtonian forces and other active principles were part of an "invisible realm" which could not definitively be said to be either corporeal or incorporeal, but was rather something in between. See McGuire, 'Force, Active Principles and Newton's Invisible Realm' (see note 18).

23. R. S. Westfall, *Force in Newton's Physics: The Science of Dynamics in the Seventeenth Century* (London and New York, 1971), p. 396. Westfall, 'The Rise of science and the Decline of Orthodox Christianity: A Study of Kepler, Descartes, and Newton' (see note 11), p. 233.
24. Newton, *Opticks* (see note 9), p. 403. B. J. T. Dobbs, 'Newton's Alchemy and his "Active Principle" of Gravitation' (see note 11), p. 74; see also idem, 'Stoic and Epicurean Doctrines in Newton's System of the World', in Margaret J. Osler (ed.), *Atoms, Pneuma, and Tranquility: Epicurean and Stoic Themes in European Thought* (Cambridge, 1991), pp. 221–38, especially pp. 232–38.
25. For a full discussion of Newton's speculations about space as the "sensorium" of God see Edward Grant, *Much ado about Nothing: Theories of Space and Vacuum from the Middle Ages to the Scientific Revolution* (Cambridge, 1981), pp. 247–54; B. P. Copenhaver, 'Jewish Theologies of Space in the Scientific Revolution: Henry More, Joseph Raphson, Isaac Newton and their Predecessors', *Annals of Science* 37 (1980): 489–548.
26. Newton, *Principia* (see note 1), p. 545.
27. Newton, *Principia* (see note 1), p. 544, 546.
28. Newton, *Principia* (see note 1), p. 546. Westfall, 'The Rise of Science and the Decline of Orthodox Christianity: A Study of Kepler, Descartes, and Newton' (see note 11), p. 233. McMullin also believes that Newton held immanentist views of God's activity in the world, *Newton on Matter and Activity* (see note 1), p. 55. For an excellent discussion of how Newton tries to arrive at putative secondary causes rather than rely on God's direct intervention see David Kubrin, 'Newton and the Cyclical Cosmos: Providence and the Mechanical Philosophy', *Journal of the History of Ideas* 28 (1967): 325–46. For further discussion of Newton's transcendentalist theology see R. H. Popkin, 'Newton's Biblical Theology and His Theological Physics', in Scheurer and Debrock (eds), *Newton's Scientific and Philosophical Legacy* (see note 11), pp. 81–97; and James E. Force, 'Newton's God of Dominion: The Unity of Newton's Theological, Scientific, and Political Thought', in J. E. Force and R. H. Popkin, *Essays on the Context, Nature, and Influence of Isaac Newton's Theology* (Dordrecht, 1990), pp. 75–102, especially pp. 85–8.
29. On the distinction between *potentia absoluta* and *potentia ordinata* see Francis Oakley, *Omnipotence, Covenant and Order: An Excursion in the History of Ideas from Abelard to Leibniz* (Ithaca and London, 1984); J. E. McGuire, 'Boyle's Conception of Nature', *Journal of the History of Ideas* 33 (1972): 523–42; and idem, 'Force, Active Principles and Newton's Invisible Realm' (see note 18).
30. Mediation does not necessarily imply an ongoing continuous process. According to the *Oxford English Dictionary* "to mediate" can mean: "To be the intermediary or medium concerned in bringing about [a result] or conveying [a gift etc.] . . . ".
31. For discussion of Leibniz's accusation and Newton's response see Koyré, 'Attraction an Occult Quality?', idem, *Newtonian Studies* (see note 3), pp. 139–48.
32. *Papers & Letters*, p. 280 (1).
33. *Papers & Letters*, p. 342 (30).
34. *Papers & Letters*, p. 341 (29). Bentley, letter to Newton, 18 February 1692/3, in H. W. Turnbull *et al.* (eds), *Correspondence of Isaac Newton* (see note 17), p. 249.
35. Koyré, 'Gravity an Essential Property of Matter?' (see note 3), p. 149, 152. Hall, *The Scientific Revolution* (see note 2), p. 275.
36. Newton, *Opticks* (see note 9), p. 339. In saying that English contemporaries would have taken Newton's voluntarism for granted I do not mean to imply that they all would have accepted it,

simply that they would have recognized it for what it was. There were, of course, English thinkers who subscribed to intellectualist forms of theology. They would, presumably, have sympathized more with the thinking of Leibniz. On Leibniz's intellectualist theology see A. O. Lovejoy, *The Great Chain of Being: A Study of the History of an Idea* (Cambridge, Mass., 1936), pp. 144–51; Oakley, *Omnipotence, Covenant, and Order* (see note 29); H. G. Alexander (ed.), *The Leibniz-Clarke Correspondence* (Manchester, 1956); and Steven Shapin, "Of Gods and Kings: Natural Philosophy and Politics in the Leibniz-Clarke Disputes", *Isis* 72 (1981): 187–215.

37. McMullin, *Newton on Matter and Activity* (see note 1), pp. 95–101.
38. Koyré, 'Gravity an Essential Property of Matter?' (see note 3), p. 149.
39. Newton, *Principia mathematica* (see note 1), Book I, Section XI, p. 164.
40. Cohen, 'Newton's Third Law' (see note 4), p. 592, and footnote 47 on the same page.
41. For further discussion of this point see Peter Dear, '*Totius in verba*: Rhetoric and Authority in the Early Royal Society', *Isis* 76 (1985): 145–61; and John Henry, 'Occult Qualities and the Experimental Philosophy: Active Principles in pre-Newtonian Matter Theory', *History of Science* 24 (1986): 335–81, especially pp. 358–68.
42. On Cohen's 'Newtonian style' see I.B. Cohen, 'The *Principia*, Universal Gravitation, and the "Newtonian Style", in relation to the Newtonian Revolution in Science: Notes on the Occasion of the 250th Anniversary of Newton's Death', in Bechler (ed.), *Contemporary Newtonian Research* (see note 1), pp. 21–108; and idem, *The Newtonian Revolution: With Illustrations of the Transformation of Scientific Ideas* (Cambridge, 1980), and 'Newton's Third Law and Universal Gravitation' (see note 4). It was not, incidentally, merely fortuitous that the Royal Society's carefully forged methodology should prove amenable to Newton's voluntarism. The leading members of the Society who were responsible for establishing its methods were all theological voluntarists themselves, and they, no less than Newton, were always conscious of the theological import of their work. For a fuller discussion see Henry, 'Occult Qualities and the Experimental Philosophy' (see note 41), especially pp. 352–58; and idem, 'The Scientific Revolution in England', in R. Porter and M. Teich (eds), *The Scientific Revolution in National Context* (Cambridge, 1992), pp. 178–210.
43. Newton, *Opticks* (see note 9), p. cxxiii.
44. Alexander (ed.), *Leibniz-Clarke Correspondence* (see note 36), Clarke's Fourth Reply, para. 45, p. 53. Leibniz's Fifth Paper, paras. 118–19, p. 94, Clarke's Fifth Reply, para. 118–23, p. 118.
45. Koyré, 'Gravity an Essential Property of Matter?' (see note 3), p. 149.
46. Miller, 'Bentley and Newton' (see note 15), p. 274, 277.
47. *Papers & Letters*, pp. 363–4 (11–12, in Part III, the Eighth Lecture).
48. Hall, *Henry More* (see note 2), p. 238.
49. Bentley, Letter to Newton, 18 February, 1693, in H. W. Turnbull *et al.* (eds), *The Correspondence of Isaac Newton* (see note 18) Vol. 3, p. 249. Miller, "Bentley and Newton" (see note 15), p. 275.
50. Cotes's letter to Bentley quoted from Koyré, 'Attraction, Newton and Cotes', in *Newtonian Studies* (see note 3), pp. 273–82, p. 281.
51. Koyré, 'Attraction, Newton and Cotes', in *Newtonian Studies* (see note 3), pp. 273–82, p. 281.
52. Cotes, *Preface* to Newton, *Principia mathematica* (see note 1), p. xxvi. See also Cotes's letter to Bentley, quoted in Koyré, 'Attraction, Newton and Cotes', in *Newtonian Studies* (see notes 3), pp. 281–2.
53. Newton, *Principia mathematica* (see note 1), pp. 399–400. For a fuller discussion see J.E. McGuire, 'Atoms and the "Analogy of Nature": Newton's Third Rule of Philosophizing', *Studies in the History and Philosophy of Science*, 1 (1970): 3–57.
54. Newton, *Principia mathematica* (see note 1), p. 400. This line and what follows, the somewhat laconic closing words of Newton's comments on Rule III, have posed severe problems for Newtonian exegetes: "Not that I affirm gravity to be essential to bodies: by their *vis insita* I mean nothing but their inertia. This is immutable. Their gravity is diminished as they recede from the earth" (p. 400). McMullin (see note 1), p. 67. calls this "a perturbed and confusing disclaimer". For a fuller discussion of the difficulties of interpreting this passage see McMullin, pp. 61–71. In spite of McMulin's and others' efforts the passage remains inscrutable. Fortunately the argument presented here does not depend upon the meaning of this passage.

55. See references in note 41 above. The best illustration of how the Royal Society's method was seen by its adherents as the securest means of arriving at truth is Steven Shapin and Simon Schaffer. *Leviathan and the Air-Pump: Hobbes, Boyle and the Experimental Life* (Princeton, 1985); but see also Henry, 'The Scientific Revolution in England' (see note 42).
56. A. Rupert Hall, in a letter to R. S. Westfall, quoted with permission in Westfall, 'Newton and Alchemy', in Brian Vickers (ed.), *Occult and Scientific Mentalities in the Renaissance* (Cambridge, 1984), pp. 315–335, p. 316.
57. *Papers & Letters*, p. 341 (29).
58. Newton, *Principia* (see note 1), p. xxvii.
59. On the "Newtonian style" see note 42 above. The reference to Newton as the last of the magi is intended, of course, to endorse John Maynard Keynes's 'Newton, the Man', in The Royal Society, *Newton Tercentenary Celebrations* (Cambridge, 1947), pp. 27–34.
60. For example, Arnold Thackray, *Atoms and Powers: An Essay on Newtonian Matter-Theory and the Development of Chemistry* (Cambridge, Mass., 1970); Robert E. Schofield, *Mechanism and Materialism: British Natural Philosophy in an Age of Reason*, (Princeton, 1970); Peter M. Heimann and J. E. McGuire, 'Newtonian Forces and Lockean Powers: Concepts of Matter in Eighteenth-Century Thought', *Historical Studies in the Physical Sciences*, 3 (1971): 233–306. For a recent essay which makes the same point see Curtis Wilson, 'Euler on action-at-a-distance and Fundamental Equations in Continuum Mechanics', in P. M. Harman and A. E. Shapiro (eds), *The Investigation of Difficult Things: Essays on Newton and the History of the Exact Sciences* (Cambridge, 1992), pp. 399–420.
61. F. H. van Lunteren, 'Gravitation and Nineteenth-Century Physical Worldviews', in P. B. Scheurer and G. Debrock (eds), *Newton's Scientific and Philosophical Legacy* (see note 4), pp. 161–73, pp. 166; see also note 38, p. 171, where he provides a sample of references.
62. These are, in case anyone does not recognise them, the closing words of F. Scott Fitzgerald's, *The Great Gatsby* (1926).

X

THE FRAGMENTATION OF RENAISSANCE OCCULTISM AND THE DECLINE OF MAGIC

THE TOUCH OF COLD PHILOSOPHY?

At a Christmas dinner party in 1817 an admittedly drunken Charles Lamb berated the famed Isaac Newton as "a Fellow who believed nothing unless it was as clear as the three sides of a triangle". He and John Keats then agreed that Newton had destroyed all the poetry of the rainbow, by reducing it to a prism.[1]

Lamb and Keats, it seemed, wished Newton in hell, as William Blake had seemed to do earlier in his long poem *Jerusalem* (1804), where he blamed the "cogs tyrannic" of the newly industrializing Britain on the "Water-wheels of Newton" which drove the "Loom of Locke, whose woof rages dire".[2] Not long after this Christmas dinner party Keats made a public statement about Newton's "cold philosophy" in his poem, *Lamia* (1820):

> Do not all charms fly
> At the mere touch of cold philosophy?
> There was an awful rainbow once in heaven:
> We know her woof, her texture; she is given
> In the dull catalogue of common things.
> Philosophy will clip an Angel's wings,
> Conquer all things by rule and line,
> Empty the haunted air, and gnomèd mine –
> Unweave a rainbow....[3]

But the Newton that Lamb and Keats denounced, and whom Blake despised as a facet of the monstrous "Urizen",[4] was merely an image of what the previous age, the "Age of Reason", thought the great scientist should be like. It was an image of Newton that was the product of Voltaire and other admirers of Newton who saw him as a symbol of what human reason might achieve. The Enlightenment image of Newton, particularly in France, served ideological purposes. For Voltaire Newton's success did not simply derive from his mathematical acuity, his natural philosophy was in some way seen as a testament to the toleration, reason and good sense which prevailed among the English and which also provided them with an enviable political system. The Enlightenment image of Newton had little to do, therefore, with the man himself.[5]

If we take a closer look, however, at what we can reconstruct of Newton's actual beliefs about the nature of the rainbow, it seems hard to resist the feeling that Keats, Lamb and Blake might have been much more inclined to embrace his ideas. Indeed, the very fact that all of us raised in the British tradition believe that there are seven colours in the rainbow — which we try to remember by means of mnemonics like "Richard Of York Gave Battle In Vain" — owes more to Newton's belief in the harmonies of the world than it does to his reputed genius as an experimental scientist (or to the visible appearance of the rainbow — has anyone ever been able to see *seven* colours? Has anyone ever seen indigo in the spectrum?).[6]

Newton's preoccupations become apparent when we look at his own account of his experimental discovery of the nature of the spectrum. In the *Opticks*, first published in 1704, he describes in detail how he came to distinguish the colours in the rainbow. Having projected the spectrum from a prism on to a piece of paper, he called upon the aid of an assistant:

> I held the paper so that the spectrum might fall upon it, whilst an Assistant, whose eyes for distinguishing colours were more critical than mine, did by Right Lines … drawn cross the Spectrum, note the Confines of the Colours, that is of the red…, of the orange…, of the yellow…, of the green…, of the blue…, of the indico…, and of the violet…. And this Operation being divers times repeated both in the same and in several Papers, I found that the observations agreed well enough with one another, and that the Rectilinear Sides [of the projected spectrum] … were by the said cross Lines divided *after the manner of a Musical Chord*.[7]

It is this last comment which leads us deep into the notion of cosmic harmonies and Newton's belief in ancient Pythagorean wisdom. The seven colours of the rainbow, according to Newton, correspond to the seven notes of the octave, and they correspond so closely that the lines drawn across the spectrum by his supposed assistant are in precisely the same place you would need to bridge a monochord of corresponding length to give you each of the seven notes in the diatonic scale. The importance of this idea for Newton can be inferred, I think, from the fact that it appears in the *Opticks* essentially unchanged from the version which he had presented in his earliest public discussions of the nature of light in one of the papers delivered before the Royal Society of London in the mid-1670s.[8]

The immediate importance of the analogy between light and music for Newton was that it enabled him to answer an anticipated objection to his claim that colours were not, as all earlier natural philosophers insisted, modifications or even corruptions of pure, white, light, but that they were in fact, each one of them, the pure forms of light.[9] If Newton's theory of light was true, it raised an obvious question for his contemporaries: Why might God have made the glorious light of the sun the mere result of a mixture of what, in the aesthetics of the day, were the less glorious kinds of light seen typically through coloured glasses?[10] Right from the outset Newton had a ready answer. White sunlight should not be seen as a corrupt form of light made by a mixture of purer lights, it should be seen as the result of a superb harmony of the coloured lights shining in unison. Furthermore, it seems safe to say that it must have seemed obvious to Newton that if God had made sunlight in this way, then the colours would show some close similarities to the notes in the octave (and so, of course, there had to be seven). That Newton was predisposed towards this view becomes clear when we look at Newton's first announcement of this idea, in the Optical Lectures which he delivered at Cambridge University from 1670. There is no mention of an assistant in this earlier account, and Newton admits that he could only distinguish five colours. Accordingly, he told his students, he added indigo and orange "in order to divide the image into parts more elegantly proportioned to one another", and to make the analogy with the musical scale possible.[11]

It seems perfectly evident that underlying this aspect of Newton's thought was a belief in the essential truth of the ancient Pythagorean doctrine of cosmic harmony. Pythagoras was renowned as one of the greatest of the ancient sages, and as the first teacher of the concept of cosmic harmonies and of the importance of number and measure for a complete understanding of the world. We have direct evidence of Newton's familiarity with this ancient tradition from a number of remarkable manuscript passages, known as the "Classical scholia", which he wrote for inclusion in an abandoned second edition of the *Principia mathematica*. Here Newton intended to draw upon what was known of ancient Pythagorean doctrine to support his views on gravitational attraction.[12]

Until quite recently, the seriousness of such ideas in Newton's thinking remained controversial. For those commentators who were more positivistically inclined, it was assumed that Newton could not have been serious about such dubious historical claims, and the fact that they remained in manuscript, withheld from publication, was cited as evidence for his lack of commitment. We now know, however, that these ideas were bound up with a much greater project that Newton was engaged in, namely, his attempted reconstruction of the original religion before it became corrupted in the generations succeeding that of Noah and his sons. There can be no doubt of the seriousness with which Newton pursued these ideas.[13]

It is important to note that the link between the spectrum and the musical scale was not confined to Newton's unpublished manuscripts. It cannot be argued, therefore, that Newton wasn't serious about this idea and thought better of publishing it. As well as announcing it to the Royal Society in the 1670s, he published it in the highly influential *Opticks*. And yet, in spite of the unambiguously clear account of the analogy between the colours of the spectrum and the notes of the diatonic scale in the second of Newton's two great books, it was evidently not an idea which resonated with Enlightenment natural philosophers. Newton's affirmation of Pythagorean cosmic harmonies was soon forgotten and by the early nineteenth century Newtonian science could be disparaged by Keats (evidently speaking for others also) as coldly and unimaginatively unweaving the rainbow. Only in the second half of the twentieth century did scholars begin to correct the Enlightenment image of Newton and to recognize the more magical aspects of his work. The result of this on-going research is to acknowledge Newton as the "last of the magicians", the "last wonder child to whom the Magi could do sincere and appropriate homage".[14] There can be little doubt that he was the last, or among the last. By the time of Newton's death in 1727, the new reformed natural philosophy, which began to emerge in the sixteenth century and which had found its first programmatist in Francis Bacon, was sufficiently well established that its promoters saw it as *sui generis*, and felt no need to acknowledge its parentage. But Newton was by no means the only natural philosopher who had drawn upon magical traditions. Indeed, Newton's own interest in various magical traditions can best be understood by locating it within a late-Renaissance movement to reform natural philosophy by paying closer attention to various magical or occult traditions.[15]

Although it is now (at last) diminishing, there is still enormous resistance among the more positivist philosophers and historians of science to any suggestion that magic might have been instrumental in the emergence of modern science. It is remarkable, for example, that the authors of two recent books on the role of alchemy in the Scientific Revolution, one introductory the other advanced, both felt the need to justify the claims they were making on behalf of alchemy because of its "associations with magic and the occult".[16] For the most part, the arguments against the possible influence of magic on science are presented *a priori*, while the historical evidence is simply ignored. So, magic is characterized as irrational and its influence upon a supremely rational pursuit like modern science is easily dismissed as inherently implausible. Similarly, magic is said to be concerned with the *supernatural* and therefore could only be antithetical to mankind's heroic intellectual endeavour to explain phenomena in entirely *naturalistic* terms.[17] What is particularly unfortunate about this approach is that, by dismissing magic at the outset, it fails to put any effort into understanding the nature and significance of magic in the pre-modern and early modern periods. But this ahistorical approach is intellectual chauvinism of the most arrogant kind, and the result is undoubtedly a diminishing of our understanding of the origins of modern science.[18] To carry on in this vein is to repeat the errors of Sir David Brewster, Isaac Newton's first biographer. Taking the opportunity to scrutinize Newton's manuscript remains, Brewster soon came across the huge mass of alchemical manuscripts. His appalled response is well known:

> ... we cannot understand how a mind of such power, and so nobly occupied with the abstractions of geometry, and the study of the material world, could stoop to be even the copyist of the most contemptible alchemical poetry, and the annotator of a work, the obvious production of a fool and a knave.[19]

When seen in the light of Brewster's overwhelming admiration for Newton this is highly significant. An observer might have expected that Brewster would be led by his otherwise slavish veneration for his great forebear to conclude that, if Newton was so interested in alchemy, then there must have been something in it. But no, evidently Brewster's conviction that alchemy was worthless rubbish outweighed even his awe of Newton's genius. This paper, however, is based on the assumption that if many of the leading figures in the Scientific Revolution (undoubtedly among the leading thinkers of their age) drew upon magical traditions, it is our job as historians to try to recover what it was that they saw in those traditions. In the process, we will not only learn more about the nature of magic in the Renaissance and in the early modern period, but also about the origins of modern science.

A positivistically inclined reader might be thinking at this point, however, that the fact that Newton's musical analogy as a way of understanding the nature of light has been all but overlooked by succeeding generations of physicists (and so forgotten that even a romantic thinker like Keats was unaware of it) shows that it wasn't of any real significance in the history of science. Knowledge of Newton's belief in such matters is only an historical nicety, helping us to gain a more complete picture of "Newton, the man", but of no relevance to our understanding of the historical development of science. In view of the fact that less than a hundred years after his death Newton was found guilty of unweaving the rainbow, of removing its awfulness (in the sense of its ability to strike awe) and putting it in the "dull catalogue of common things", this point perhaps should be conceded. But what about those other aspects of Newton's physics which were also influenced by magical traditions? We cannot dismiss Newton's belief in actions at a distance as equally irrelevant to the history of science. Although, in this particular case, it is interesting to note that Newton's acceptance of *actio in distans*, clear as it is to see in his writings, has nonetheless been vigorously denied, firstly by scientists and then by historians of science. [20]

The concept of *actio in distans* had been rejected by Aristotle and therefore excluded from mainstream natural philosophy from its beginnings in the Latin West in the twelfth century. It was always a prominent feature of the magical tradition, however, being a mainstay of notions of sympathy and antipathy, and therefore appearing in many of the occult arts and sciences, from astrology to alchemy and beyond. It is now generally acknowledged that Newton's easy acceptance of actions at a distance (manifested most obviously in his concept of gravity, but also in the micro-matter theory discussed in the "Queries" appended to the *Opticks*) derived from his own work in alchemy, where he can be seen to make assumptions about particles operating on one another across a distance.[21] Furthermore, it cannot be denied that these ideas were immensely influential on succeeding generations of chemists and natural philosophers. The history of eighteenth-century natural philosophy, especially in Britain, can be seen in terms of those who accept the Newtonian claim that all phenomena can be explained in terms of attractive and repulsive forces operating between particles, or in terms of those who accept Newton's idea that all phenomena might be explained by a highly rarefied yet highly transmissive aether, consisting of particles widely separated as a result of strong repulsive forces operating between the particles.[22]

It seems perfectly clear from Newton's example, therefore, that some aspects of the magical tradition were recognized by early modern thinkers as useful, and by implication, valid or true, while other aspects of the tradition were either ignored or rejected, and were by implication held to be invalid or false. The main aim of this article is to demonstrate that this was not just a feature of magical ideas as far as Newton was concerned, but was in fact the more general fate of the magical tradition. Some aspects of the tradition were taken up by practitioners and became absorbed into reformed versions of natural philosophy, while other aspects of the tradition were rejected.

Indeed, it seems perfectly clear that something recognizably like modern science first emerged as a direct result of the absorption of various aspects of the magical tradition into traditional contemplative natural philosophy. Both the experimental method, and the concern that knowledge of the natural world should be put to use for the benefit of mankind, can be seen to have been long-established aspects of the magical tradition which came increasingly to be embraced by students of nature, who thereby turned traditional natural philosophy into one or other of the so-called new philosophies of the early modern period.[23] It hardly seems necessary to repeat the arguments in support of this claim here. This does not mean, however, that there is nothing more to be said. If it is true that major aspects of the magical tradition became absorbed into what we might call (if we are allowed a bit of anachronistic leeway) modern science, then this has implications for claims that have been made about a perceived decline of magic at the end of the seventeenth century. The 'decline' in question, of course, is the decline of magic as a topic for serious scholarly investigation and discussion. While magical ideas continued to flourish in popular culture, they declined dramatically among the highly educated élite, and came to be regarded as well beyond the intellectual pale. The reasons for this are undoubtedly manifold, and a number of reasons for the decline have already been discussed, most notably, of course, in Keith Thomas's *Religion and the decline of magic*.[24] The aim of this paper is to suggest another major reason for the decline of magic; a reason which has not been discussed before, and which has significant implications for our understanding of early modern intellectual history.

In the rest of this paper, therefore, I want to offer a new perspective to add to previous attempts to explain the so-called decline of magic. I suggest that there was a *fragmentation* of the occult arts and sciences during the Renaissance and early modern periods, as some aspects of the magic tradition became appropriated into the new philosophy, or new science. To a large extent it was the input from magic that made the new philosophies what they were, not only with regard to the experimental method and the new ethos that natural knowledge should be pragmatically useful, but also with regard to the substantive content of those new philosophies. At the same time, however, other aspects of the magical tradition were firmly rejected. These historical changes are perhaps best understood in terms of what sociologists of science have called 'boundary work', the process of demarcating supposed legitimate and valid procedures and presuppositions in establishing natural knowledge from those that are deemed invalid and illegitimate.[25] From the Renaissance through the period known as the Scientific Revolution there was a complete rearrangement of the boundaries of what was magic or occult and what was not, which in turn involved a redrawing of the boundaries which determined what was natural philosophy and what was not. Furthermore, it is my contention that this led to a decline in the fortunes (among orthodox thinkers at least) of what was left behind in the realms of magic.

My account also helps us to see, therefore, what is in fact a major historical and historiographical irony. The reason why positivistically inclined commentators on the development of science have refused to acknowledge the relevance of magic to the history of science is because they mistake the rejected left-overs of the magical tradition — the pathetic rump of the tradition remaining after early modern natural philosophers had taken what they wanted from it — as the whole of the tradition. I said earlier that such historians, like Brewster confronted with Newton's alchemy, refuse to make any attempt to understand the nature of the magical tradition. Just as Brewster, writing in the 1850s, knew the current reputation of alchemy and didn't try to assess its reputation in Newton's day, so certain modern commentators of science have relied upon their current understanding of what magic is (and by implication always has been), and have refused to accept the claims of other historians that magic was once so different that, properly understood, it is easy to see how it might have, and indeed did, influence the development of modern science. The currently prevailing conception of the magical tradition began to be forged in the eighteenth century and has continued into our own times, no less than the image of Newton as "a Fellow who believed nothing unless it was as clear as the three sides of a triangle", began to be forged in the Enlightenment and has been reinforced ever since.[26]

Before proceeding it is important to note that I frequently talk here of a magical tradition as if there was indeed a unified tradition. This is in fact almost certainly not the case, and I am fully aware that I am imposing a unity on numerous different aspects of occult thinking in order to then say this unity is fragmented! I do this simply to make my overall argument clear and avoid tedious circumlocutions at every turn. My argument by no means relies on the false claim that there was a unified tradition; on the contrary, I am arguing here that, during the Renaissance and in some cases persisting through to the late seventeenth century, various different ways of understanding natural phenomena which were excluded from mainstream scholastic natural philosophy, and which were all to some extent based on occult or magical assumptions, began to be considered more seriously by the learned, and were considered as possible ways of reforming the traditional natural philosophy which came to be seen as increasingly inadequate for a proper understanding of the natural world. My claim is not that the upshot of these fresh looks at occult arts or sciences led natural philosophers to conclude that there was a unified tradition here which they then proceeded to break up. I simply wish to assert that some aspects of this diverse set of arts and sciences were seen as useful for contemporary reformist ambitions and were absorbed into natural philosophy, completely transforming that philosophy in the process, while other aspects were rejected. The fragmentation of Renaissance occultism should not be seen as similar to the fragmentation of western Christianity after the Reformation, therefore, because there was no previous occult tradition comparable in its monolithic nature to Roman Catholicism. My title should perhaps have indicated a discussion of the selective take-up of various occult arts and sciences and the rejection of others, but I hope the more succinct phrasing of the first part of my title is not too misleading. It fits in better, anyway, with the historiographically commonplace (and equally misleading) talk of the "decline of magic".[27]

Having said that, let us begin by taking as broad an overview as we can of the nature of what I have called in my title "Renaissance occultism" and its would-be practitioners.

THE NATURE OF 'MAGIC' AND OF 'MAGICIANS'

One major reason for the prevailing mistaken conception (by positivist historians and others) of the nature of magic in the Renaissance is the lack of any understanding of what was known as *natural* magic. Lack of awareness of the natural magic tradition is due to the fact that it was to a large extent completely absorbed into what we now think of as science, while other, lesser, aspects of the tradition have remained in what should be regarded as merely a rump of the magical tradition — what was left over after parts of the tradition had been absorbed into natural philosophy. Today, we tend to identify magic with the supernatural (if we leave aside the stage trickery of 'show-business' magic), but in the period we are looking at, to describe an event or a phenomenon as supernatural was to say that it had been brought about miraculously by God — only God was above nature, and only God could perform a supernatural act.[28] Magic, by contrast, exploited the *natural* properties of things and the successful magician was believed to be highly knowledgeable about the different occult qualities of things. As Giovanni Battista Della Porta wrote:

> Magic is nothing else but the knowledge of the whole course of Nature. For whilst we consider the Heavens, the Stars, the Elements, how they are moved, and how they are changed, by this means we find out the hidden secrecies of living creatures, of plants, of metals, and of their generation and corruption; so that this whole science seems merely to depend upon the view of Nature.... This Art, I say, is full of much vertue, of many secret mysteries; it openeth unto us the properties and qualities of hidden things, and the knowledge of the whole course of Nature; and it teacheth us by the agreement and the disagreement of things, either so to sunder them, or else to lay them so together by the mutual and fit applying of one thing to another, as thereby we do strange works, such as the vulgar sort call miracles, and such as men can neither well conceive, nor sufficiently admire.... Wherefore, as many of you as come to behold Magick, must be persuaded that the works of Magick are nothing else but the works of Nature, whose dutiful hand-maid magick is.[29]

This definition appears in the most popular textbook of magic of its day, simply titled *Magia naturalis* (quoted from London edition of 1658, but first published in 1589), but the same claims about the nature of magic are repeated time and again. Cornelius Agrippa, a leading Renaissance contributor to the magical tradition, insisted that "magicians are careful explorers of nature only directing what nature has formerly prepared, uniting actives to passives and often succeeding in anticipating results so that these things are popularly held to be miracles when they are really no more than anticipations of natural operations".[30]

The major assumption of natural magic, then, was that all bodies have occult qualities which make them capable of acting upon other bodies in various ways, though in many cases the working of these occult qualities are supposed to be highly specific. The main method of putting magical knowledge to use, therefore, is to bring together a body known to have a specific action and the body upon which it is known to act, or else to separate such reactants for a negative effect. This is what Della Porta meant by sundering or laying together things in accordance with their "agreement and disagreement", and what Agrippa meant by "uniting actives to passives". This doctrine made a major impression on the great reformer of natural philosophy, Francis Bacon, who stated it in the fourth Aphorism of his influential *New organon*: "Towards the effecting of works, all that man can do is put together or part asunder natural bodies. The rest is done by nature working within."[31] Such occult interactions were often described in terms of sympathies and antipathies between bodies, a notoriously magical way of talking which nevertheless was employed with minor changes by such leading exponents of the new philosophy as Robert Hooke, who spoke of congruities and incongruities between bodies, and Isaac Newton, who explained certain chemical phenomena in terms of principles of sociability and unsociability.[32]

When compared with natural magic, other aspects of the tradition, aspects which today are all too often held to be characterizing features of magic, were distinctly subordinate. The truly learned magus was held to be a man (it was always a man of course[33]) with a vast knowledge of "how to effect things worthy of the highest admiration ... by the mutual application of natural actives and passives".[34] The great magician, in other words, knew by experience many of the operations of the occult qualities of things and knew how to put that knowledge to use. Lesser magicians, however, might have to resort to one or other of two alternative aspects of the tradition, as a substitute for their lack of knowledge. Both of these aspects of the magical tradition were seen as means of cutting corners, or of taking a short cut, to the knowledge of the occult qualities which the real magus would learn by experience (in principle at least — though in practice more usually by relying on magical lore, increasingly *printed* magical lore).[35]

I am referring here to sorcery (which includes necromancy, theurgy, witchcraft and all other arts of summoning spiritual beings), and semeiology or symbolic magic (which relies upon the power of signs, words and other symbols, and includes numerology, gematria, spellbinding, incantation and so on). These are the very things which many think of today as definitive of magic, but this is largely thanks to the re-drawing of boundaries which took place in the early modern period. In the pre-modern period demonology and symbolic magic were seen primarily as subordinate to natural magic. An important element of symbolic magic, for example, involved the reading of the 'signatures' of things. It was supposed that God, at the Creation, had left physical clues about the secret workings of things, and these were the signatures. So, as one commentator wrote: "besides the manifest and occult qualities of plants, from which their uses may be inferred, [Nature] has marked those which are most useful to us with certain signs and characters."[36] God and Nature, after all, did nothing in vain, and so there must be a reason for every characteristic feature that a thing might have. A favourite example among historians is the walnut: crack open the shell and the flesh of the walnut resembles the human brain sitting in the skull, being divided down the middle and having a surface made up of convolutions. What else could this be but a sign from God that the walnut bears some relationship to the human head or brain? The usual assumption was that the signature indicated some curative power, and so walnuts were assumed to offer a cure, perhaps for headaches, or for mental disturbance. Needless to say, precise determination of the efficacy of walnuts would require empirical research of a trial and error kind. In this respect at least, then, symbolic magic can be seen as a short-cut to the knowledge of occult qualities required by the natural magician.[37]

The link between natural magic and the signatures of things is easy to see, but to link *sorcery* to natural magic seems, on the face of it, bizarre in the extreme. Natural magic, as we've seen, in spite of its name, seems rather close to what we think of as science. It depends, after all, on the assumption that particular bodies have a power of interacting with others to create new bodies, or have a power of easing pain, curing specific diseases, and so on — assumptions which are no different from those of modern chemistry or pharmacology. Demonology, by contrast, seems to us to be completely divorced from any naturalistic conceptions. There is a lesson to be learned here, however, about the astonishing flexibility of our cultural categories, and of our ability to redefine conceptual or disciplinary boundaries.

Surprising though it may seem, sorcery too was seen as little more than a way of avoiding the painstaking gathering of knowledge of occult qualities from experience. If we wish to understand this we must once again be aware of the shift in meaning of the notion of the 'supernatural'. From our perspective it seems a reasonable assumption that for pre-modern thinkers demons were capable of performing supernatural acts to bring about some miraculous event. This fits *our* assumptions about demons. *Our assumptions, however, are historically misguided.* It is something of an irony that present-day notions of demons with comic-book superpowers (as seen on TV in *Buffy the vampire slayer* and other such shows) are the products of secular imaginations. In the pre-modern and early modern intellectual cosmology only God could do supernatural things. Demons, even the Devil himself, were God's creatures and as such were subject to natural law just like the rest of us.[38] As John Cotta wrote in *The triall of witch-craft* (1616): "Nature is nothing else but the ordinary power of God in all things created, among which the Divell being a creature, is contained, and therefore subject to that universal power."[39] In so far as the Devil *could* perform marvellous feats it was only by virtue of the fact that he was a consummate natural magician. The Devil knew the occult qualities of things, and how to apply appropriate actives to passives to accomplish whatever might be required. William Perkins put it rather well in his *Discourse of the damned art of witch-craft* (1618):

> [The Devil has] exquisite knowledge of all natural things, as of the influences of the starres, the constitutions of men and other creatures, the kinds, vertues, and operation of plantes, rootes, hearbes, stones etc., which knowledge of his goeth many degrees beyond the skill of all men, yea even those that are most excellent in this kind, as Philosophers and Physicians are.[40]

King James VI of Scotland likewise believed that the Devil was "farre cunningner then man in the knowledge of all the occult proprieties of nature".[41] Accordingly, if a would-be practitioner of natural magic was at a loss about how to accomplish a particular outcome he might indulge in necromancy to summon one of the dead, on the assumption that the dead person might know how to bring about the desired end.[42] The more ambitious magus, or one with a bigger problem might similarly decide to summon a demon, or even the Devil himself. It is important to note, however, that if the Devil did succeed in performing what the magus wanted it would be because, as William Perkins wrote:

> ... in nature there be some properties, causes, and effects ... most familiar unto him [the Devil], because in themselves they be no wonders, but only mysteries and secrets, the vertue and effect whereof he hath sometime observed since his creation.[43]

The Devil, in other words, does these things in the same way that the natural magician does, but with greater success because of his greater experience — the Devil, after all, has been around for a very long time. The vulgar might think that the Devil and the magus are capable of producing miracles, but, as Cornelius Agrippa insisted, both merely anticipate and exploit natural operations.[44] "Demons operate nothing", wrote Francesco Giuntini, "except by natural application of active forces to the appropriate and proportionate passive objects, which is the work of nature".[45]

So, although the logic of symbolic magic and the logic of demonology were closely linked to the logic of natural magic in the pre-modern period, by the end of the sixteenth century a re-alignment was under way. By the end of the seventeenth century major aspects of the natural magic tradition had been appropriated by the new philosophies or redefined in order to fit more easily with the new kinds of naturalism. But symbolic magic, demonology and some aspects of natural magic, such as astrology, and the chrysopoeic aspects of alchemy, were left aside in what was effectively a new, differently defined, category of magic.

An important aspect of the re-designation of natural magic as a set of assumptions that could be more fruitfully exploited in natural philosophy was, effectively, a denial that natural magic was magic. Thanks principally to the power of the Church, magic had always had what today would be called a 'bad press'. As if the escapades of frauds and charlatans claiming to be alchemists, astrologers, and magicians were not enough to damage the reputation of magic, the Roman Catholic Church tended to emphasize its demonological aspects in order to present it as dangerous and irreligious. It seems clear that the Church wanted to avoid confusion between the miraculous and the kind of marvellous things which were achievable through natural though occult means. Natural magic seemed to suggest, to the uneducated at least, that miraculous things could be accomplished by laymen without supernatural aid. This implicit threat to the authority of the Church could be neatly turned around by insisting that all magic was accomplished by demonic aid, and so condemning it in the most vigorous terms. For the Church, every magus was a Doctor Faustus (and during the witch-crazes every village 'cunning man' or more especially 'cunning woman' was accused of deriving their knowledge not from local lore but directly from Satan).[46] It is this attitude of the Church which underlies the comments we have already seen, in which Della Porta, Agrippa, and others insist that magic is merely the study of nature and so, by implication, no more dangerous to the faith than natural philosophy.

It is hardly surprising, therefore, that although we come across many reputed magicians in the historical record, we do not come across many who declare *themselves* to be magicians; on the contrary, they usually deny it. Nobody was reputed a greater magician than Roger Bacon and yet Bacon himself vigorously denied that he did anything by magic. If we were to take Bacon, and other magicians in denial, at their word, however, we might have to conclude that there was no such thing as a magical tradition, and that nobody ever was a magician. In a sense the latter is true, because there never was a Merlin, or a Faust, there were only mathematicians, alchemists, cabbalists, natural philosophers of a more mystical bent than usual, humanist scholars enthralled by Neoplatonic theurgy, and so forth.[47] But we need to bear in mind the historical actors' categories, not our own. From the point of view of his contemporaries, Roger Bacon was, as the late George Molland pointed out, "a full-blooded magician", and in Molland's estimation this was hardly surprising since, in spite of his protestations to the contrary, Bacon "went some way to meriting his later classification as a magician".[48]

As a result of religious condemnation of magic, then, it wasn't possible simply to appropriate occult traditions in an open way into natural philosophy. Accordingly, an important aspect of the absorption of natural magic into reformed versions of natural philosophy was the defence of those past thinkers who were alleged to be magicians from all charges that they were magicians. Again, as George Molland has pointed out, reputed medieval magicians like Al-Kindi, Albertus Magnus, Roger Bacon, Arnald of Villanova and Michael Scot were transformed in early modern scholarly literature from magicians into heroes of experimental science. The major contribution to this new enterprise was Gabriel Naudé's *Apologie pour tous les Grands Personages qui ont esté faussement soupçonnez de Magie* (Paris, 1625), but John Dee evidently defended Roger Bacon, in a work now lost, from charges of sorcery.[49] Similarly, Robert Hooke later took it upon himself to defend Dee. Having acquired increased notoriety from a newly published account of his supposed converse with various angels, Dee was defended by Hooke as a cryptographer rather than a sorcerer. According to Hooke, these angelic conversations were in fact a "concealed History of Nature and Art". In taking this line, Hooke was simply re-using the same defence which had been used to protect the reputation of Johannes Trithemius — another magus who reported his conversations with angels, but which were later claimed to be merely exercises in cryptography (the point of the exercises being to find what was really being said under the guise of these conversations with angels).[50]

Other exploiters of the magical tradition chose to obscure their indebtedness to the tradition, or to confuse contemporaries as to their commitment to magic. Francis Bacon vigorously criticized magic even as he appropriated many of its precepts and doctrines.[51] Cornelius Agrippa made the status of his *De occulta philosophia* (Cologne, 1533) somewhat ambiguous by publishing what looked like a retraction of it three years before publishing the work itself (although the supposed retraction, *De incertitudine et vanitate scientiarum*, has recently been shown to be far from straightforward).[52] Isaac Newton would never have declared himself to be the last of the magi, but it is significant that he responded to Leibniz's charge that his principle of gravity was a "scholastic occult quality" not by denying that it was occult, but by denying that it was 'scholastic', which is what he meant by rejecting occult qualities which were supposed to depend upon specific forms:

> These Principles [gravity and other "active principles" in matter] I consider, not as occult Qualities, supposed to result from the specifick Forms of Things, but as general Laws of Nature, by which the Things themselves are form'd; their Truth appearing to us by Phaenomena, though their Causes be not yet discover'd. For these are manifest Qualities, and their Causes only are occult.[53]

Although Newton wants rhetorically to present the obvious fact that bodies fall to the ground as 'manifest', gravity was never a manifest quality in the scholastic sense; if its causes were occult then it was occult. Newton knew this, just as he knew that the actions at a distance he invoked in the Queries at the end of the *Opticks*, and in the Preface to the *Principia*, were as much a part of the magical tradition as the alchemy he had so fervently pursued.[54]

Neither silence about magical influence, nor even explicit denial of magic, should be taken as evidence that magical traditions did not play a role in the origins of modern science. We need to be constantly aware of the process which sociologists of science refer to as 'boundary-drawing'. As we shall see, early modern thinkers re-constituted symbolic magic as beyond the intellectual pale, for example, while continuing to accept natural magic; others reasserted the untenability of sorcery (whether on sceptical or religious grounds[55]), while claiming other facets of the magical tradition as defining aspects of natural philosophy. Those positivistic historians and philosophers of science who have regarded magic as antithetical to science have made the mistake, in my view, of neglecting such changes in what constituted magic. They have tended to assume that magic in the early modern period was essentially the same as it is now. In fact, magic has changed radically, chiefly because significant parts of the original tradition have been absorbed into natural philosophy, and redefined by the historical actors themselves (all too conscious of religious opposition to magic) as though they were always aspects of natural philosophy or other legitimate attempts to understand the natural world.[56] I disagree, therefore, with the suggestion of Frank L. Borchardt that, sooner or later, magicians themselves expressed a "disappointment in magic", recognizing that it led inexorably to demonolatry, and repudiated it as they all turned back to religious orthodoxy. It seems to me that the story is rather one of negotiating with the faith, their own as much as that of leading Churchmen, and appropriating certain aspects of magic into their own philosophical systems, while leaving the more religiously dangerous aspects to remain in what became an increasingly demonologically defined (as opposed to the former more *naturally* defined) magic.[57] One of the major reasons why the influence of magic on science (if we can speak anachronistically for the sake of an historiographical argument) has been denied is precisely because those aspects of magic which clearly did influence science are now simply regarded as part of the history of science, and so no longer recognized to be part of the history of magic. Meanwhile, those aspects of magic which were not absorbed into science, and to a large extent were seen in the early modern period as antithetical to a proper understanding of natural phenomena, have come to be regarded as entirely representative of magic, not just as it was after the end of the seventeenth century, but as it was throughout the whole of its career through Western culture. This is simply a very misleading mistake.

CLIPPING ANGELS' WINGS: THE CHANGING STATUS OF DEMONOLOGY

The separation of natural magic from demonic magic is such an important part of this story that it is worth considering the background in more detail. We have suggested that natural magic and demonology were always closely linked, so why didn't these linkages persist after the absorption of various aspects of the magical tradition into the new philosophies? After all, we do know that the various new philosophies continued to be closely affiliated to religion,[58] and so we might expect to see the precepts of demonology being carried over into the new philosophies.[59] As Stuart Clark has shown, however, it is possible to see why demonology was separated from both natural magic and natural philosophy in the early modern period, in spite of (or maybe even because of) the otherwise friendly relations between science and religion. In what follows, I rely entirely upon Clark's analysis.

The sixteenth century was, of course, a period of intense religious turmoil, not only by virtue of the factionalism of the Reformation, but also as a result of major efforts (not unconnected) to increase the levels of spirituality and religious observance among the laity. It was almost inevitable in this atmosphere that the detailed scrutiny of what occult qualities can and cannot do in natural philosophy was bound to have repercussions in demonology, especially as this was also the age of the witch craze.[60]

Since accusations of witchcraft always began with notions of occult influence — the evil eye, or laying a curse or some such — it became important for the demonologist to be able to distinguish between what was a natural effect and what was not. If a villager was able to make a neighbour's milk-cow go dry by natural, even though occult, means, the Church, in principle at least, would not be so concerned about it (because it could be regarded as simply a criminal matter, equivalent to an assault, and a matter for the secular courts). The Church's main concern was with those who were believed to have made a pact, a bargain, with the Devil. Accordingly, it was important to be able to distinguish what could be accomplished without demonic aid, from what could not. The decision as to whether a particular malfeasance was brought about by natural or unnatural means depended of course upon what could be said to be natural (even though occult). This does not mean, however, that demonologists had now departed from tradition and orthodoxy and had decided that the Devil could, after all, perform supernatural phenomena (remember, we saw earlier that the pre-modern view was that the Devil could not do anything supernatural but could only exploit his knowledge of natural magic). No, the unanimous assumption was still that the Devil could perform his deeds only by natural means, but that he was so clever at exploiting occult qualities, that he could deceive onlookers into mistaking just how a particular accomplishment was achieved. The witch might have thought that covering herself with an ointment made from the fat of birds enabled her to fly, but in fact the Devil enabled her to fly by some other natural means, or more likely, simply used natural means to give the witch the illusion of flying (the ointment, in other words, was not an anti-gravity substance, but merely a hallucinogen). In this case, following Thomas Aquinas, the theologian will argue that the ointment is effectively a sign of the witch's compact with the Devil.[61]

So, natural philosophers and theologians were both concerned to decide what was natural and what was not. Needless to say, there were numerous disputes. The up-shot, in demonology, was that if an effect was brought about by spurious means, that is to say, by means which could not be shown to bring about that effect in a natural way, then the conclusion was that the devil was affecting the outcome, and so the human agent was guilty of a satanic pact. A witch charged with poisoning a neighbour who said she had administered the poison by contaminating his well, would be considered differently from one who said she had walked through his wall and dropped the poison into his mouth as he slept. The Devil, being a spirit, can walk through walls, but he cannot arrange for a human being to walk through a wall. He can, however, make the witch *believe* she has walked through a wall, but if she did believe that then she is guilty of colluding with the Devil in some way.

This aspect of demonology was not confined to witch trials, nor to those occasions when an accused was supposedly successful in bringing about a magical outcome. It is easy to see that for the Churchman concerned to improve the general spirituality of his flock, even a *false* belief that a charm, or a particular incantation, or even a particular herb, will bring about a desired outcome, is a sign of lack of trust in God. Popular beliefs about the efficacy of various techniques and rituals for bringing about good health or good fortune, for foretelling the future, or for making the right decision, ignored the divine aspect of Providence and all that went with it (such as an awareness of the need for prayer and repentance). It seemed to the theologian to be at best an idolatry, placing faith in God's creatures rather than God himself, and at worst to be a form of paying homage to the Devil — since these procedures are not thought to work by natural means, the persons who perform them must expect the Devil to intercede for them.[62]

In this way, then, demonology extended itself to embrace the study of superstition. But superstition at this time was regarded by Churchmen with great seriousness; like witchcraft it was seen as "religion's opposite", and it therefore became of the utmost importance to eradicate or to suppress the superstitious magic of popular culture. We can't pursue here the various ways in which the Church tried (often unsuccessfully) to eradicate popular magical beliefs, but it is important for us to note that one essential ingredient in all such efforts was an insistence upon the spuriousness of the causal links between the supposed magical procedure and the desired outcome. Churchmen were concerned, therefore, first of all to understand the *real* causal linkages available in nature (even if they may be occult) and secondly to use that knowledge to draw a clear distinction between legitimate natural philosophy and illicit magic. The category of magic was again redefined in this process; the intellectual boundaries around it redrawn. For Churchmen, magic became interchangeable with witchcraft and superstition, all being seen as attempts to use the power of the devil to bring about a desired outcome.[63] The appalling enthusiasm with which Churchmen redefined magic — changing it from a knowledge of the natural powers of things to a commerce with the Devil — provided the intellectual underpinning for the European witch-crazes.[64] Such religious excesses made it all the more important for those natural philosophers who recognized the usefulness of the magical tradition to extract what they needed from that tradition and to incorporate it into the safe intellectual haven of natural philosophy, denying that it had anything to do with magic. Here again, we can see that what was left behind by the new philosophers, what was *not* taken up by them and incorporated into natural philosophy, was in fact just a patchwork of the earlier magical tradition, but it soon came to be seen as the full picture.

It can be seen, then, that the Churches were vigorously re-asserting what had always been their dominant view, that all magic is sorcery, at the same time that natural philosophers were absorbing much of the tradition of natural magic into their new philosophies. The result was a major shift in perceptions of what was magic and what was not. But such a sea-change in the categorization of magic did not take place at the throwing of a switch. Clearly it was a rather more piecemeal and complex process than I've been able to indicate here. Throughout the second half of the seventeenth century, on the eve of the Enlightenment, it was still possible for a natural philosopher like Boyle, or Newton (or any number of others), to draw upon the old natural magic tradition to provide them with theories of matter, or methodological justifications for occultist explanations, and they can clearly be seen to have done so. It was no longer possible, however, for them to speak meaningfully about the importance of the magical tradition in their work: even by then such a pronouncement would have been misunderstood. Here, for example, we can discern a clear difference between Boyle and Newton on the one hand, and Francis Bacon on the other. Bacon, writing at the beginning of the century, could explicitly discuss the validity of magic, or simply invoke precepts which he overtly affirmed to be appropriated from the magical tradition, but Boyle and Newton never did this.

This is not to say that Boyle, Newton and others at the end of the seventeenth century would have *self-consciously* kept quiet about their indebtedness to the magical tradition — using it, but deliberately avoiding any acknowledgement that they were using it for fear of the Church. That's not how the social process of re-defining disciplinary boundaries works. Participating in the boundary-drawing process themselves, Boyle and his contemporaries almost certainly would have simply denied magical influence (as Roger Bacon had done centuries before). Boyle himself, we know, was highly troubled in his conscience by his attempts to succeed at the old alchemical dream of transmuting lead into gold. Shortly before his death Boyle consulted his close friend Gilbert Burnet, Bishop of Salisbury, for advice and reassurance on some matters that troubled his conscience. Burnet took a record of the discussion and Michael Hunter has recently analysed these notes.[65] It is important to realize that the discussion between Boyle and Burnet does not involve concerns about the matter theory of alchemy, nor are there any doubts that alchemical transmutation is possible. What does concern Boyle, however, is the fact that every successful transmutation which Boyle knew about, including one he allegedly performed himself, was brought about by the use of a ready-made powder which came from a mysterious source. If such an alchemical powder had been made by Boyle himself, he could be sure that it was produced by natural means. There was an obvious danger however if the ready-made powder was simply given to him (as indeed it was) — the powder might not be a natural cause of transmutation but merely a sign from the Devil. Perhaps alchemical transmutation has only ever succeeded with the help of the Devil, whom even Boyle unwittingly invoked when he used the powder given to him by a stranger.[66]

Hunter provides us with another example, this time told by John Flamsteed, first Astronomer Royal. A local Greenwich washerwoman who had had a parcel of linen stolen asked Flamsteed if he could tell her where it was by divination (note that she saw no distinction between an astronomer and an astrologer). Flamsteed was evidently in a flippant mood and he drew circles and squares at his desk before telling her where she might find the linen. You can no doubt guess what happened. She *did* find the linen exactly where Flamsteed said it would be. She returned to the Royal Observatory to give Flamsteed half-a-crown, but Flamsteed was horrified:

> Good woman, I am heartily glad you have found your Linen; but I assure you I know nothing of it, and intended only to joke with you.... But I see the Devil has in mind I should deal with him: I am determined I will not. Never come or send any one to me any more, on such Occasions, for I will never attempt such an Affair again while I live.[67]

It seems clear from these stories that the new concept of magic, as the locus of activity for the Devil, was gaining ground with natural philosophers as well as theologians. It does not follow, however, that these natural philosophers rejected the possibility of transmuting lead into gold by natural means, much less all the other occult phenomena which until recently had been routinely seen as part of the natural magical tradition.

If what was left of magic became increasingly identified with superstition throughout the seventeenth century, it was forced to undergo another shift in the eighteenth century. The intellectual leaders of the succeeding age, the would-be "Age of Reason", had their own agenda, and their own very different reasons for dismissing all superstition, and in their case, all serious talk of demons. In that further shift, during the Enlightenment, the old idea that demons (thanks to the long experience they gained as immortals) were merely adepts of natural magic was completely lost. The secularists of the Enlightenment tended on the one hand to imagine that those benighted individuals who believed in demons believed them to be supernatural beings, while on the other hand they were hardly aware of the tradition of natural magic, since much of it had by then become absorbed into the new philosophies.

THE SELECTIVE ABSORPTION OF ASPECTS OF THE MAGICAL TRADITION

The foregoing should not be taken to mean that natural philosophers simply decided in a deliberate way to look into the magical tradition to see if there was anything they could incorporate into their natural philosophies. This was no more the case than that natural philosophers in the sixteenth century deliberately decided to go and see what artisans and craftsmen were doing, on the chance there might be something they could use. Nevertheless, it is just as true to say that natural philosophers began to become more and more familiar with occult arts and sciences as it is to say that scholars and craftsmen began to interact during the Renaissance as they never had before.[68] In some cases, of course, reformers did extol the deliberate appropriation of knowledge from craft or magical traditions, and no doubt some of their readers did follow suit. Certainly, Juan Luis Vives did urge his scholarly readers in 1531 "to enter into shops and factories, and to ask questions from craftsmen, and to get to know about the details of their work"; and Francis Bacon urged readers of his *Novum organum* (1620) to systematically search through magical lore,

> for although such things lie buried deep beneath a mass of falsehood and fable, yet they should be looked into ... for it may be that in some of them natural operations lie at the bottom; as in fascination, strengthening of the imagination, sympathy of things at a distance, transmission of impressions from spirit to spirit no less than from body to body and the like.[69]

For the most part, however, what we are dealing with is a diffuse movement, throughout Europe and spread over the sixteenth and seventeenth centuries, of thinkers adopting, or adapting, theories, assumptions, and techniques which previously would have been seen as too occult or too susceptible to the charge of being demonic, into what came to be accepted as the new philosophy. Again, different aspects of this diffuse movement need to be understood differently. William Gilbert, who developed an explanation of the perpetual movement of the Earth (demanded by Copernican theory) based on the occult properties of magnets, may not have been looking for a way to explain the motion of the Earth, but realized he could offer an explanation after reading Maricourt's *Epistle on the magnet*. Robert Boyle did not turn to alchemy as a result of his dissatisfaction with Cartesian mechanical philosophy — he already was an alchemist, and may well have recognized inadequacies in Cartesianism precisely because of his alchemical knowledge.[70] In lots of different ways magical ideas became incorporated into the mainstream of philosophical thought, but only in a few cases was this the result of a self-conscious effort to plunder magical traditions.

The precise way in which different areas of the occult arts and sciences were taken up by natural philosophers can be seen, therefore, to be complex and affected by many historical contingencies. It is not possible to provide a model which reveals how the occult was absorbed into mainstream philosophy because each case was very different. The process was not systematic, and may not even have been fully comprehensive, embracing every aspect of the occult, but it was undeniably extensive. What follows in this section is not intended to be a complete account, but merely a preliminary attempt to show how aspects of at least some of the occult sciences came to be incorporated into the new philosophies of the early modern period, while others were considered for inclusion but ultimately rejected.

So-called mathematical magic, for example, was concerned with the demonstration of what could be accomplished by machinery. Machines, after all, were intended to perform marvellous feats which could not be done by normal means, and they did so in ways that were by no means manifest to a casual observer. Their operations were, therefore, by definition *occult*. In part this can be seen as an example of Arthur C. Clarke's 'law', that "any sufficiently advanced technology is indistinguishable from magic". But it would be a mistake to assume that this meant that Renaissance thinkers believed that machinery was worked by hidden demons.[71] The ill-educated were superstitious, of course, and sometimes might well have thought this way, but among the educated it was perfectly well known that machines worked by means of cunningly arranged mechanical contrivances. Consider, for example, Salluste du Bartas's description of the "iron fly", allegedly built by the mathematician Regiomontanus and capable of flying around a room:

> O devine wit, that in the narrow wombe
> Of a small Flie, could finde sufficient roome
> For all those springs, wheels, counterpoise, & chaines,
> Which stood in stead of life, and spurre, and raines.[72]

This iron fly would still have been held to work by occult means. Because the mechanical arrangements which "stood in stead of life" were hidden, and their mode of operation was not obvious to the senses, and moreover could not be explained in the terms of Aristotelian natural philosophy, they were regarded as occult powers, analogous to the workings of occult qualities in natural bodies.[73]

The mathematization of the world picture has always been regarded as an important element in the Scientific Revolution, but scholarly analysis of this crucial historical process has failed to pay sufficient attention to the undeniable associations between mathematics and magic in the Renaissance. Historians of mathematics have looked to humanists, astronomers, Jesuit mathematicians, and mathematical practitioners of the more pragmatic kind (artillerymen, surveyors, merchant book-keepers, and engineers of various kinds), but have largely ignored those Renaissance intellectuals who were more concerned with the magic of mathematics.[74] This is undoubtedly another result of the positivist tendencies among historians of science, tending to dismiss anything which smacks of magic. Where magical mathematics has been discussed it has been seen purely as an aspect of Renaissance culture, and its possible relevance to the subsequent development of mathematics is left unconsidered. J. Peter Zetterberg, for example, takes it for granted that there was something called "the mathematicks" which was unfortunately all too often *mistaken* for magic. It evidently never occurred to him that mathematics could have been, as indeed it was, regarded by pre-modern thinkers as a major part of the magical tradition.[75] As far as most pre-modern thinkers were concerned, to describe a man as a mathematician was to describe him as a wizard; this was certainly true, for example, in the cases of John Napier and John Dee.[76]

By the time John Wilkins, one of the major contributors to the new philosophy in England, came to publish his *Mathematical magick* (1648), he felt it necessary to apologize for the title. By now, educated men were getting used to the idea that mathematics had an important place in natural philosophy (something which had always been denied by Aristotle), and mechanics was increasingly being seen as a science which depended upon natural phenomena, and so the workings of machinery could be seen as part of natural philosophy, and could shake off its old association with magic.[77] The process of incorporating mechanics into natural philosophy began with the Renaissance discovery of the *Mechanical questions*, attributed (wrongly) to Aristotle and first translated into Latin by Vittore Fausto in 1517. It required the re-casting of mechanics from an art to one of the mixed mathematical sciences, before Descartes could insist, in 1644, that "there really are no reasonings in Mechanics which do not also pertain to Physics, of which it is a part or species".[78] Undoubtedly one of the most important contributors to this transformation in attitudes to mechanics was the sixteenth-century Girolamo Cardano, a mathematician who was prominent in the occult tradition.

Cardano has still not attracted the scholarly attention he undoubtedly deserves and so his historical importance has yet to be properly assessed. It is clear, however, that his "complete account of the universe in a single volume", *De subtilitate* (1550), is one of the earliest attempts to provide a comprehensive system of philosophy intended as a substitute for the Aristotelian system. One of the striking features of this work, most evident in its opening book on the principles of natural philosophy, is the way Cardano easily moves back and forth between explaining natural phenomena and the way machines work. Indeed, he uses his accounts of machinery to explain the principles of natural motion, matter, the void, and so forth.[79] This approach, mixing the natural and the artificial, together with his emphasis in his mathematical works on the relevance of mathematics to the understanding of nature, is radically different from the traditional Aristotelian approach but contributed strongly to the "epistemological optimism with emphasis on the utility of knowledge" which the historian of mathematics, Jens Høyrup, has seen as characteristic of Renaissance occultism.[80]

Although historians of mathematics have tended to be even less interested in assessing the possible influence of occultism than historians of science have been, there seems to be a *prima facie* case for assuming that changing attitudes to mathematics, and changes in the intellectual status of mathematics, owed something to the reassessment of occult traditions which took place during the Renaissance. There is even a case to be made for the role of numerology in the Scientific Revolution, albeit a case that hinges almost entirely on Kepler and Newton. In the so-called "Classical scholia", in which Newton sought support for universal gravitation in Pythagorean doctrine, he seemed to take seriously the belief that the number seven has some cosmic significance — as revealed by its appearance in music, optics, and the number of heavenly bodies in our system. These facts of nature, together with what Newton thought of as clear historical evidence of the significance of the number seven in ancient wisdom (that is to say, wisdom closer in time to that of Adam), revealed something about the mind of God.[81] Similarly, Kepler was at least as much concerned to know why God created only seven heavenly bodies in our system, as he was to understand what force actually moved the planets. Kepler believed that he had found a *natural* explanation for the fact that God created only six planets (instead of 10, 100, or even an infinite number) to circle the Sun in the details of what he called the geometrical archetype.[82] It is important to note, however, that for both Kepler and Newton numerology had to be firmly grounded on empirical facts about the physical world. This set their numerology apart from those whose opinions about what a particular number might signify were based on mysterious traditions, or their own whims. This is precisely why Kepler objected so vigorously to the writings of Robert Fludd, and tried to ensure that his own work was not seen in the same light as Fludd's. Kepler found numerical significances in the Creation, while Fludd found them in his own imagination and imposed them on the natural world (at least, this is how Kepler saw the situation).[83]

It might be suggested that the latter kind of numerologist was the more common, and so Neoplatonising fantasists like Robert Fludd, for example, are more representative than Kepler or Newton. But this could easily be an illusion created by the prevailing historiography (or perhaps by the lack of an adequate historiography). Although Kepler's geometrical archetype is well known, it isn't always stressed in the literature that his deployment of the five Platonic solids in between the planetary spheres enabled him to answer the question as to why God created only six planets (the point being that, if God spaced the planets between these solids, He had to stop at six because He'd run out of solids, and the rules of geometry make it clear that not even God could create a sixth regular solid). Similarly, although Newton's analogy between the colours in the spectrum and the notes in the octave are plain for all to see in the *Opticks*, it is hardly ever mentioned, even in scholarly works dealing with Newton's optical theories. It is perfectly possible, therefore, that numerological concerns which are closer to Kepler's and Newton's than to Fludd's are waiting to be found in the writings of other contributors to the historical development of science.

The entirely undeniable role of alchemy in the development of modern science also needs to be understood in a carefully nuanced way. In an important article Lawrence Principe and William Newman have recently shown that our modern view of the nature of alchemy is severely distorted by various reconstructions of it which derive from nineteenth-century occultist movements, and have no real historical basis in the alchemy of the pre-modern period.[84] The historiographical rot set in when

Enlightenment thinkers drew a spurious distinction between chemistry, in something like the modern sense, and alchemy, which was presented as being concerned solely with transmutation of base metal into gold.[85] This in itself can be seen as part of the trend, still active in the eighteenth century, to separate the new natural philosophy from magic. The need for this kind of separation of alchemy from chemistry became even more urgent for later spokesmen on science because of what Principe and Newman see as three "residues of Victorian occultism".[86] Firstly, "spiritual alchemy", which extended Jacob Boehme's ecstatic use of alchemical imagery in his religion of self-purification and self-transmutation, implied that all alchemy was to be understood in this kind of mystical light. This view of alchemy was in turn transformed on the one hand into an exclusive concern with psychic transformation by the psychologist Carl Jung; and on the other into what they call a "panpsychic interpretation" of alchemy (in which alchemy was equated with an organic, vitalist, and generally enchanted worldview which was, so the story continued, inimical to modern science) by the doyen of comparative religious studies, Mircea Eliade. Newman and Principe are in the forefront of on-going efforts to recover the real history of alchemy from these obfuscations, and to show precisely how alchemy was absorbed into modern science, and what was left out (or, in this case, what came to be interpolated subsequently into the bogus history of alchemy). In connection with this, Principe and Newman point out that not all alchemists subscribed to an animist, or even a vitalist, view of matter, and that such differences might have resulted in a different kind of take-up of alchemical ideas by reforming natural philosophers. Similarly, they point to recent work which has shown that alchemical matter theory was often corpuscularian, and even mechanistic, and which certainly played a part in the new matter theories of the Scientific Revolution.[87]

If alchemy was broken down into a general concern with chemical interactions and processes on the one hand and a concern with metallic transmutation on the other, and only the former made it into the new science, we can see a similar process with regard to the absorption of herbalism and the lore of medieval bestiaries. Studies of flora and fauna in the pre-modern period were overlaid with assumptions about the religious, moral and symbolic significance of all God's creatures, as well as their potential for providing *materia medica*. Many of these assumptions, in what historians have referred to as the "emblematic worldview", derived from the belief in correspondences within the Great Chain of Being, and included various occult associations, based upon what were considered to be God-given signatures. As botany and zoology came to be included in the new science, however, much of this magical and mythological lore, once considered to provide essential information about the plant or animal in question, was excluded.

To understand this change in attitude about knowledge of natural things, we need to consider the effect of the discovery of the New World. Plants and animals from the New World came to the West devoid of any symbolic associations — they had no religious or moral significance deriving from either historical and religious legends, or from humbler folklore. Naturalists had no choice but to confine themselves to known facts about this new flora and fauna. In subsequent compendiums of natural history, therefore, there was a clear shift towards treating all plants and animals in the same strictly descriptive way. Just as creatures unknown to European culture had to be recorded merely in terms of what could actually be observed, so the old folkloric associations were stripped away from familiar plants and animals. To a large extent this resulted in a less magical world picture, but it would be wrong to see this as a steady triumph of science over magic.[88]

Belief in the occult qualities of those plants and animals used in *materia medica*, supposedly based upon past experience anyway, was not affected by this increased emphasis upon observation. Furthermore, the tendency of explorers was to bring back just those plants which were deemed by native populations to be most useful in curing disease. More often than not, European doctors could not decide how, or even whether, such unknown drugs worked on the four humours of the body. Without so much as an *obiter dictum* from Aristotle, or Dioscorides, or some other ancient authority, it was often impossible to tell whether a plant worked through heat, cold or one of the other manifest qualities. Increasingly, therefore, medical thinkers declared new drugs to work by means of the alternative to the manifest qualities: occult qualities. This even led to a major re-working of medical and therapeutic theory, in which some diseases were held to be the result not of an imbalance in the humoral constitution, but a corruption of the whole substance of the body. The only drugs capable of curing these diseases were those that operated, likewise, on the 'total substance' of the body, not merely on a particular humour.[89]

So, while the occult qualities of *materia medica* were increasingly recognized and accepted into the new natural philosophy, there was a marked rejection of belief in the supposed symbolic significance of natural objects. This exclusion of symbolism as a factor relevant to the workings of the physical world is a general trend that can be discerned elsewhere in Renaissance and early modern intellectual life. Again, it is easy to understand changing attitudes here in terms of a gradual fragmentation. The use of amulets or talismans was, of course, a prominent aspect of the magical tradition. In his investigation of the *Hidden causes of things* (*De abditis rerum causis*, 1548), Jean Fernel separated amulets which were supposed to work by the occult powers of natural objects from those supposed to work by magic spells — the power of words, pictures, or other symbols. The former seemed perfectly acceptable. Even much later Francis Bacon accepted the use of bloodstone, hung around the neck, to prevent nosebleeds. The colour of bloodstone was its signature, revealing that it must have an occult but natural power as a styptic.[90] Fernel and many others, however, had severe doubts that words or pictures could have any natural power. Although the Roman Catholic Church insisted upon the power of the priest to turn bread and wine into the Eucharist by reciting the words of the Mass, this was always regarded as a supernatural event. As Johann Wier insisted: "Words are brought forth from the mouth of the priest. But they are consecrated by the power and grace of God." When God said "Let there be light", there was light not because it had been said, but because *God* had said it.[91]

According to the orthodox Christian view, if words or other symbols seemed to have any power it must have been due to the secret intervention of demons, accomplishing what on-lookers thought the symbols themselves had accomplished. As Thomas Aquinas pointed out, words have no power except over intellects. If a magician achieves power over inanimate nature by chanting words it must be because a demon has intervened, for its own purposes, and is making the magician believe he has succeeded.[92] Within the natural magic tradition, however, there was a belief that words *did* have their own natural power, and so could be used to accomplish natural events or processes. This derived from the suggestion in Genesis 2, 19 that Adam named all things. It was assumed that Adam knew the true names of things, names which somehow captured the essence or true nature of the things and therefore had power over them. To us (and to Thomas) this simply looks like a *non sequitur*, but it seems clear that for many pre-modern thinkers there could be a power linking the "real", God-given, name of a thing to its essence. In this essentially Neoplatonic view, language was not just used for communication, but was bound up with knowledge of things. If God had given names to things, they would not have been arbitrary signs, since God does nothing arbitrarily, and so knowledge of the names constitutes knowledge of Creation.

The revival of magic in the Renaissance led to increased hopes, therefore, that perhaps this Adamic language could be rediscovered, and Adam's dominion over all things, lost at the Fall, could be restored. The major searchers after such 'universal' languages, therefore, were leading players in the reform of natural philosophy.[93] In his *Valerius Terminus* of 1603 Francis Bacon expressed his belief that man's sovereignty and power over nature will only be restored "whensoever he shall be able to call the creatures by their true names". By the time he came to write his *Great instauration* (1620) he was still thinking of restoring "that commerce between the mind of man and the nature of things", but it was clear that he saw this as an exercise in classificatory natural history rather than in linguistics. Similarly, the leading light of the Royal Society, John Wilkins, denied the existence of an Adamic language with a natural magical power over things, but held out the hope that a language might be constructed which, by being rationally based upon the true nature of things, would prove to be a genuinely universal language capable of repairing "the ruins of Babel".[94] The distinction is a subtle one. Words in the supposed Adamic language are held to have power over things because they somehow capture the essence of the thing. Words in Wilkins's 'philosophical language' capture the essence of the thing by being based on (and somehow linguistically reflecting) a sound classificatory scheme which reveals the relationships between all things. What is clear is that Wilkins's language has power only over the intellects of its users, not over the objects being discussed. It is equally clear, however, that Wilkins's interest in a universal language first appeared in an early work which was undeniably magical in tone: *Mercury: Or the secret and swift messenger. Shewing how a man may with privacy and speed communicate his thoughts to a friend at any distance* (1641).

As in the case of mathematical magic, Wilkins can be seen to have contributed to the transformation of universal language schemes from an ambition of the magicians to one of natural philosophers. Another important aspect of this story can be seen in the Renaissance interest in the Jewish Kabbalah, a mystical system which invested words with real power. Christian cabbalism arguably originated with Giovanni Pico della Mirandola's *Conclusiones philosophicae, cabalasticae et theologicae* of 1486, but it soon became an important feature of Renaissance magic in its own right. In the end, it can be seen that cabbalism was not absorbed into the reformed natural philosophies, but this was not through lack of trying. It certainly figured in the universal language schemes, even in one of the latest, and potentially the greatest, of them, that of G. W. Leibniz. In the end, however, not even Leibniz could bring these universal language schemes to fruition and so they failed to find their way into the new science.[95] Nevertheless, historically these schemes can be seen as another aspect of the magical tradition which for a while were seriously discussed by leading shapers of the new philosophy, as they tried to incorporate what they recognized to be useful into their new philosophy.

A similar tale is to be told about the symbolic value of the supposed signatures. We have already noted how compendia of natural history began to jettison much of the earlier symbolic paraphernalia that were previously regarded as important pieces of information about God's creatures. As Brian Vickers has shown, the doctrine of signatures also began to attract sceptical scrutiny in the early modern period. Guy de la Brosse, for example, believed that the supposed resemblances were too uncertain, "like clouds, which can be made to look like anything that fantasy can project". John Ray, by contrast, could clearly see the signatures but refused to see any meaning in them. He noted that between some plants and various natural or artificial objects "so great a similarity exists that no one could fail to recognize it immediately nor could he persuade himself that it had ever come about by chance". Such resemblances were, on the one hand, "clear proof of intention on the part of nature" (Ray, like all his contemporaries, believed that nature, or God, does nothing in vain). On the other hand, as if to contradict this "clear proof", he suggested that the number of similarities was not "so great nor the signatures they bear so obvious and plain to everybody that they suggest a pointer or a deliberate plan on the part of Nature". The contradictions in Ray's account surely reflect a conflict between his belief in natural theology, with its assumption that the world is full of examples of God's design in the natural world, together with his religiously-inspired distaste of anything that would encourage the practice of magic.[96] The concept of signatures was not simply discarded, it was broken up, and partially incorporated into the new natural philosophy, appearing especially in the tradition of natural theology, which sought to prove God's wisdom and beneficence by pointing to seemingly obvious cases of intelligent design in nature.

Although astrology is now very far beyond the pale as far as most scientists are concerned (a few extreme empiricist statisticians notwithstanding), there was a time when it was a mainstream subject in the universities, particularly in Medical Faculties. What's more, during the re-drawing of the boundaries around and between natural philosophy and magic, there was a time when astrology looked as though it might continue to be included even in the new science. Astrology had always attracted detractors but the evidence in its favour seemed to override all arguments.[97] The effect of the Moon on the tides seemed undeniable and the Sun's movement through the Zodiac defined the seasons. It was also hard to believe that God had placed the myriad fixed stars, as well as the Sun, Moon and planets, without having a very good reason for doing so. Kepler was in many ways an idiosyncratic thinker but he spoke for many devout believers when he insisted that we had to consider God's reasons:

> For if we do not, we shall be driven to admit that God acted arbitrarily in the universe, even though perfectly good rational procedures were open to Him. And this is a conclusion I will not accept on anyone's authority....[98]

Even so, Kepler recognized the problems with traditional astrology and jettisoned much of its old lore when he came to write his treatise *On the more certain foundations of astrology* (*De fundamentis astrologiae certioribus*, 1602).[99]

Certainty seems to have been an important issue in the fragmentation of astrology. The rise in intellectual status of mathematics, which was such an important aspect of the mathematiziation of the world picture, depended to a large extent upon arguments about the greater certainty of mathematics. So-called judicial astrology was an early casualty. Notoriously unreliable, and attracting many cogent arguments against its validity, the personal divinatory aspects of astrology were already being rejected by the early seventeenth century.[100] Once again, however, we can see that the story was one of fragmentation, and partial absorption, not wholesale rejection. Kepler's own attempt to establish more certain principles of astrology came to nothing, but even at the end of the seventeenth century we can see continuing attempts to establish, on empirical grounds, the links between the heavenly movements and the weather.[101]

For a while it even looked as though Newtonian science, the apogee of certainty in mathematical physics, would embrace one major aspect of astrology. As Maupertuis wrote, even as late as 1742, "one of the greatest astronomers of the century [Newton] has spoken of comets in a manner which re-establishes them in all the reputation of terror where once they were".[102] Newton and some of his close followers, including the down-to-earth Edmund Halley as well as the more religiously fanciful William Whiston, saw comets as God's way of continuing to adjust and reshape the nature of our planetary system. The gravitational pull of a passing comet was considered as the likely cause of the Noachian Deluge, since the Almighty prefers to make use of "Natural Means to bring about his Will". Bearing in mind that God does nothing in vain, it seemed likely that comets might well be once again, "the Instruments of Divine Vengeance".[103]

It would be absurd to suppose that Newton and his followers were determined to rescue, by hook or by crook, some aspect of astrology. Clearly, they were not. Newton's principal concern was to allow continued discourse about God within natural philosophy. For Newton the Cartesian system, which required God only at the Creation, after which His intervention in the mechanistic system of the world was never required, "was made on purpose to be the foundations of infidelity".[104] Newton's belief that God had continually to tinker with his system to keep it running smoothly outraged his continental rival, G. W. Leibniz, who was appalled that Newton could imagine God to be such a poor artificer. As far as Newton was concerned, however, the use of comets as the natural means by which God winds up and replenishes the system allowed him to avoid atheistic interpretations of his philosophy, without actually having to suppose direct supernatural interventions by God.[105] Here then we can see the old astrological belief in comets as portents of great change, and instruments of God's wrath, being absorbed for a while into the new natural philosophy. By the late eighteenth century, however, thanks to increasing secularization no less than to the increased success of Newtonian mechanics in explaining our solar system without recourse to inputs from comets, even this aspect of astrology was excluded from science.[106]

Another significant example of the adoption, and adaption, of magical thinking concerns the transformation of the magnet from one of the most occult of occult objects to a body with a unique but easily demonstrated power, which could even be invoked to explain various non-occult phenomena. A well-known magical manuscript on the magnet, written by Pierre de Maricourt in 1269, was first printed in 1558, when the value of magical works was becoming widely recognized. It must have been after reading this that William Gilbert, so-called father of magnetic science, hit upon the idea of using magnetism to explain the Copernican rotations of the Earth. Magnets are capable of spontaneous movement, Gilbert reasoned, and therefore must be endowed with souls, because (according to Aristotle) only creatures with souls are capable of self-movement. Using experiments and ideas highly reminiscent of those found in Maricourt's *Epistle on the magnet*, Gilbert then went on to show that the Earth was a giant magnet, and must have a soul, and be capable of making itself move in the ways required by Copernican astronomy.[107] Subsequent writers, particularly in England, recognized the value of Gilbert's attempt to provide a physical explanation of the Earth's movement (something which Copernicus never provided), but evidently found his emphasis upon a magnetic soul too animistic for their tastes, and focused instead on Gilbert's talk of a magnetic *orbis virtutis*, surrounding a planet. Where Gilbert had used the magnetic soul to suggest planets moved themselves, later natural philosophers, starting with Kepler, developed a dynamic account of planetary movements based on tangential motions past the Sun and an attractive force towards it. Talk of magnetic souls was replaced by talk of "magnetic attractions", "magnetic virtues", and eventually "attractive principles" and "gravitating powers"; such talk was less animistic, but it was not less occult, even though for us it seems definitely more 'scientific'.[108] The culmination of this trend was the natural philosophical essay by Sir William Petty, *A new hypothesis of springing or elastique motions* (1674), in which he actually sought to explain phenomena like elasticity, solidity and fluidity, expansion and contraction, and more, by assuming that, just as all planets were spherical magnets in Gilbert's cosmological scheme, so all atoms are spherical magnets. Although magnetic attractions and repulsions continue to be occult qualities, magnetic phenomena can be clearly and easily demonstrated on a laboratory workbench, and so magnets can be used to provide legitimate explanations in natural philosophy.[109]

There were even some thinkers, although mostly on the fringes of natural philosophy, who tried to absorb demonology into the new natural philosophy. Henry More, the Cambridge Platonist divine who introduced the teaching of Cartesianism into Cambridge, Joseph Glanvill, FRS and apologist for the new philosophy, and even for a while Robert Boyle, were keen to make the study of demons, witchcraft, and ghosts an important part of the new experimental philosophy. The motivation was perfectly clear: to combat the uptake of Cartesian and other versions of the mechanical philosophy by irreligious thinkers who were seeking to promote an entirely materialist worldview. The would-be experimentalist demonologists wanted to demonstrate that there were phenomena which could not be explained in terms of the mechanical philosophy, and so immaterial beings and immaterial principles had to be acknowledged.[110]

I believe it would be possible to continue in this vein, showing how different aspects of the magical tradition were partially taken up by natural philosophers, while other aspects were entirely excluded. In all cases, as with the examples briefly discussed here, the story will be one where the historical complexities derive from a host of contingencies. It is fairly obvious, for example, that the precise way in which alchemy, or part of it, became absorbed into natural philosophy was different from the way that so-called 'mathematical magic' — the use of machinery to improve human faculties — was adopted.[111] Furthermore, some prominent aspects of the natural magic tradition, such as a belief in the natural power of words, were not included within the new boundaries at all. In other cases, aspects of the tradition were taken up only in a very restricted sense, as in the case of numerology, which seems to have led Kepler and Newton to draw specific conclusions about the natural world, but can hardly be said to have been a general influence.[112] If we bear in mind, also, that the magical tradition was undoubtedly a major source for the experimentalism of the new philosophies, and for the idea that knowledge of the natural world should be useful for the benefit of mankind, it seems hard to deny the claim that the new philosophies were greatly indebted to the magical tradition.[113]

WHY THE CHANGE?

The question arises, therefore, as to why this sea-change occurred. Why was the map of knowledge redrawn between the end of the Renaissance and the beginning of the Enlightenment? Why were the boundaries redefined so that natural magic lost its identity by becoming largely absorbed into the new 'natural philosophy' (which now, thanks to the experimental method, the integral use of mathematics, and the concern with pragmatism was closer to our modern concept of science than it was to the earlier tradition of contemplative natural philosophy), while symbolic magic came to be seen, by the educated at least, as superstitious nonsense,[114] and demonology, formerly a borderline category linking religion and natural philosophy, became first of all an entirely religious category, and with increasing secularization was thoroughly rejected?[115]

As with so many other problems in history, the answers to this question are no doubt legion. We have already considered a few reasons along the way, such as the change in status of mathematics, helped by the discovery of the supposedly Aristotelian *Quaestiones mechanicae*, but it is beyond the scope of this article to survey all the other possible factors. It is clear, for example, that the full story could not be told without paying careful attention to the social and political context of Renaissance and Reformation Europe, and how developments in these spheres affected intellectual life.[116] What I want to do here is simply to consider briefly some of the more immediate reasons why the boundaries of magic and natural philosophy were redrawn in just the way they were. By "immediate" I mean those reasons which arose directly out of the efforts of Renaissance and early modern thinkers to improve their understanding of the natural world.

Perhaps the first thing to mention in this connection was the change in the intellectual status of magic as a result of the discovery of the essentially religious writings attributed to Hermes Trismegistus. Thanks to a generally accepted belief in the wisdom of Adam, it was usually assumed in the pre-modern period that knowledge was something that needed to be recovered from the past. Adam knew all things, but thanks to the Fall, this wisdom had been successively forgotten. For the pre-moderns, therefore, thinkers of great antiquity were more likely to know more — to have forgotten less of the Adamic wisdom — than a contemporary thinker. This belief is most familiar to historians of science through the common designation of the Copernican theory as the Pythagorean theory.[117] The Copernicans knew that if they were to have any chance of persuading their contemporaries of the truth of the Copernican theory they had to show that it had been believed in the past. The ancient sage, Hermes Trismegistus, was regarded as a contemporary of Moses, and he was seen to be responsible for transmitting the Adamic wisdom to the pagan Greeks, as Moses had transmitted it to the Jews. This belief was easy to sustain in the light of the fact that the newly discovered Hermetic writings showed clear foreshadowings of Christian belief, including its trinitarianism. As we now know, these supposed foreshadowings were in fact echoes of Christian belief, since these writings were actually compiled by Neoplatonists in the early centuries of the Christian era.

The first translator of the Hermetic Corpus (a substantial part of it, at least) was Marsilio Ficino who was clearly fascinated by Neoplatonic theurgical beliefs and who began to develop his own theory of what D. P. Walker called "spiritual magic" in his *De vita coelitus comparanda* (1489). Ficino's work proved immensely influential and helped to promote the view that Hermes Trismegistus was above all else a magician.[118] This identification of Hermes as a magus was helped by the fact that, as well as the Neoplatonic theistic writings attributed to him, there were also a considerable number of astrological, alchemical and natural magical texts attributed to this great sage. Given the belief in Adamic wisdom, and the belief in the great antiquity of the Hermetic writings, magic came to be seen as one of the oldest forms of knowledge, and was therefore newly invested with great respectability. After centuries of being disparaged by the Church, magic came to be seen as a major aspect of Adamic wisdom. Accordingly, as Eugenio Garin has suggested, there was a brief time, shortly after the discovery of the Hermetic Corpus, when it was acceptable to be called a magus, and to acknowledge oneself to be a magus (remember, we have already seen that it was much more usual to deny that one was a magician).[119] It is hardly surprising therefore that reforming natural philosophers of the Renaissance and early modern periods should look with fresh eyes at the magical tradition, and consider more seriously than before what it had to offer.[120]

Another reason why the boundaries of magic came to be redrawn arose from rapid developments in the understanding of earlier magical traditions as a result of the humanist scholarship of the Renaissance. In particular, the recovery of the works of Ancient Neoplatonists, such as Plotinus, Proclus, Iamblichus and others, revealed a theory of magic in which "spiritual and demonic magic" played a greater role than natural magic. This alternative to the Aristotelian tradition first became known as a result of the work of the famous Florentine philosopher and translator, Marsilio Ficino, whose *De vita coelitus comparanda* of 1489 was a full exposition of the theory of magic which drew not only upon Aristotelian traditions of occult qualities but also upon the more theurgical theorizing of the later Neoplatonists. Although Ficino himself seems to have managed to stay within the confines of natural magic as it was traditionally conceived (with the emphasis on the *natural*), his exposition drew attention to the fact that later Neoplatonists seemed to believe that occult qualities in matter were clear signs of divine or demonic presence within the matter. For the pagan Neoplatonists, in other words, the occult but *natural* effects discussed in traditional natural magic were, in fact, *supernatural* effects brought about directly by gods or demons. Although such ideas were pagan and could easily be shown not to fit in with Christian Aristotelianism, they ensured that the boundary demarcation between natural and supernatural and the abilities of demons were placed firmly on the agenda of scholarly discussions. In this way, occult qualities, formerly hardly discussed within the scholastic tradition, became important items for discussion.[121]

One of the most important of such discussions was that of the notorious secular Aristotelian philosopher, Pietro Pomponazzi. Pomponazzi was, as Brian Copenhaver has recently said, "entirely and aggressively naturalist" and in 1520 he wrote a treatise *On the causes of marvelous natural effects and on spells*, in which he explicitly intended to make demons redundant for any understanding of the natural world. One of his arguments was that, even if demons were capable of knowing better than men all the occult qualities of things and how to accomplish things by bringing together sympathetic actives and passives, because demons were spiritual, incorporeal, beings they were completely incapable of manipulating matter to accomplish anything by their knowledge.[122] Once again, Pomponazzi's book, like Ficino's, stimulated debate about occult qualities, demons, and the demarcations between and around them. By rejecting the possibility of demonic intervention, Pomponazzi greatly expanded the role of the supposed occult qualities of matter. Such qualities could stimulate thinking about the nature of matter itself, and could subsequently be absorbed into the new natural philosophy. Meanwhile, sceptical philosophers had a new set of arguments for dismissing demons as ineffectual, and ultimately as nothing more than the result of superstitious beliefs.[123]

Another important stimulus towards a new and detailed consideration of occult qualities arose out of university medical faculties. We have already noted that developments in botany and other subjects affiliated to the production of *materia medica* resulted in increased numbers of new drugs being designated as "occult" in their operation, because their efficacy did not depend upon their effects upon the manifest qualities of the patients' humours. This coincided with awareness of the need for reform of medical theory from another quarter. The increased prevalence of pestilential diseases in a Europe where bubonic plague was endemic and where syphilis was cutting a swathe through all classes of society presented problems for traditional medical theory. Galen saw all disease in terms of a disturbance of the four humours such that the normal healthy temperament (the balance of the humours in the body) was disrupted. It follows from this that diseases do not have a separate existence in their own right. The difficulty with this individualistic physiological approach to sickness, seeing every illness as the special problem of one patient, is that it cannot easily explain conditions which seem to be infectious. Why should one patient's physiological disruption be capable in some cases of being passed on to others with different constitutions or temperaments? The difficulty is especially severe for Galenic theory in the case of epidemic diseases, such as plague. Epidemic pestilences strongly suggest, contrary to ancient authority, that diseases have a kind of life of their own; they are real, distinct entities, which can pass from one person to another, or can simultaneously attack great numbers of people irrespective of their individual temperaments. A new understanding of the nature of diseases demands new ways of dealing with them, and the sixteenth century saw three major attempts to reform medical theory. The three would-be reformers, Paracelsus, Girolamo Fracastoro and Jean Fernel, all drew upon occult traditions in their suggested reforms. Paracelsus looked to alchemy, not just as a way of producing new medicines, but as a way of understanding the nature of the physiology, and the nature of disease. Fracastoro developed the idea of "seeds" of disease, seminal principles capable of growing in the body and disrupting it, while Fernel believed that pestilences acted not on the humours, but on the substantial form of the body, which Fernel called the "total substance" of the body, and they did so, not by affecting the manifest qualities but by means of some occult power over the total substance.[124]

The works of Ficino, Pomponazzi, Paracelsus, Fracastoro, and Fernel are, of course, just salient points in a rapidly changing landscape of intellectual discussion. Throughout the sixteenth century the nature and the role of occult qualities become increasingly prominent in natural philosophizing. This was bound to have unfortunate repercussions for traditional Aristotelianism, since, although occult qualities were allowed for in scholasticism, and traditional natural magic throughout the Middle Ages was loosely premised upon Aristotelian assumptions (alchemy, for example, although going far beyond anything to be found in Aristotle's writings still assumed the truth of the four elements and four qualities), in fact there was very little in Aristotle himself about occult qualities.[125] Indeed, it became increasingly obvious during the Renaissance, when Aristotle came to be studied in the original Greek, that the natural magic tradition owed a great deal to medieval and Arabic interpolations, for example, from Thomas Aquinas, Albertus Magnus, Avicenna, and Alkindi.[126] More to the point, the ideal of science in the Aristotelian tradition was based on the form of the logical syllogism (deductive reasoning), but the premises, the starting points upon which the reasoning was based, had to be uncontentious, evident truths to which all could freely assent. Small wonder, therefore, that the main emphasis in natural philosophical argument was on the supposedly *manifest* qualities, which could fulfil the criteria of being undeniable and evident to all.[127] It was one thing, within this system to occasionally have to resort to occult qualities in one's explanations; it was quite another, however, to see occult qualities playing an increasingly prominent role in a widespread range of natural phenomena. There was a real crisis in Aristotelianism, therefore, concerning the very possibility of dealing with *insensible* properties and entities in a philosophy that was supposedly grounded on human sensation. When writers like Pomponazzi, Jean Fernel, and Daniel Sennert can be seen to be elevating the role of occult qualities in Aristotelianism it seems legitimate to ask whether they are best seen as eclectic Aristotelians or as contributors to the demise of Aristotelianism. Whatever the truth of that, we cannot help but conclude, I think, that Renaissance developments in the notion of occult qualities resulted in a major re-arrangement, in which these qualities came to play a much greater role in reformed versions of natural philosophy, and eventually became absorbed into the mainstream of the new philosophies which completely displaced Aristotelianism.[128]

Further difficulty for Aristotelian natural philosophy was caused by emerging problems with the theory of substantial forms. These arose especially as a result of considering the nature of what we would call chemical compounds, compared with mere mixtures, and why new substantial forms seem to be created in the one case, but not in the other. Just as the medical reformers turned to the occult to develop new theories of medicine, so innovative scholastics and anti-Aristotelians alike turned to alchemy to help understand the production of new substances. At the leading edge of this work, as performed by alchemists like Daniel Sennert and Robert Boyle, the notion that the substantial form defined a body became untenable when it was realized that the original constituents of a 'mixt' (what we would call a chemical compound) could be recovered. According to scholastic theory this required the recreation of the constituents' substantial forms after they had been replaced by the new substantial form of the mixt. Alternative explanations in terms of the conjunction of unchanging atoms or corpuscles began to replace Aristotelian accounts but these explanations, as William R. Newman has shown, owed more to alchemical theorizing than they did to the emerging mechanical philosophy.[129]

Finally, unlikely as it may seem to us (or to those of us who remain, like David Brewster, recalcitrant in their belief that all magic was merely the production of knaves and fools), there is every reason to suppose that, as far as Renaissance thinkers were concerned, the occult sciences were the most likely source for the reform of natural philosophy, and for the establishment of a true understanding of God's Creation. Indeed, to a large extent they were the *only* alternative sources of natural knowledge.

It has been suggested by Brian Vickers that modern science emerged as the scientific mentality overcame the occult mentality that had been dominant among earlier thinkers. The assumption here is that Paracelsus, say, representing the occult mentality, chose the wrong path to reform natural knowledge, while Descartes, representing the scientific mentality, chose the right path.[130] In a sense, of course, this is true — Cartesianism has proved more influential on subsequent scientific thinking than Paracelsianism, but we cannot infer from this that Descartes was more scientific in his thinking than Paracelsus, unless we choose to make it a matter of our definitions of who is a scientific thinker and who is not. Relying on our perfect hindsight we can easily declare Descartes to be a more scientific thinker than Paracelsus, but at the time of Descartes's death in 1650 judgements as to the relative merits of these two thinkers could not have been couched in these terms.

Furthermore, for the sake of the argument being presented here, Paracelsus is by no means the best representative of magical thinking to choose. His thinking seems to be at best confused and at worst wilfully obscure, and was clearly recognized as such by many contemporaries. But not every thinker in the occult tradition was as poor a communicator (or as weak a thinker, if that was his problem) as Paracelsus. Many, on the contrary, must be counted among the greatest thinkers of their age, and they believed, like Descartes, that they had discovered the key to understanding, and explaining natural phenomena. Although Jean Fernel chose very deliberately to argue for the increased importance of occult qualities in natural philosophy and in the theory and practice of medicine, he did so in accordance with the best canons of logic and rhetoric of the day, and by marshalling powerful evidence and argument in support of his views. To dismiss Fernel as a man with an "occult mentality", with implications of woolly-minded inconsequentiality if not downright irrationality, would be entirely unjust. The same is true, for the most part, of other thinkers in the magical tradition. Girolamo Cardano, and Cornelius Agrippa, to mention but two, should be recognized as leading shapers of European thought, not simply dismissed as deluded contributors to a worthless pseudo-intellectual tradition.

This is not to say that all magicians should *ipso facto* be honoured as great thinkers. We need to exercise the same care in making judgements about the historical significance of these thinkers as we would in the case of those who are supposed contributors to the history of science. Lauren Kassell has recently shown that the London-based astrologer Simon Forman was an autodidact who was barely capable of understanding the tradition he wished to exploit for his own personal gain. If we want to understand how practitioners in the occult tradition, even more than medical practitioners, came to be seen as frauds and charlatans, we could start with Forman. Similarly, the historical significance of Robert Fludd has not yet been established beyond doubt, in spite of sympathetic treatments of his work by a number of scholars.[131]

The leading thinkers in the magical tradition, at any rate, should be seen as thinkers who were trying to find solutions to problems which the Aristotelian and Galenic traditions could not resolve, and who turned to the occult, not because they were befuddled fools with an occult mentality, but quite simply because occult traditions seemed to offer the most likely source of help. When the Aristotelian theory of substantial forms, and the associated hylomorphic matter theory, began increasingly to seem inadequate, natural philosophers turned to alchemy as a likely pointer to alternative ways of understanding the relationship between bodies and their properties.[132] When Galenic medical theory, which relied almost exclusively on the balance (or imbalance) of the four qualities in the body for understanding disease and treating it, came to be seen as increasingly inadequate, medical reformers like Fernel, Fracastoro, and the Paracelsians all turned in one way or another to occult qualities as an alternative.[133] Similarly, every one of the Renaissance thinkers who tried to develop new systems of philosophy, intended to completely replace Aristotelianism, relied to a large extent on aspects of the magical tradition. So much so, in fact, that each of these system-builders can be seen as contributors themselves to the magical tradition. The title of Girolamo Fracastoro's *De sympathia et antipathia rerum* (1550) reveals its magical nature, but Girolamo Cardano's more cryptically titled *De subtilitate* (1554), which includes a substantial book "On Marvels, and the way to represent diverse things beyond belief", is equally indebted to the occult tradition.[134] Bernardino Telesio's *De rerum natura iuxta propria principia* (1587) assumes that all things are sentient and relies heavily on ideas of spirit which derive from the earlier magical philosophy of Ficino. Furthermore, Telesio was a major influence on at least two other thinkers who were indebted to the magical tradition, Tommaso Campanella, author of *De sensu rerum et magia* (1620) and other magical works, and Francis Bacon, whose own system of philosophy combined Telesian ideas on spirit with alchemical ideas to make what has been described as a semi-Paracelsian cosmology.[135] The inventive group of Italian system-building "nature philosophers", as they are usually known, also included Francesco Patrizi and Giordano Bruno. The magical nature of Bruno's world-view is so well known that it can be taken for granted, but Patrizi's views are indicated, if not by the title of his great system, *Nova de universis philosophia* (1593) then by the fact that he published it alongside his own translation of Chaldean and Hermetic works, under the title *Magia philosophia* (1593).[136]

This situation, in which would-be reformers of natural philosophy turned to the occult tradition as the most likely way out of all difficulties, continued even late into the seventeenth century, to the period seen as witnessing the dramatic decline of magic. Robert Boyle, recognizing the inadequacy of strict versions of the mechanical philosophy, including Cartesian claims that there were no new motions generated in nature, only transfers of motion from one part of the system to another via collisions, turned once again, as the opponents of Aristotelianism had before him, to alchemy. Newton, recognizing the absurdity of Cartesian vortex theory as an explanation for planetary movements and for gravity, preferred to rely instead upon the assumption that bodies could attract one another across empty space.

The example set by Newton makes it hard to deny that, if reformers of natural philosophy believed the occult sciences offered the most likely source for a viable alternative to Aristotelianism, they were right. Descartes was proud of the fact that he had eschewed all occult qualities from his system, and so in a sense believed he had succeeded where Aristotelianism had failed (since it had never quite managed to dispose of occult qualities). But for many, the Cartesian system could be seen to be ultimately unworkable, and rather than eschewing occult qualities, they embraced them as the only realistic alternative.[137] In so doing, occult qualities became absorbed into mainstream reformed natural philosophy. The triumph of Newtonianism, then, with its basic premise that all phenomena could be explained in terms of attractive and repulsive particles capable of acting at a distance, showed not only that Newton was right, but that earlier would-be reformers of natural philosophy who had tried to draw upon the magical tradition were not too far wrong.

ACKNOWLEDGEMENTS

This paper grew out of talks given to the History Society at the University of Lancaster in 1995, the Third Annual Arizona Center for Medieval and Renaissance Studies Conference in 1997, and a symposium on early modern thought at the University of Western Sydney in 2002. I am grateful to my hosts and other participants on these occasions for their encouragement. I also wish to thank the anonymous referees for their positive reactions. Above all, however, I wish to thank the Editor for allowing me the space for such an inordinately long paper.

REFERENCES

1. *The diary of Benjamin Robert Haydon*, ed. by W. B. Pope (5 vols, Cambridge, MA, 1960–63), ii, 173.
2. William Blake, *Jerusalem: The emanation of the giant Albion* (1804), Plate 15, lines 15–20; see Donald D. Ault, *Visionary physics: Blake's response to Newton* (Chicago, 1974).
3. John Keats, *Lamia* (1820), Part II, lines 229–37. On Keats's location in Enlightenment science, see Donald C. Goellnicht, *The poet-physician: Keats and medical science* (Pittsburgh, 1984).
4. William Blake, *The [first] book of Urizen* (1794).
5. See, for example, Gerd Buchdahl, *The image of Newton and Locke in the Age of Reason* (London, 1961); P. M. Rattansi, "Voltaire and the Enlightenment image of Newton", in H. Lloyd-Jones, V. Pearl and B. Worden (eds), *History and imagination: Essays in honour of H. R. Trevor-Roper* (London, 1981), 218–31; and I. B. Cohen, "The eighteenth-century origins of the concept of scientific revolution", *Journal of the history of ideas*, xxxvii (1976), 257–88.
6. Alan E. Shapiro, in his "Artists' colours and Newton's colours", *Isis*, lxxxv (1994), 600–30, suggests that Newton discovered that white light was made of an infinite number of spectral colours (p. 600). This is just loose talk, however, written in the light of our knowledge of the electromagnetic spectrum. Newton himself clearly believed that there were only seven primary colours.
7. Isaac Newton, *Opticks, based on the fourth edition London, 1730* (New York, 1952), Bk I, Pt II, Prop. III, Problem I, 126, my emphasis. The ellipses indicate where I have omitted Newton's

labelling of the lines in accordance with his accompanying diagram.

8. Isaac Newton, "An hypothesis explaining the properties of light, discoursed of in my several papers", in I. B. Cohen (ed.), *Isaac Newton's papers and letters on natural philosophy* (Cambridge, MA, 1978), 192–3.
9. Newton had shown, by further prismatic experiments, that the colours of the spectrum could not themselves be broken into further colours. Repeated refractions merely separated the colours further apart, but did not alter their colours. *Opticks*, Bk I, Pt I, Prop. V, Theorem IV, *ed. cit.* (ref. 7), 73–75.
10. See, for example, A. I. Sabra, *Theories of light from Descartes to Newton* (Cambridge, 1981).
11. *The optical papers of Isaac Newton*, ed. by Alan E. Shapiro (Cambridge, 1984), 543. Newton defended this musical analogy in a letter to an Oxford undergraduate, John Harrington, in May 1698, where he wrote: "I am inclined to believe some general laws of the Creator prevailed with respect to the agreeable or unpleasing affections of all our senses." I. Newton, *Correspondence*, ed. by H. W. Turnbull *et al.* (7 vols, Cambridge, 1959–77), iv, 275, quoted by Shapiro, *ed. cit.*, 547.
12. The classic account is still J. E. McGuire and P. M. Rattansi, "Newton and the 'Pipes of Pan'", *Notes and records of the Royal Society of London*, xxi (1966), 108–43, pp. 115–17. For the Latin text see Paolo Casini, "Newton: The classical scholia", *History of science*, xxii (1984), 1–58. A complete edition, including translations, of these proposed scholia is now available: Volkmar Schuller, "Newton's *Scholia* from David Gregory's estate on the Propositions IV through IX Book III of his *Principia*", in W. Lefevre (ed.), *Between Leibniz, Newton and Kant* (Dordrecht, 2001), 213–65. For Newton's work on music see Penelope Gouk, "The harmonic roots of Newtonian science", in J. Fauvel, R. Flood, M. Shortland and R. Wilson (eds), *Let Newton be!* (Oxford, 1988), 101–25; and *idem*, *Music, science and natural magic in seventeenth-century England* (New Haven, 1999), 224–57.
13. See R. S. Westfall, *Never at rest: A biography of Isaac Newton* (Cambridge, 1980), 351–9; and *idem*, "Isaac Newton's *Theologiae gentilis origines philosophicae*", in W. Warren Wagar (ed.), *The secular mind: Transformations of faith in modern Europe* (New York, 1982), 15–34. See also Matt Goldish, "Newton's *Of the Church*: Its contents and implications", in J. E. Force and R. H. Popkin (eds), *Newton and religion: Context, nature and influence* (Dordrecht, 1999), 145–64.
14. John Maynard Keynes, "Newton, the man", in The Royal Society, *Newton tercentenary celebrations* (Cambridge, 1947), 27–34. Keynes was one of the first scholars to accept that Newton's undeniable interest in alchemy was sincere and highly positive. For other treatments of Newton's occult interests see, for example, McGuire and Rattansi, "Newton and the 'Pipes of Pan'" (ref. 12); R. S. Westfall, "Newton and alchemy", in Brian Vickers (ed.), *Occult and scientific mentalities in the Renaissance* (Cambridge, 1984), 315–35; B. J. T. Dobbs, *The foundations of Newton's alchemy: Or, "The hunting of the Greene Lyon"* (Cambridge, 1975); *idem*, *The Janus faces of genius: The role of alchemy in Newton's thought* (Cambridge, 1991); Carolyn Merchant, *The death of nature: Women, ecology and the Scientific Revolution* (San Francisco, 1980); and John Henry, "Newton, matter and magic", in Fauvel *et al.* (eds), *Let Newton be!* (ref. 12), 127–45.
15. Paolo Rossi, *Francis Bacon: From magic to science* (London, 1968); Charles Webster, *From Paracelsus to Newton: Magic and the making of modern science* (Cambridge, 1982); Brian Copenhaver, "Astrology and magic", in C. B. Schmitt and Q. Skinner (eds), *The Cambridge history of Renaissance philosophy* (Cambridge, 1988), 264–300; *idem*, "Natural magic, hermetism, and occultism in early modern science", in D. C. Lindberg and R. S. Westman (eds), *Reappraisals of the Scientific Revolution* (Cambridge, 1990), 261–302; Ron Millen, "The manifestation of occult qualities in the Scientific Revolution", in M. J. Osler and P. L. Farber (eds), *Religion, science and worldview: Essays in honor of Richard S. Westfall* (Cambridge, 1985), 185–216; John Henry, "Occult qualities and the experimental philosophy: Active principles in pre-Newtonian

matter theory", *History of science*, xxiv (1986), 335–81; and Paul Richard Blum, "Qualitates occultae: Zur philosophischen Vorgeschichte eines Schlüsselbegriffs zwischen Okkultismus und Wissenschaft", in August Buck (ed.), *Die 'okkulten Wissenschaften in der Renaissance* (Wiesbaden, 1992), 45–64.

16. Bruce Moran, *Distilling knowledge: Alchemy, chemistry, and the Scientific Revolution* (Cambridge, MA, 2005), 1–7 and 185–9; and William R. Newman, *Atoms and alchemy: Chymistry and the experimental origins of the Scientific Revolution* (Chicago, 2006), 1–20 and 224–5.
17. Examples of this approach can be seen in a number of papers in M. L. Rhigini Bonelli and W. R. Shea (eds), *Reason, experiment, and mysticism in the Scientific Revolution* (London, 1975), particularly: Paolo Casini, "Newton, a sceptical alchemist?", 233–8; A. Rupert Hall, "Magic, metaphysics and mysticism in the Scientific Revolution", 275–82; and perhaps even Paolo Rossi, "Hermeticism, rationality and the Scientific Revolution", 247–73. But see also Bronislaw Malinowski, *Magic, science and religion and other essays* (Boston, 1948); and Mary B. Hesse, "Reason and evaluation in the history of science", in M. Teich and R. M. Young (eds), *Changing perspectives in the history of science* (London, 1973), 129–47. For further discussion see David Katz, *The occult tradition from the Renaissance to the present day* (London, 2005), 11–16.
18. It is important to exonerate Brian Vickers from these charges. Professor Vickers's extensive, careful and thoughtful scholarship has done much to increase our understanding of the history of magic. Even so, I have to say that I find his distinction between two 'mentalities', occult and scientific, in the pre-modern period, is a clear case of putting the positivist cart before the historical horse. See his Introduction to Vickers (ed.), *Occult and scientific mentalities in the Renaissance* (ref. 14), 1–55.
19. Sir David Brewster, *Memoirs of the life, writings and discoveries of Sir Isaac Newton* (2 vols, Edinburgh, 1855), ii, 374–5.
20. The denials are based on a misreading of a single comment by Newton in a letter to Richard Bentley written in February 1693. Although this comment, on a superficial glance, might look like a denial of action at a distance, it is not. The misreading was pointed out long ago by Emile Meyerson, "Leibniz, Newton, and action at a distance", in *idem*, *Identity and reality* (London, 1930), 447–56; and reiterated more recently in John Henry, "'Pray do not ascribe that notion to me': God and Newton's gravity", in James E. Force and Richard H. Popkin (eds), *The books of nature and Scripture: Recent essays on natural philosophy, theology and Biblical criticism in the Netherlands of Spinoza's time and the British Isles of Newton's time* (Dordrecht, 1994), 123–47. For the quotation in question, see Cohen (ed.), *Isaac Newton's papers and letters on natural philosophy* (ref. 8), 302.
21. Newton, *Opticks* (ref. 7), Queries, 1, 4, 21, 29, and 31, pp. 339, 352, 371, 375–6. The role of Newton's alchemy in his concept of force was first suggested by R. S. Westfall, "Newton and the Hermetic tradition", in Allen G. Debus (ed.), *Science, medicine and society in the Renaissance* (2 vols, New York, 1972), ii, 183–98. See also *idem*, "Newton and alchemy", in Vickers (ed.), *Occult and scientific mentalities* (ref. 14), 315–35; P. M. Rattansi, "Newton's alchemical studies", in Allen G. Debus (ed.), *Science, medicine and society in the Renaissance* (this ref.), ii, 167–82; and Dobbs, *The foundations of Newton's alchemy* (ref. 14).
22. The major study is Robert E. Schofield, *Mechanism and materialism: British natural philosophy in an age of reason* (Princeton, 1970). But see also Arnold Thackray, *Atoms and powers: An essay on Newtonian matter-theory and the development of chemistry* (Cambridge, MA, 1970); and P. M. Heimann and J. E. McGuire, "Newtonian forces and Lockean powers: Concepts of matter in eighteenth-century thought", *Historical studies in the physical sciences*, iii (1971), 233–306.
23. I have argued this elsewhere: John Henry, "Magic and science in the sixteenth and seventeenth centuries", in G. N. Cantor, J. R. R. Christie, J. Hodge, and R. C. Olby (eds), *Companion to the history of modern science* (London and New York, 1990), 583–96; and *The Scientific Revolution*

and the origins of modern science, 2nd edn (Basingstoke and New York, 2002), 54–67. See also Rossi, *Francis Bacon* (ref. 15); and Webster, *From Paracelsus to Newton* (ref. 15). See also Ann Blair, "Natural philosophy", in Katharine Park and Lorraine Daston (eds), *The Cambridge history of science*, iii: *Early modern science* (Cambridge, 2006), 365–406, who writes of the "Transformation of natural philosophy by empirical and mathematical methods" but does not explicitly mention the magical tradition as a source of these methods.

24. Keith Thomas, *Religion and the decline of magic* (London, 1971). For a convenient summary of criticisms of Thomas's theses, and an alternative view, see Alan Macfarlane, "Civility and the decline of magic", in Peter Burke, Brian Harrison, and Paul Slack (eds), *Civil histories: Essays in honour of Sir Keith Thomas* (Oxford, 2000), 145–60. On the continued fortune of magic in popular culture see, for example, Robert Muchembled, *Popular culture and elite culture in France, 1400–1750* (Baton Rouge and London, 1985); and Owen Davies, *Cunning folk: Popular magic in English history* (New York and London, 2003).
25. For studies of the importance of drawing intellectual or disciplinary boundaries in order to support knowledge claims, see T. F. Gieryn, "Boundary-work and the demarcation of science from non-science: Strains and interests in professional ideologies of scientists", *American sociological review*, xlviii (1983), 781–95; and B. Barnes, D. Bloor and J. Henry, *Scientific knowledge: A sociological analysis* (London and Chicago, 1996), 140–68.
26. I am encouraged in this line of argument by the similar claim made brilliantly with regard to alchemy in Lawrence M. Principe and William R. Newman, "Some problems with the historiography of alchemy", in W. R. Newman and A. Grafton (eds), *Secrets of nature: Astrology and alchemy in early modern Europe* (Cambridge, MA, 2001), 385–431.
27. For a discussion of the clear separation, and distance, between two major aspects of what we might think of as a unified magical tradition, astrology and alchemy, see Newman and Grafton (eds), *Secrets of nature* (ref. 26), 14–27.
28. Stuart Clark, "The scientific status of demonology", in Vickers (ed.), *Occult and scientific mentalities* (ref. 14), 351–74; and *idem*, *Thinking with demons: The idea of witchcraft in early modern Europe* (Oxford, 1997), 161–79.
29. Giambattista della Porta, *Natural magick … in twenty books* (London, 1658), Bk I, chap. 2, p. 2. This popular manual of natural magic was originally published in four books in Naples, 1554, and in twenty books in 1589. For a brief discussion, see Millen, "Manifestation of occult qualities" (ref. 15).
30. Cornelius Agrippa, *De incertitudine et vanitate omnium scientiarum et artium* (n.p., 1531), chap. 42.
31. Francis Bacon, *Novum organum*, Pt I, Aphorism IV. On magic in the work of Francis Bacon see Rossi, *Francis Bacon* (ref. 15); and John Henry, *Knowledge is power: Francis Bacon and the method of science* (Cambridge, 2002), 42–81.
32. Robert Hooke, *Micrographia* (London, 1665), 12, 15, 16, etc.; and Isaac Newton, Letter to Robert Boyle, February 28, 1679, in Cohen (ed.), *Newton's papers and letters* (ref. 8), 251. See John Henry, "Robert Hooke, the incongruous mechanist", in Michael Hunter and Simon Schaffer (eds), *Robert Hooke: New studies* (Woodbridge, Suffolk, 1989), 149–80.
33. Women and, to a lesser extent, uneducated men were held to be capable of magical knowledge, of course, but usually only to a limited extent. Prejudiced assumptions by the élite about the limits of the knowledge of such "cunning" men and women were to have appalling consequences during the witch crazes. Supposed success in magical operations was assumed to have been achieved thanks to the Devil's help (rather than by knowledge of natural occult qualities and powers), and so the witch was presumed guilty of commerce with Satan. The educated magus would always have been able to defend himself from similar charges by insisting that he used only natural magic, and by demonstrating a clear understanding of the distinction between natural and demonic

magic. Uneducated witches were not always able to make such clear distinctions, and much less so during inquisitorial proceedings. On popular knowledge of medicine see John Henry, "Doctors and healers: Popular culture and the medical profession", in Stephen Pumfrey, Paolo Rossi, and Maurice Slawinski (eds), *Science, culture and popular belief in Renaissance Europe* (Manchester, 1991), 191–221. Literature on the witch crazes is vastly extensive, but see, for example, Norman Cohn, *Europe's inner demons: The demonization of Christians in medieval Christendom* (Chicago, 1993); Hugh Trevor-Roper, *The European witch-craze of the sixteenth and seventeenth centuries* (Harmondsworth, 1969); and Bengt Ankarloo, Stuart Clark, and William Monter (eds), *Witchcraft and magic in Europe: The period of the witch trials* (Philadelphia, 2002).

34. Agrippa, *De incertitudine et vanitate omnium scientiarum* (ref. 30), chap. 42. See also Lauren Kassell, "'All was that land full fill'd of faerie', or magic and the past in early modern England", *Journal of the history of ideas*, lxvii (2006), 107–22, p. 112.
35. William Eamon, *Science and the secrets of nature: Books of secrets in medieval and early modern culture* (Princeton, 1994); and Bert Hansen, "Science and magic", in D. C. Lindberg (ed.), *Science in the Middle Ages* (Chicago, 1975), 483–506, pp. 493–5.
36. Recorded during one of the weekly conferences conducted at Theophraste Renaudot's Bureau d'Adresse from 1633 to 1642. Quoted from Kathleen Wellman, "Talismans, incubi, divination and the Book of M*: The Bureau d'Adresse confronts the occult", in A. G. Debus and Michael T. Walton (eds), *Reading the book of nature: The other side of the Scientific Revolution* (St Louis, 1998), 215–38, p. 228.
37. Brian Copenhaver, "Did science have a Renaissance?", *Isis*, lxxxiii (1992), 387–407; Michel Foucault, *The order of things: An archaeology of the human sciences* (London, 1974); and E. M. W. Tillyard, *The Elizabethan world picture* (London, 1943).
38. Clark, "Scientific status of demonology" (ref. 28); and *idem*, *Thinking with demons* (ref. 28), 161–78.
39. John Cotta, *The triall of witch-craft* (London, 1616), 34.
40. William Perkins, *Discourse of the damned art of witch-craft* (Cambridge, 1610), 59. The fact that Perkins does not list magicians among those "that are most excellent" in the knowledge of natural magic reflects the Church's reluctance to acknowledge anything worthy in magicians (Perkins was an Anglican clergyman). This is discussed more fully below.
41. James VI, *Daemonologie* (Edinburgh, 1597), 44.
42. The most notorious example of this kind of short cut to wisdom is provided by John Dee, who thanks to his "scrying stone" and a supposedly psychic "medium", Edward Kelly, who was all too willing to please, held many conversations with angels from 1582 to 1587, with a view to being able to make the philosopher's stone. Note that Dee, for obvious religious reasons, always claimed he was summoning angels, not demons. See Deborah E. Harkness, *John Dee's conversations with angels: Cabala, alchemy, and the end of nature* (Cambridge, 1999).
43. Perkins, *Discourse of the damned art of witch-craft* (ref. 40), 20.
44. Agrippa, *De incertitudine et vanitate omnium scientiarum* (ref. 30), chap. 42.
45. F. Giuntini, *Speculum astronomiae* (Paris, 1573), quoted from Lynn Thorndike, *A history of magic and experimental science* (8 vols, New York, 1923–58), vi, 132.
46. See D. P. Walker, *Spiritual and demonic magic from Ficino to Campanella* (London, 1958), 36, 83; and Hansen, "Science and magic" (ref. 35), 488–9. As in so many other cases, the Church's attitude forged popular consciousness. Hence, Christopher Marlowe, *The tragical history of Doctor Faustus* (1604), or the less well known Robert Greene, *The honourable historie of Frier Bacon and Frier Bungay* (*c.* 1592). On the legend of Faust, see E. M. Butler, *The myth of the magus*, *Ritual magic*, and *The fortunes of Faust* (Cambridge, 1948, 1949, and 1952). Literature on the witch-crazes is vastly extensive, but see, for example, the works cited in ref. 33 above.

47. Or some who were more than one of these things — magical traditions featured prominently in Renaissance eclecticism. John Napier, for example, inventor of logarithms, also devoted much of his time to alchemy, and was known locally, in Edinburgh, as a wizard. See Francis Shennan, *Flesh and bones: The life, passions and legacies of John Napier* (Edinburgh, 1989).
48. A. G. Molland, "Roger Bacon as magician", *Traditio*, xxx (1974), 445–60, p. 459–60. Consider also the revealingly contradictory title of Roger Bacon's *Epistola de secretis operibus artis et naturae, et de nullitate magiae*, written between 1248 and 1267.
49. For a recent discussion of Naudé's book see Kassell, *op. cit.* (ref. 34); and Molland, "Roger Bacon as magician" (ref. 48), 448.
50. Robert Hooke, "Of Dr Dee's book of spirits" (1690), in Richard Waller (ed.), *The posthumous works of Robert Hooke* (London, 1705), 203–10. See Henry, "Robert Hooke" (ref. 32), 176–8; and Richard Deacon, *John Dee, scientist, geographer, astrologer and secret agent to Elizabeth I* (London, 1968). Dee's séances with angels had been published in 1659: *A true & faithful relation of what passed for many yeers between dr. John Dee ... and some spirits*, ed. by Meric Casuabon (London, 1659). See Harkness, *John Dee's conversations with angels* (ref. 42). On Trithemius, see Klaus Arnold, *Johannes Trithemius (1462–1516)* (Würtzburg, 1971); and Wayne Shumaker, *Renaissance curiosa* (Binghampton, 1982).
51. On Bacon's criticisms of magic see Rossi, *Francis Bacon* (ref. 15), 31–35.
52. The retraction appeared in a general attack on all human knowledge, and an affirmation of Christian fideism, Cornelius Agrippa, *De incertitudine & vanitate scientiarum declamatio invectiva...* (n.p., 1530), but has been shown to share the same magico-religious foundations as the *De occulta philosophia* — so much so that the two works are said to share a basic unity. See Michael H. Keefer, "Agrippa's dilemma: Hermetic rebirth and the ambivalence of *De vanitate* and *De occulta philosophia*", *Renaissance quarterly*, xli (1988), 614–53.
53. Newton, *Opticks* (ref. 7), Bk III, Pt I, Query 31, p. 401.
54. For example, *Opticks* (ref. 7), Queries, 1, 4, 21, 29, and 31, pp. 339, 352, 371, 375–6; and Isaac Newton, *The Principia: Mathematical principles of natural philosophy*, a new transl. by I. B. Cohen and Anne Whitman (Berkeley, 1999), 382–3. On Newton's alchemy see the works cited in ref. 15 above. On action at a distance in Newton see Henry, "'Pray do not ascribe that notion to me'" (ref. 20).
55. On the sociology of 'boundary demarcation' see the works cited in ref. 25 above. We have mentioned the continuing and constant religious objection to demonology, but there was a new development in the late Renaissance and early modern periods, namely a sceptical denial of the existence of demons and other spiritual beings that went hand-in-hand with the rise of atheism, irreligion and secularism. For a general survey of such trends see Richard H. Popkin, *The history of scepticism from Erasmus to Spinoza* (Berkeley, 1979). For the relevance of these developments to the decline of magic, especially demonology, see Michael Hunter, "Witchcraft and the decline of belief", *Eighteenth-century life*, xxii (1998), 139–47.
56. See Vickers, "Introduction", in *idem* (ed.), *Occult and scientific mentalities* (ref. 14), 1–55. Professor Vickers is more sophisticated than the label "positivist" might imply. In a recent essay he has distinguished between "occult1", "a neutral distinction between visible and invisible causes", and "occult2", which is not so clearly defined (he refers to it as a synthesis of "magic and astrology", a "philosophy in its own right", and an "aggressive formulation" which led to "an increasingly bitter denunciation of occultism by both the Church and secular scholars"). I can agree that in the process of redefining boundaries there was this kind of bifurcation of the occult, but I see it as a process hinging upon many historical contingencies, while Professor Vickers seems to see it merely as a matter of an ahistorical entity called science, triumphing over an ahistorical entity called magic. See Brian Vickers, "The occult in the Renaissance", *Annals of science*, lii (1995), 77–84.

57. Frank L. Borchardt, "The magus as Renaissance man", *Sixteenth century journal*, xxi (1990), 57–76, pp. 59, 72.
58. Literature on the relations between the new philosophies and religion is vast, but for recent provocative insights see Amos Funkenstein, *Theology and the scientific imagination from the Middle Ages to the seventeenth century* (Princeton, 1986); Peter Harrison, *The Bible, Protestantism, and the rise of natural science* (Cambridge, 1998); and Stephen Gaukroger, *The emergence of a scientific culture: Science and the shaping of modernity, 1250–1685* (Oxford, 2006).
59. In fact there was a short-lived attempt to do just this, and this will be briefly discussed later (see ref. 110 and text at that point). But this was a minor historical episode that serves to highlight the more general trend.
60. Stuart Clark, "The rational witchfinder: Conscience, demonological naturalism and popular superstitions", in Pumfrey, Rossi and Slawinski (eds), *Science, culture and popular belief* (ref. 33), 222–48; and *idem, Thinking with demons* (ref. 28).
61. Clark, "Rational witchfinder" (ref. 60); and Hansen, "Science and magic" (ref. 35), 488–9. See also Caroline Oates, "Metamorphosis and lycanthropy in Franche-Comté, 1521–1643", in M. Feher (ed.), *Fragments for a history of the human body, Part One* (New York, 1989), 305–63, especially p. 320.
62. Clark, "Rational witchfinder" (ref. 60).
63. *Ibid.*
64. Literature on the witch-crazes is dauntingly vast, but see, for example, Christina Larner, *Witchcraft and religion: The politics of popular belief* (Oxford, 1984); Joseph Klaits, *Servants of Satan: The age of the witch hunts* (Bloomington, 1985); Jonathan Barry, Marianne Hester and Gareth Roberts (eds), *Witchcraft in early modern Europe: Studies in culture and belief* (Cambridge, 1996); Robin Briggs, *Witches and neighbours: The social and cultural context of European witchcraft* (London, 1996); and the works cited in ref. 33 above.
65. Michael Hunter, "Alchemy, magic and moralism in the thought of Robert Boyle", in *idem, Robert Boyle (1627–91): Scrupulosity and science* (Woodbridge, Suffolk, 2000), 93–118. See also *idem* (ed.), *Robert Boyle by himself and his friends* (London, 1994).
66. See Hunter, *Boyle by himself* (ref. 65), 29–31.
67. Quoted from Michael Hunter, "Alchemy, magic and moralism" (ref. 65), 116.
68. For detailed expositions of what has come to be called the scholar and craftsman thesis, see Edgar Zilsel, *The social origins of modern science* (Dordrecht, 2000); Pamela H. Smith, *The body of the artisan: Art and experience in the Scientific Revolution* (Chicago, 2004); and *idem*, "Laboratories", in Park and Daston (eds), *Cambridge history of science*, iii (ref. 23), 290–319.
69. Foster Watson, *Vives: On education. A translation of the De tradendis disciplinis, together with an introduction* (Cambridge, 1913), 209. On Vives, see Carlos G. Noreña, *Juan Luis Vives* (The Hague, 1970). Francis Bacon, *New organon*, Pt II, Aphorism 31.
70. On Gilbert, see John Henry, "Animism and empiricism: Copernican physics and the origins of William Gilbert's experimental method", *Journal of the history of ideas*, lxii (2001), 99–119. On Boyle as a thoroughgoing alchemist, rather than one who turned to it later, see Lawrence Principe, *The aspiring adept: Robert Boyle and his alchemical quest* (Princeton, 1998); and William Newman, *Atoms and alchemy* (ref. 16).
71. Arthur C. Clarke, *Profiles of the future: An inquiry into the limits of the possible* (London, 1962).
72. Salluste du Bartas, *His divine weekes and workes*, transl. by J. Sylvester (London, 1606), 221; quoted from J. Peter Zetterberg, "The mistaking of 'the mathematicks' for magic in Tudor and Stuart England", *Sixteenth century journal*, xi (1980), 83–97, p. 93. On superstitious beliefs about mathematics and magic and demonology, see Mordechai Feingold, "The occult tradition in the English universities of the Renaissance: A reassessment", in Vickers (ed.), *Occult and*

scientific mentalities (ref. 14), 73–94; and Katherine Neal, "The rhetoric of utility: Avoiding occult associations for mathematics through profitability and pleasure", *History of science*, xxxvii (1999), 151–78.

73. See Thorndike, *History of magic and experimental science* (ref. 45); M. Sherwood, "Magic and mechanics in medieval fiction", *Studies in philology*, xli (1947), 567–92; Zetterberg, "The mistaking of 'the mathematicks' for magic" (ref. 72); William Eamon, "Technology as magic in the late Middle Ages and the Renaissance", *Janus*, 1983, 171–212; A. George Molland, "Cornelius Agrippa's mathematical magic", in Cynthia Hay (ed.), *Mathematics from manuscript to print* (Oxford, 1988), 209–19; and Henry, "Magic and science" (ref. 23).

74. See, for example, Paul Lawrence Rose, *The Italian Renaissance of mathematics: Studies on humanists and mathematicians from Petrarch to Galileo* (Geneva, 1975); Robert S. Westman, "The astronomer's role in the sixteenth century: A preliminary survey", *History of science*, xviii (1980), 105–47; Nicholas Jardine, "Epistemology of the sciences", in Schmitt and Skinner (eds), *The Cambridge history of Renaissance philosophy* (ref. 15), 685–711; Mario Biagioli, "The social status of Italian mathematicians, 1450–1600", *History of science*, xxvii (1989), 41–95; J. A. Bennett, "The challenge of practical mathematics", in Pumfrey, Rossi and Slawinski (eds), *Science, culture and popular belief* (ref. 33), 176–90; Anna De Pace, *Le matematiche e il mondo: Ricerche su un dibattito in Italia nella seconda metà del cinquecento* (Milan, 1993); James M. Lattis, *Between Copernicus and Galileo: Christoph Clavius and the collapse of Ptolemaic cosmology* (Chicago, 1994); and Peter Dear, *Discipline and experience: The mathematical way in the Scientific Revolution* (Chicago, 1995).

75. Zetterberg, "The mistaking of 'the mathematicks' for magic" (ref. 72). For more positivist tendencies in the history of Renaissance mathematics see Edward W. Strong, *Procedures and metaphysics: A study in the philosophy of mathematical-physical science in the sixteenth and seventeenth centuries* (Berkeley, 1936).

76. On Napier, see Mark Napier, *Memoirs of John Napier of Merchiston: His lineage, life, and times, with a history of the invention of logarithms* (Edinburgh, 1834); and Shennan, *Flesh and bones* (ref. 47). Napier's alchemy remains largely unstudied, but see J. Small, "Sketches of later Scottish alchemists: John Napier of Merchiston, Robert Napier, Sir David Lindsay, first earl of Balcarres, Patrick Ruthven, Alexander Seton, and Patrick Scot", *Proceedings of the Society of Antiquaries of Scotland*, xi (1874–76), 410–38. His alchemical papers are in Edinburgh University Library. Literature on Dee is extensive. See Stephen Clucas, "Recent Works on John Dee (1988–2005): A select bibliography", in *idem* (ed.), *John Dee: Interdisciplinary studies in English Renaissance thought* (Dordrecht, 2006), 345–50. The best single study is Nicholas H. Clulee, *John Dee's natural philosophy: Between science and religion* (London and New York, 1988).

77. John Wilkins, *Mathematical magick, or, the wonders that may be performed by mechanical geometry* (London, 1648). The major study is still Barbara Shapiro, *John Wilkins: An intellectual biography* (Berkeley, 1969), but see also Ana Maria Alfonso-Goldfarb, "An 'older' view about matter in John Wilkins' 'modern' mathematical magick", in Debus and Walton (eds), *Reading the book of nature* (ref. 36), 133–46.

78. René Descartes, *Principia philosophiae*, IV, §203. On the impact of the pseudo-Aristotelian *Quaestiones mechanicae*, see Helen Hattab, "From mechanics to mechanism: The *Quaestiones mechanicae* and Descartes' physics", in Peter R. Anstey and John A. Schuster (eds), *The science of nature in the seventeenth century: Patterns of change in early modern natural philosophy* (Dordrecht, 2005), 99–129.

79. This book has been translated into English: Girolamo Cardano, *De subtilitate, Book 1*, translated, with introduction and notes, by Myrtle Marguerite Cass (Williamsport, PA, 1934). The best edition of the *De subtilitate* (currently only the first seven books) is *De subtilitate libri XXI (Libri I–VII)*, ed. by Elio Nenci (Milan, 2004). For studies of Cardano see Angelo Bellini, *Girolamo Cardano*

e il suo tempo (Milan, 1947); Nancy G. Siraisi, *The clock and the mirror: Girolamo Cardano and Renaissance medicine* (Princeton, 1997); and Anthony Grafton, *Cardano's cosmos: The worlds and works of a Renaissance astrologer* (Cambridge, MA, 1999). See also Girolamo Cardano, *De libris propriis: The editions of 1544, 1550, 1557, 1562*, ed. by Ian Maclean (Milan, 2004).

80. Jens Høyrup, "Philosophy: Accident, epiphenomenon, or contributory cause of the changing trends of mathematics — a sketch of the development from the twelfth through the sixteenth century", in *idem, In measure, number, and weight: Studies in mathematics and culture* (Albany, 1994), 123–71, p. 165, see also pp. 154–6. See also William R. Newman, *Promethean ambitions: Alchemy and the quest to perfect nature* (Chicago, 2004).
81. See McGuire and Rattansi, "Newton and the 'Pipes of Pan'" (ref. 12). Consider also, in this connection, Ayval Leshem, *Newton on mathematics and spiritual purity* (Dordrecht, 2003).
82. For the details of Kepler's geometrical archetype see Alexandre Koyré, *The astronomical revolution: Copernicus, Kepler, Borelli*, transl. by R. E. W. Maddison (London, 1973); J. V. Field, *Kepler's geometrical cosmology* (London, 1988); and Rhonda Martens, *Kepler's philosophy and the new astronomy* (Princeton, 2000).
83. Kepler and Fludd engaged in a dispute, played out in dauntingly long publications, on the issue of numerology. See Robert S. Westman, "Nature, art, and psyche: Jung, Pauli, and the Kepler–Fludd polemic"; and J. V. Field, "Kepler's rejection of numerology", both in Vickers (ed.), *Occult and scientific mentalities* (ref. 14), 177–229, and 273–96, respectively.
84. Principe and Newman, "Some problems with the historiography of alchemy" (ref. 26). See also Lawrence M. Principe, "Reflections on Newton's alchemy in light of the new historiography of alchemy", in J. E. Force and S. Hutton (eds), *Newton and Newtonianism: New studies* (Dordrecht, 2004), 205–19.
85. *Ibid.*, 386. See also W. R. Newman and L. M. Principe, "Alchemy vs. chemistry: The etymological origins of a historiographic mistake", *Early science and medicine*, iii (1998), 32–65.
86. Principe and Newman, "Some problems with the historiography of alchemy" (ref. 26), 418.
87. *Ibid.*, 413–15. The principle contributor to this new scholarship is William Newman himself. See, for a recent example, Newman, *Atoms and alchemy* (ref. 16).
88. William B. Ashworth Jr, "Natural history and the emblematic world view", in David C. Lindberg and Robert S. Westman (eds), *Reappraisals of the Scientific Revolution* (Cambridge, 1980), 303–32. See also James J. Bono, *The word of God and the languages of man: Interpreting nature in early modern science and medicine* (Madison, 1995); and Harrison, *The Bible, Protestantism, and the rise of natural science* (ref. 58). But see also Ann Blair, "*Historia* in Zwinger's *Theatrum humanae vitae*", in Gianna Pomata and Nancy G. Siraisi (eds), *Historia: Empiricism and erudition in early modern Europe* (Cambridge, MA, 2005), 269–96.
89. On these new occultist developments in medicine see Linda Deer Richardson, "The generation of disease: Occult causes and diseases of the total substance", in A. Wear, R. K. French and I. M. Lonie (eds), *The medical renaissance of the sixteenth century* (Cambridge, 1985), 175–94; and John Henry and John M. Forrester, "Jean Fernel and the importance of his *De abditis rerum causis*", in *idem, Jean Fernel's On the hidden causes of things: Forms, souls and occult diseases in Renaissance medicine* (Leiden, 2005), 37–44.
90. Jean Fernel, *De abditis rerum causis* (Paris, 1648), Bk II, chap. 16. For a new edition and translation of this important work see Forrester and Henry, *Jean Fernel's On the hidden causes of things* (ref. 89); on amulets, see pp. 646–73.
91. Johann Wier, *De praestigiis daemonum*... (Basel, 1583), Bk V, chap. 8, column 535, quoted from Walker, *Spiritual and demonic magic* (ref. 46), 154; and, of course, Genesis, 1, 3.
92. See Hansen, "Science and magic" (ref. 35), 488.
93. M. M. Slaughter, *Universal languages and scientific taxonomy in the seventeenth century* (Cambridge,

1982); and Paolo Rossi, *Logic and the art of memory: The quest for a universal language*, transl. by Stephen Clucas (London, 2000).

94. Francis Bacon, *Valerius Terminus*, and 'Proemium' to *Great instauration*, in John M. Robertson (ed.), *Philosophical works of Francis Bacon* (London, 1905), 188, and 241 respectively. John Wilkins, *Essay towards a real character and a philosophical language* (London, 1668).

95. On the historical phenomenon of Christian cabbalism, see Joseph Leon Blau, *The Christian interpretation of the Cabala in the Renaissance* (New York, 1944); and Moshe Idel, "The magical and Neoplatonist interpretation of the Kabbalah in the Renaissance", in B. D. Cooperman (ed.), *Jewish thought in the sixteenth century* (Cambridge, MA, 1983), 186–242. On Giovanni Pico see S. A. Farmer, *Syncretism in the West. Pico's 900 theses (1486): The evolution of traditional, religious, and philosophical systems*, with text, translation, and commentary (Medieval & Renaissance Texts & Studies; Tempe, AZ, 1998). On Leibniz's efforts to develop a universal language and the role of cabbalism in these schemes, see Allison P. Coudert, "Some theories of a natural language from the Renaissance to the seventeenth century", *Studia Leibnitiana*, vii (1978), 119–41; *idem*, "Forgotten ways of knowing: The Kabbalah, language, and science in the seventeenth century", in Donald R. Kelley and R. H. Popkin (eds), *The shapes of knowledge: From the Renaissance to the Enlightenment* (Dordrecht, 1991), 83–99; and Rossi, *Logic and the art of memory* (ref. 93).

96. Guy de la Brosse (1628) and John Ray (1660), both quoted from Brian Vickers, "Critical reactions to the occult sciences during the Renaissance", in E. Ullmann-Margalit (ed.), *The scientific enterprise* (Dordrecht, 1992), 43–92, pp. 77–79.

97. For a recent convenient summary of the Renaissance arguments against astrology, see Vickers, "Critical reactions to the occult sciences during the Renaissance" (ref. 96). But see also, Jim Tester, *A history of Western astrology* (Woodbridge, 1987); Eugenio Garin, *Astrology in the Renaissance: The zodiac of life* (London, 1983); and H. Darrel Rutkin, "Astrology", in Park and Daston (eds), *The Cambridge history of science*, iii (ref. 23), 542–61.

98. Johannes Kepler, *Mysterium cosmographicum* (1597), in *Gesammelte Werke*, i (Munich, 1938), 21. Galileo, always loath to accept magical influence, did deny the effect of the Moon on the tides. On this issue, however, the magicians were correct and Galileo was embarrassingly wrong. Galileo Galilei, *Dialogue on the two chief world systems*, transl. by Stillman Drake (New York, 2001), Fourth Day, 487.

99. See J. V. Field, "Astrology in Kepler's cosmology", in Patrick Curry (ed.), *Astrology, science and society* (Woodbridge, 1987), 143–70; and *idem*, "A Lutheran astrologer: Johannes Kepler", *Archive for history of exact sciences*, xxxi (1984), 189–272 (which includes a complete translation of *De fundamentis astrologiae certioribus*).

100. Vickers, "Critical reactions to the occult sciences" (ref. 96); and Tester, *A history of Western astrology* (ref. 97).

101. On empirical investigations of astrology and the weather, see Michael Hunter and Annabel Gregory, *An astrological diary of the seventeenth century: Samuel Jeake of Rye, 1652–1699* (Oxford, 1988); and Michael Hunter, "Science and astrology in seventeenth-century England: An unpublished polemic by John Flamsteed", in Curry (ed.), *Astrology, science and society* (ref. 99), 261–300.

102. P. de Maupertuis, "Lettre sur la comète", in *Oeuvres* (Lyons, 1768), iii, 209–56, p. 240. Quoted from Simon Schaffer, "Newton's comets and the transformation of astrology", in Curry (ed.), *Astrology, science and society* (ref. 99), 219–43, p. 237.

103. Edmund Halley, "Some considerations about the cause of the universal deluge", *Philosophical transactions*, xxxiii (1724), 118–25; and William Whiston, *Vindication of the new theory of the Earth* (London, 1698), Preface; both quoted from Schaffer, "Newton's comets" (ref. 103), 233.

104. Newton's belief recorded by his friend John Craig, Cambridge University Library, Keynes MS 130.7,

f. 1r. See also Isaac Newton, "De gravitatione et aequipondio fluidorum", in A. R. Hall and M. B. Hall (eds), *Unpublished scientific papers of Isaac Newton* (Cambridge, 1962), 139.

105. The rivalry on this issue between Newton and Leibniz is brought out most starkly in the so-called Leibniz–Clarke correspondence. See H. G. Alexander (ed.), *The Leibniz–Clarke correspondence* (Manchester, 1956); and Ezio Vailati, *Leibniz and Clarke: A study of their correspondence* (Oxford, 1997). For an account of the role of comets in Newton's system, see David Kubrin, "Newton and the cyclical cosmos", *Journal of the history of ideas*, xxviii (1967), 325–46; Sara Schechner Genuth, "Comets, teleology and the relationship of chemistry to cosmology in Newton's thought", *Annali dell'Istituto e Museo di Storia della Scienza di Firenze*, x (1985), 31–65; and Schaffer, "Newton's comets" (ref. 103).

106. Two developments summed up by Pierre Simon Laplace, author of *Traité du mécanique céleste* (5 vols, 1799–1825), who used Newtonian mechanics to show that the planetary system was self-regulating, without the need for comets, and who also told Napoleon that God was not mentioned in his book because he "had no need of that hypothesis". See C. C. Gillispie, *Pierre-Simon Laplace, 1749–1827: A life in exact science* (Princeton, 1997).

107. Petrus Peregrinus Maricurtensis, *De magnete, seu rota perpetui motus, libellus ... nunc primum promulgatus* (Augsburg, 1558). See S. P. Thompson, *The epistle of Peter Peregrinus de Maricourt to Sygerus de Foucaucourt, soldier, concerning the magnet* (London, 1902); and H. D. Harradon, "Some early contributions to the history of geomagnetism — I: The letter of Peter Peregrinus de Maricourt to Sygerus de Foucaucourt, soldier, concerning the magnet", *Terrestrial magnetism and atmospheric electricity*, xlviii (1943), 3–17. William Gilbert, *De magnete* [1600], transl. by P. Fleury Mottelay (New York, 1958). For details see Henry, "Animism and empiricism" (ref. 70). For a general study of the concept of self-movement in Aristotelianism see Mary Louise Gill and James G. Lennox (eds), *Self-movement: From Aristotle to Newton* (Princeton, 1994).

108. J. A. Bennett, "Magnetical philosophy and astronomy from Wilkins to Hooke," in R. Taton and C. Wilson (eds), *Planetary astronomy from the Renaissance to the rise of astrophysics*, Part A: *Tycho Brahe to Newton* (Cambridge, 1989), 222–30. On Kepler's use of Gilbert see, for example, Koyré, *The astronomical revolution* (ref. 82), 185–214.

109. Sir William Petty, *The discourse made before the Royal Society ... together with a new hypothesis of springing or elastique motions* (London, 1674). See Henry, "Occult qualities and the experimental philosophy (ref. 15), 350–1.

110. Moody E. Prior, "Joseph Glanvill, witchcraft and seventeenth-century science", *Modern philology*, xxx (1932), 167–93; George Edelin, "Joseph Glanvill, Henry More, and the phantom drummer of Tedworth", *Harvard Library bulletin*, x (1956), 186–92; Thomas Harmon Jobe, "The devil in Restoration science: The Glanvill–Webster debate", *Isis*, lxxii (1981), 343–56; and Ian Bostridge, *Witchcraft and its transformations c. 1650–c. 1750* (Oxford, 1997). On similar efforts in the Netherlands see Andrew Fix, *Fallen angels: Balthasar Bekker, spirit belief and confessionalism in the seventeenth-century Dutch Republic* (Dordrecht, 1999).

111. It is like comparing Wilkins, *Mathematical magick* (ref. 77), with, say, Petrus Severinus, *Idea medicinae philosophicae: Fundamenta continens totius doctrinae Paracelsicae, Hippocraticae, & Galenicae* (Basel, 1571).

112. But remember what was said earlier about the possibility that the work of other numerologists on the Keplerian or Newtonian model might simply be being overlooked.

113. The main support for the claim that experimentalism and the utilitarianism of the new philosophies derives from the occult tradition is to be found in Francis Bacon. See Rossi, *Francis Bacon: From magic to science* (ref. 15); and Henry, *Knowledge is power* (ref. 31). But see also, for example, Henry, "Magic and science" (ref. 23); *idem*, "Occult qualities and the experimental philosophy" (ref. 15); and *idem*, "Animism and empiricism" (ref. 70).

114. On the decline of belief in symbolic magic, see Brian Vickers, "Analogy versus identity: The rejection

of occult symbolism, 1580–1680", in *idem* (ed.), *Occult and scientific mentalities* (ref. 14); and *idem*, "Critical reactions to the occult sciences" (ref. 96). See also, Brian Copenhaver, "The occultist tradition and its critics", in D. Garber and M. Ayers (eds), *The Cambridge history of seventeenth-century philosophy* (Cambridge, 1998), 454–512.

115. Walker, *Spiritual and demonic magic* (ref. 46), 84; Thomas, *Religion and the decline of magic* (ref. 24); and Clark, "The rational witchfinder" (ref. 60).

116. The literature here is vast but an obvious starting point is Thomas, *Religion and the decline of magic* (ref. 24). Consider also Clark, *Thinking with demons* (ref. 28). On the socio-political dimensions of alchemy, which has been well served in the literature, see for example P. M. Rattansi, "Paracelsus and the Puritan revolution", *Ambix*, xi (1963), 24–32; Hugh Trevor-Roper, "The Paracelsian movement", in *idem*, *Renaissance essays* (London, 1985), 149–99; J. Andrew Mendelsohn, "Alchemy and politics in England", *Past and present*, cxxxv (1992), 30–78; and William Newman, "From alchemy to 'chymistry'", in Park and Daston (eds), *The Cambridge history of science*, iii (ref. 23), 497–517. On the political dimension to astrology, see for example, Patrick Curry, *Prophecy and power: Astrology in early modern England* (Oxford, 1989). See also, for another example, Paola Zambelli, "Magic and radical reformation in Agrippa of Nettesheim", *Journal of the Warburg and Courtauld Institutes*, xxxix (1976), 69–103.

117. Consider, for example, Thomas Digges, *A perfit description of the caelestiall orbes, according to the most auncient doctrine of the Pythagoreans* (London, 1576). On this tradition, usually called the *prisca theologia* tradition, consult Charles B. Schmitt, "*Prisca theologia* e *philosophia perennis*: Due temi del Rinascimento italiano e la loro fortuna", in *Atti del V Convegno Internazionale del Centro di Studi Umanistici: Il pensiero italiano del Rinascimento e il tempo nostro* (Florence, 1970), 211–36; and D. P. Walker, *The ancient theology: Studies in Christian Platonism from the fifteenth to the eighteenth century* (London, 1972).

118. *De vita coelitus comparanda* was the third part of Ficino's *De vita triplici* (Florence, 1489). The meaning of the title is obscure, possibly something like: "On arranging one's life in accordance with the heavens". See Walker, *Spiritual and demonic magic* (ref. 46), 58, where Walker makes it clear that 'spiritual magic' is his own phrase to describe Ficino's ideas. For an indication of Ficino's immense influence see, for example, Hiro Hirai, *Le concept de semence dans les theories de la matière à la Renaissance: De Marsile Ficin à Pierre Gasendi* (Turnhout, 2005).

119. See Eugenio Garin, "The philosopher and the magus", in *idem* (ed.), *Renaissance characters*, transl. by Lydia G. Cochrane (Chicago, 1997), 123–53. See also Borchardt, "The magus as Renaissance man" (ref. 57). We should also remember, however, that even a highly learned magician such as Agrippa might call himself a magus and at the same time be anxious about the religious implications of being a magus. See Keefer, "Agrippa's dilemma" (ref. 52).

120. The most convenient source on the religious *Corpus Hermeticum*, which includes an invaluable introduction, is Brian P. Copenhaver, *Hermetica: The Greek Corpus Hermeticum and the Latin Asclepius in English translation* (Cambridge, 1991). On the so-called "technical" writings see A.-J. Festugière, *La révélation d'Hermès Trismégiste*, i: *L'Astrologie et les sciences occultes* (Paris, 1950). For brief assessments of the historical significance of the Hermetic writings see Cophenhaver, "Astrology and magic" (ref. 15); *idem*, "Natural magic, Hermetism, and occultism in early modern science" (ref. 15); and *idem*, "Magic", in Park and Daston (eds), *Cambridge history of science*, iii (ref. 23), 518–40.

121. Copenhaver, "Astrology and magic" (ref. 15); *idem*, "Did science have a Renaissance?" (ref. 37); and Walker, *Spiritual and demonic magic* (ref. 46).

122. Copenhaver, "Astrology and magic" (ref. 15), 273.

123. For an important study of the role of discussions of natural powers and the abilities of demons, see Newman, *Promethean ambitions* (ref. 80).

124. On Paracelsus see, for example, Walter Pagel, *Paracelsus: An introduction to philosophical medicine*

in the era of the Renaissance (Basel, 1958); and A. Weeks, *Paracelsus: Speculative theory and the crisis of the early Reformation* (Albany, 1997). On Fracastoro, see Vivian Nutton, "The seeds of disease: An explanation of contagion and infection from the Greeks to the Renaissance", *Medical history,* xxvii (1983), 1–34; and *idem,* "The reception of Fracastoro's theory of contagion: The seed that fell among thorns?", *Osiris,* n.s, vi (1990), 196–234. On Fernel, see Richardson, "Generation of disease" (ref. 89); Laurence Brockliss and Colin Jones, *The medical world of early modern France* (Oxford, 1997); Hiroshi Hirai, "Humanisme, néoplatonisme et *prisca theologia* dans le concept de semence de Jean Fernel", *Corpus,* xli (2002), 43–69; D. P. Walker, "The astral body in Renaissance medicine", *Journal of the Warburg and Courtauld Institutes,* xxi (1958), 119–33; *idem,* "Medical spirits in philosophy from Ficino to Newton", in *Arts du spectacle et histoire des idées* (Tours, 1984), 287–300; and Henry and Forrester, "Jean Fernel and the importance of his *De abditis rerum causis*" (ref. 89).

125. For brief accounts of the nature of alchemy see Bruce T. Moran, *Distilling knowledge: Alchemy, chemistry, and the Scientific Revolution* (Cambridge, 2005); Newman, *Atoms and alchemy* (ref. 16); and *idem,* "From alchemy to 'chymistry'" (ref. 116). On occult qualities see Keith Hutchison, "What happened to occult qualities in the Renaissance?", *Isis,* lxxiii (1982), 233–53; and Millen, "Manifestation of occult qualities" (ref. 15).

126. Copenhaver, "Astrology and magic" (ref. 15).

127. On this aspect of Aristotelianism see, for example, Peter Dear, *Discipline and experience* (ref. 74), 11–31; and *idem,* "The meanings of experience", in Park and Daston (eds), *Cambridge history of science,* iii (ref. 23), 106–31.

128. On Pomponazzi see Walker, *Spiritual and demonic magic* (ref. 46), 107–11. On Fernel: Henry and Forrester, "Jean Fernel and the importance of his *De abditis rerum causis*" (ref. 89). On Sennert: Newman, *Atoms and alchemy* (ref. 16). For general studies of this theme see Millen, "Manifestation of occult qualities" (ref. 15); and Henry, "Magic and science" (ref. 23).

129. For more details see Newman, *Atoms and alchemy* (ref. 16). See also Lynn S. Joy, "Scientific explanation from formal causes to laws of nature", in Park and Daston (eds), *Cambridge history of science,* iii (ref. 23), 70–105; and Andrew G. van Melsen, *From atomos to atom: The history of the concept atom* (New York, 1960).

130. Brian Vickers, "Introduction", in *idem* (ed.), *Occult and scientific mentalities* (ref. 14), 1–55.

131. Lauren Kassell, *Medicine and magic in Elizabethan London: Simon Forman, astrologer, alchemist, and physician* (Oxford, 2005). On Fludd, consider, for example, William H. Huffman, *Robert Fludd and the end of the Renaissance* (London and New York, 1988); and Allen G. Debus, *Man and nature in the Renaissance* (Cambridge, 1978), 121–6. Fludd's work has even been considered alongside that of Agrippa and Cardano: Silvia Parigi, *La magia naturale nel Rinascimento: Testi di Agrippa, Cardano, Fludd* (Turin, 1989).

132. Both Daniel Garber and Lynn S. Joy acknowledge the role of chemical ideas in early modern changes in natural philosophy. See Daniel Garber, "Physics and foundations", and Lynn S. Joy, "Scientific explanation from formal causes to laws of nature", in Park and Daston (eds), *Cambridge history of science,* iii (ref. 23), 29–33, and 70–105, respectively.

133. See Henry and Forrester, "Jean Fernel and the importance of his *De abditis rerum causis*" (ref. 89); Sylvain Matton, "Fernel et les alchimistes", *Corpus,* xli (2002), 135–97; Nutton, "The seeds of disease" (ref. 124); and *idem,* "The reception of Fracastoro's theory of contagion" (ref. 124). As representative of the Paracelsians, consider Severinus: Jole Shackelford, *A philosophical path for Paracelsian medicine: The ideas, intellectual context, and influence of Petrus Severinus (1540/2–1602)* (Copenhagen, 2004).

134. Girolamo Fracastoro, *De sympathia et antipathia rerum* (Lyons, 1554). Girolamo Cardano, *De subtilitate libri XXI* (Basle, 1554). The book mentioned is Book XVIII, but is by no means the only book where Cardano draws upon occult ideas and attitudes.

135. Bernardino Telesio, *De rerum natura iuxta propria pricipia libri IX* (Naples, 1587). For Telesio's, Campanella's and even Bacon's credentials as magicians, see Walker, *Spiritual and demonic magic* (ref. 46), 189–93, 199–203, and 203–36. On Bacon's semi-Paracelsian cosmology, see Graham Rees, "Francis Bacon's semi-Paracelsian cosmology", *Ambix*, xxii (1975), 81–101; and *idem*, "Francis Bacon's semi-Paracelsian cosmology and the Great Instauration", *Ambix*, xxii (1975), 161–73.

136. On Bruno, see (of course) Frances Yates, *Giordano Bruno and the Hermetic tradition* (London, 1964). See also Francesco Patrizi, *Nova de universis philosophia libris quinquaginta comprehensa* (Venice, 1593); and *Magia philosophia ... Zoroaster et eius cccxx Oracula Chaldaica ... Hermetis Trismegisti Poemander...* (Hamburg, 1593).

137. Henry, "Occult qualities" (ref. 15).

INDEX

For Product Safety Concerns and Information please contact our EU representative GPSR@taylorandfrancis.com Taylor & Francis Verlag GmbH, Kaufingerstraße 24, 80331 München, Germany

Batch number: 10725979

Printed by Printforce, the Netherlands